Data Spaces

Edward Curry • Simon Scerri • Tuomo Tuikka
Editors

Data Spaces

Design, Deployment and Future Directions

 Springer

Editors
Edward Curry
Insight SFI Research Centre
for Data Analytics
University of Galway
Galway, Ireland

Simon Scerri
metaphacts GmbH
Walldorf, Germany

Tuomo Tuikka
VTT Technical Research Centre of Finland
Oulu, Finland

ISBN 978-3-030-98638-4   ISBN 978-3-030-98636-0   (eBook)
https://doi.org/10.1007/978-3-030-98636-0

This Springer imprint is published by the registered company Springer Nature Switzerland AG
The registered company address is: Gewerbestrasse 11, 6330 Cham, Switzerland

Preface

"The future is already here—it's just not evenly distributed" is a quote widely attributed to William Gibson, the science fiction author who has provoked much debate on the manifestation of our future society. This particular quote has many thought-provoking interpretations. My interpretation (and others') of this quote has been the challenge of technology adoption; the technology is already invented—it's just not widely adopted. Many world-changing technologies have taken years or decades to be widely adopted, while others may never see widespread adoption. For example, in the early part of the 2010s, data-driven innovation powered by Big Data was a clear competitive advantage for many organizations. However, European organizations were lagging in the adoption of data-driven innovation. Evenly distributing the adoption of data-driven innovation was a key motivation for the Big Data Value Public-Private Partnership.

Another interpretation of Gibson's quote takes the perspective of inequality. The future is here, but only a few can access it. Or the future is here, and someone else will define what it is for you. These are profound questions on the type of society we aspire to create and its equality. As we look to the evolution of data-driven innovation, this is a pertinent perspective to consider. Artificial Intelligence (AI) revolutionizes many industries and society, including transportation and logistics, security, manufacturing, energy, healthcare, and agriculture, by providing intelligence to improve efficiency, quality, and flexibility. Data, in particular large quantities of high-quality data, is critical to creating competitive AI solutions. However, a significant barrier to adopting data-driven AI solutions is the high upfront costs associated with data collection, integration, and sharing activities. This limits large-scale data-driven projects to those with the necessary expertise and resources. The future is here, but only if you have the scale and expertise to manage the data.

With few exceptions, our current large-scale data infrastructures are beyond the reach of small organizations that cannot deal with the complexity of data management and the high costs associated with data infrastructure. This situation needs to change to enable everyone to engage and leverage the value available from data-driven AI solutions. I believe that forward-thinking societies need to see the provision of digital infrastructure as a shared societal service in the same way as water, sanitation, education, and healthcare. Luckily, this is a vision shared by the European Commission within their data strategy and their goal to establish common European Data Spaces as a mechanism to support the sharing and exchange of data.

The Big Data Value Association views Data Spaces as an ecosystem of data models, datasets, ontologies, data sharing contracts, and specialized data management services together with soft competencies including governance, social interactions, and business processes. The data space concept has gained traction with several groups exploring its usefulness for managing data from different domains and regions within a global data ecosystem. Data Spaces offer the digital foundations to design a new Digital Society where individuals and organizations can share data in a trusted and controlled environment to create data-driven AI solutions.

This book aims to educate data space designers to understand what is required to create a successful data space. What specialized techniques, methods, and platforms are needed for data sharing and exchange in a data space? What business processes need to be re-engineered? How do we design and nurture the ecosystem of stakeholders around a data space? This book explores the cutting-edge theory, technologies, methodologies, and best practices for Data Spaces for both industrial and personal data. It provides the reader with a basis for understanding the design, deployment, and future directions of Data Spaces.

The book's contributions emanate from the Big Data Value Public-Private Partnership and the Big Data Value Association, which have acted as the European data community's nucleus to bring together businesses with leading researchers to harness the value of data to benefit society, business, science, and industry. The technological basis established in the BDV PPP is enabling the creation of embryonic Data Spaces across Europe.

The book is of interest to two primary audiences: (1) researchers interested in data management and data sharing and (2) practitioners and industry experts engaged in data-driven systems where the sharing and exchange of data within an ecosystem are critical. This book is arranged in three parts. The first part contains design contributions of technologies and methods which are needed to implement a data space. The second part includes contributions detailing the deployment of existing Data Spaces. Finally, the third part outlines future directions for Data Spaces. Chapter "Data Spaces: Design, Deployment and Future Directions" provides an overview of Data Spaces and the vision of common European data space. Then, it positions the contributions related to the different aspects of Data Spaces, including value, data, technology, organization, people, governance, and trust.

Part I: Design details critical technical contributions which are needed in the design of effective Data Spaces. Chapter "An Organisational Maturity Model for Data Spaces: A Data Sharing Wheel Approach" looks at the necessary organizational design required for Data Spaces and presents a maturity model to capture organizational best practices. Chapter "Data Platforms for Data Spaces" provides an overview of the different designs of data platforms for Data Spaces. Data Governance within a data space ecosystem is the focus of chapter "Technological

Perspective of Data Governance in Data Space Ecosystems". Trust and federated learning within Data Spaces is the topic of chapter "Increasing Trust for Data Spaces with Federated Learning". A secure, trusted, regulatory-compliant, and privacy-preserving data sharing platform is detailed in chapter "KRAKEN: A Secure, Trusted, Regulatory Compliant and Privacy-Preserving Data Sharing Platform", with chapter "Connecting Data Spaces and Data Marketplaces and the Progress Towards the European Single Digital Market with Open Source Software" investigating the link between Data Spaces and data marketplaces. Chapter "AI-based Hybrid Data Platforms" explores the design of an AI-based hybrid data platform.

Part II: Deployment details experience reports and lessons from using the data space deployments within different sectors. Chapters are co-authored with industry experts and cover domains including manufacturing, food, energy, and health. Chapter "A Digital Twin Platform for Industrie 4.0" details a platform for Digital Twins within Industrie 4.0, and chapter "A Framework for Big Data Sovereignty: The European Industrial Data Space" presents a framework for industrial data sovereignty. A food safety data space is explored in chapter "Deploying a Scalable Big Data Platform to Enable a Food Safety Data Space", with Data Spaces and FinTech the focus of chapter "Data Space Best Practices for Data Interoperability in FinTechs". Trusted sharing of sensitive data with a focus on health data is detailed in chapter "TIKD: A Trusted Integrated Knowledge Dataspace For Sensitive Data Sharing and Collaboration". Chapter "Towards an Energy Data Platform Design: Challenges and Perspectives from the SYNERGY Big Data Platform and AI Analytics Marketplace" introduced an AI analytics marketplace for an energy data space.

Part III: Future Directions details research challenges for Data Spaces and data ecosystems. The focus of chapter "Privacy Preserving Techniques for Trustworthy Data Sharing: Opportunities and Challenges for Future Research" is on the opportunities and future research for privacy-preserving techniques for trustworthy data sharing. The book closes with chapter "Common European Data Spaces: Challenges and Opportunities" identifying the challenges and opportunities for common European Data Spaces.

Gibson's quote is still inspiring and provoking debate and discussion. Data Spaces are already here—we hope this book will evenly distribute them.

Galway, Ireland   Edward Curry
May 2022

Acknowledgments

The editors thank Ralf Gerstner and all at Springer for their professionalism and assistance throughout the journey of this book. This book was made possible through funding from the European Union's Horizon 2020 research and innovation program under grant agreement no. 732630 (BDVe). This work was supported by the Science Foundation Ireland, co-funded by the European Regional Development Fund under Grant SFI/12/RC/2289_P2.

We would like to thank the team at BDVA, which guided the community on its journey to value since 2014: Thomas Hahn, Ana Garcia Robles, Laure Le Bars, Milan Petkovic, Nuria De Lama, and Jean-Christophe Pazzaglia. Thanks to Andreas Metzger as the co-Chair of the BDV Technical Committee and Sonja Zillner as the Strategic Research and Innovation Agenda leader. We appreciate the support of the BDVA Office, including Jaakko Karhu, Sinna Rissanen, Mattia Trino, and Martina Barbero. Thanks to Claire Browne for her support in the final stages of book perpetration.

We thank all the authors for sharing their work through our book. A special thanks to all the reviewers who gave their time, effort, and constructive comments that enhanced the overall quality of the chapters. We particularly recognize the dedication and commitments of Juha-Pekka Soininen, Caj Södergård, Juha Kortelainen, Amin Anjomshoaa, Felipe Arruda Pontes, Praneet Dhingra, and Atiya Usmani as reviewers.

Finally, we would like to thank our partners at the European Commission, in particular Commissioner Breton, Commissioner Gabriel, Commissioner Kroes, and the Director-General of DG CONNECT Roberto Viola, who had the vision and conviction to develop the European data economy. We thank the current and past members of the European Commission's Unit for Data Policy and Innovation (Unit G.1) Yvo Volman, Márta Nagy-Rothengass, Kimmo Rossi, Beatrice Covassi, Stefano Bertolo, Francesco Barbato, Wolfgang Treinen, Federico Milani,

Daniele Rizzi, and Malte Beyer-Katzenberger. Together they have represented the public side of the Big Data Value Partnership and were instrumental in its success.

Galway, Ireland   Edward Curry
Walldorf, Germany   Simon Scerri
Oulu, Finland   Tuomo Tuikka
May 2022

Contents

Part III Future Directions

About the Editors and Contributors

About the Editors

Edward Curry is the Established Professor of Data Science and Director of the Insight SFI Research Centre for Data Analytics and the Data Science Institute at the University of Galway. Edward has made substantial contributions to semantic technologies, incremental data management, event processing middleware, software engineering, as well as distributed systems and information systems. He combines strong theoretical results with high-impact practical applications. The excellence and impact of his research have been acknowledged by numerous awards, including best paper awards and the NUIG President's Award for Societal Impact in 2017. His team's technology enables intelligent systems for innovative environments in collaboration with his industrial partners. He is organizer and program co-chair of major international conferences, including CIKM 2020, ECML 2018, IEEE Big Data Congress, and European Big Data Value Forum. Edward is co-founder and elected Vice President of the Big Data Value Association, an industry-led European Big Data community, has built consensus on a joint European Big Data research and innovation agenda, and influenced European data innovation policy to deliver on the agenda.

Simon Scerri has over 15 years of research experience exploring Knowledge Graph methodologies and applications, with the objective of strengthening and diversifying data value chains to realize various data infrastructures. Currently, Simon fulfils the role of a senior technical consultant at metaphacts GmbH. Previously, Simon fronted the Knowledge Graph competency area at the Enterprise Information Systems department in Fraunhofer IAIS (2014–2020). He has served as a re-elected member of the Board of Directors for the Big Data Value Association (2015–2021) where he led the activity group promoting advances in data sharing ecosystems. Simon received his Ph.D. from the National University of Ireland, Galway (2011), where he also contributed to research efforts (2005–2013) at the affiliated Insight Centre for Data Analytics (formerly the Digital Enterprise Research Institute).

Tuomo Tuikka is a Lead of Data Space Solutions at VTT Technical Research Centre of Finland. He is involved in many European and national research and innovation activities especially on Data Spaces, data infrastructures, Big Data, and Artificial Intelligence. In the 1990s, his academic career at the University of Oulu included position as assistant professor mainly focusing on CSCW and HCI research. Starting from 2002, he was responsible on R&D of a product simulation software product line for international market at CCC Group. In 2007, he started at VTT by coordinating very large EUREKA projects on Near-Field Communication (NFC) technology, eventually becoming principal scientist, research team leader, and research manager. He is now a member of the Board of Directors for the Big Data Value Association, while also co-chairing Data Sharing Spaces task force. In Finland, Tuomo is a member of Gaia-X Finland hub steering group and leads the data analytics program at the city of Oulu. He received his Ph.D. from the University of Oulu (2002).

Contributors

Jesús Alonso Asociación de Empresas Tecnológicas Innovalia, Bilbao, Spain

Amin Anjomshoaa Maynooth University, Maynooth, Ireland

Spiros Athanasiou Athena Research Center, Athens, Greece

Gernot Böge FIWARE Foundation e.V, Berlin, Germany

Juan Carlos Pérez Baún ATOS Spain S.A., Madrid, Spain

Alan Barnett OCTO Research Office, Dell Technologies, Ovens, County Cork, Ireland

Pilar Pérez Berganza ATOS Spain S.A., Madrid, Spain

Evmorfia Biliri Suite5 Data Intelligence Solutions, Limassol, Cyprus

Susanna Bonura Engineering Ingegneria Informatica SpA - Piazzale dell' Agricoltura, Rome, Italy

Rob Brennan ADAPT Centre, School of Computing, Dublin City University, Dublin, Ireland

Davide dalle Carbonare Engineering Ingegneria Informatica SpA - Piazzale dell' Agricoltura, Rome, Italy

Aitor Celaya Software Quality Systems (SQS), Oswego, NY, USA

Ruben Costa UNINOVA, Guimarães, Portugal

Edward Curry Insight SFI Research Centre for Data Analytics, University of Galway, Galway, Ireland

Josu Díaz-de-Arcaya TECNALIA, Basque Research & Technology Alliance (BRTA), Bilbao, Spain

Roberto Díaz-Morales Tree Technology - Parque Tecnológico de Asturias, Llanera, Asturias, Spain

Lidia Dutkiewicz KU Leuven Centre for IT & IP Law – imec, Leuven, Belgium

Santiago Cáceres Elvira ITI – Instituto Tecnológico de Informática, Valencia, Spain

Maurizio Ferraris CU Innovation Unit – GFT Italia Srl., Milan, Italy

Paulo Figueiras UNINOVA, Guimarães, Portugal

Silvia Gabrielli Fondazione Bruno Kessler, Trento, Italy

Alexandra Garatzogianni Leibniz University Hannover, Hannover, Germany

Gisela Garcia Volkswagen Autoeuropa (VWAE), Lisbon, Portugal

Guillermo Gil TECNALIA, Basque Research & Technology Alliance (BRTA), Bilbao, Spain

Yury Glikman Fraunhofer FOKUS, Berlin, Germany

Diogo Graça Volkswagen Autoeuropa (VWAE), Lisbon, Portugal

Lukas Helminger Graz University of Technology, Graz, Austria

Know-Center GmbH, Graz, Austria

Julio Hernandez ADAPT Centre, School of Computing, Dublin City University, Dublin, Ireland

Marlène Hildebrand EPFL, Lausanne, Switzerland

Sylvia Ilieva GATE Institute, Sofia University, Sofia, Bulgaria

Dimosthenis Ioannidis Centre of Research and Technology Hellas (CERTH), Thessaloniki, Greece

Manos Karvounis Agroknow, Athens, Greece

Asterios Katsifodimos TU Delft, Delft, Netherlands

Iva Krasteva GATE Institute, Sofia University, Sofia, Bulgaria

Stephan Krenn AIT Austrian Institute of Technology, Vienna, Austria

Oscar Lázaro Asociación de Empresas Tecnológicas Innovalia, Bilbao, Spain

Begoña Laibarra Software Quality Systems (SQS), Oswego, NY, USA

Fenareti Lampathaki Suite5 Data Intelligence Solutions, Limassol, Cyprus

Christoph Lange Fraunhofer Institute for Applied Information Technology FIT, RWTH Aachen University, Aachen, Germany

Chi-Hung Le Insight SFI Research Centre for Data Analytics, Data Science Institute, University of Galway, Galway, Ireland

Stefanie Lindstaedt Graz University of Technology, Graz, Austria

Know-Center GmbH, Graz, Austria

Tomasz Luniewski CAPVIDIA, Houston, TX, USA

Diego Mallada GESTAMP, Pillaipakkam, Tamil Nadu, India

Nikos Manouselis Agroknow, Maroussi, Greece

Lucy McKenna ADAPT Centre, School of Computing, Dublin City University, Dublin, Ireland

Marco Mellia Politecnico di Torino, Torino, Italy

Christoph Mertens International Data Spaces e.V. (IDSA), Berlin, Germany

Raúl Miñón TECNALIA, Basque Research & Technology Alliance (BRTA), Bilbao, Spain

Yuliya Miadzvetskaya KU Leuven Centre for IT & IP Law – imec, Leuven, Belgium

Dimitris Miltiadou UBITECH, Thessalias 8 & Etolias, Chalandri, Greece

Simona Mladenova Agroknow, Maroussi, Greece

Alberto Mozo Universidad Politécnica de Madrid (UPM), Madrid, Spain

Azzam Naeem United Technology Research Centre Ireland, Cork, Ireland

Ángel Navia-Vázquez University Carlos III of Madrid, Madrid, Spain

Alexandros Nizamis Centre of Research and Technology Hellas (CERTH), Thessaloniki, Greece

Hosea Ofe TU Delft, Faculty of Technology, Policy and Management, Delft, The Netherlands

Charaka Palansuriya EPCC, The University of Edinburgh, Edinburgh, UK

Mihalis Papakonstantinou Agroknow, Maroussi, Greece

Antonio Pastor Telefonica I+D, Madrid, Spain

Irena Pavlova GATE Institute, Sofia University, Sofia, Bulgaria

Donato Pellegrino TX - Technology Exploration Oy, Helsinki, Finland

Konstantinos Perakis UBITECH, Thessalias 8 & Etolias, Chalandri, Greece

Florian Pethig Fraunhofer IOSB, IOSB-INA Lemgo, Fraunhofer Institute of Optronics, System Technologies and Image Exploitation, Lemgo, Germany

Dessisslava Petrova-Antonova GATE Institute, Sofia University, Sofia, Bulgaria

Athanasios Poulakidas INTRASOFT, Mumbai, Maharashtra, India

Mark Purcell IBM Research Europe, Dublin, Ireland

Gavin Purtill Banking and Payments Federation of Ireland – BPFI Ireland, Dublin, Ireland

Sebastian Ramacher AIT Austrian Institute of Technology, Vienna, Austria

Magnus Redeker Fraunhofer IOSB, IOSB-INA Lemgo, Fraunhofer Institute of Optronics, System Technologies and Image Exploitation, Lemgo, Germany

Juan Rodriguez Telefonica I+D, Madrid, Spain

Bastian Rössl Fraunhofer IOSB, IOSB-INA Lemgo, Fraunhofer Institute of Optronics, System Technologies and Image Exploitation, Lemgo, Germany

Stephanie Rossello KU Leuven Centre for IT & IP Law – imec, Leuven, Belgium

Vaia Rousopoulou Centre of Research and Technology Hellas (CERTH), Thessaloniki, Greece

Simon Scerri metaphacts, Walldorf, Germany

Martín Serrano Insight SFI Research Centre for Data Analytics, University of Galway, Galway, Ireland

J. Enrique Sierra-García Universidad de Burgos, JRU ASTI-UBU, Burgos, Spain

Konstantinos Sipsas INTRASOFT, Mumbai, Maharashtra, India

Piotr Sobonski United Technology Research Centre Ireland, Cork, Ireland

John Soldatos Athens Information Technology – AIT Athens, Athens, Greece

Wiktor Sowinski-Mydlarz Cyber Security Research Centre, London Metropolitan University, London, UK

GATE Institute, Sofia University, Sofia, Bulgaria

Giannis Stoitsis Agroknow, Maroussi, Greece

Miquel Tarzán Fundació i2CAT, Barcelona, Spain

Ana I. Torre-Bastida TECNALIA, Basque Research & Technology Alliance (BRTA), Bilbao, Spain

Rizkallah Touma Fundació i2CAT, Barcelona, Spain

Andreas Trügler Know-Center GmbH, Graz, Austria

Graz University of Technology, Graz, Austria

Ernesto Troiano CU Innovation Unit – GFT Italia Srl., Milan, Italy

Kostas Tsatsakis Suite5 Data Intelligence Solutions, Limassol, Cyprus

Tuomo Tuikka VTT Technical Research Centre of Finland, Oulu, Finland

Dimitrios Tzovaras Centre of Research and Technology Hellas (CERTH), Thessaloniki, Greece

Stanislav Vakaruk Universidad Politécnica de Madrid (UPM), Madrid, Spain

Irene Lopez de Vallejo DisCO.coop, Bilbao, Spain

Wim Vandevelde katholieke Universiteit Leuven, Leuven, Belgium

Vassil Vassilev Cyber Security Research Centre, London Metropolitan University, London, UK

GATE Institute, Sofia University, Sofia, Bulgaria

Richards Walsh Banking and Payments Federation of Ireland – BPFI Ireland, Dublin, Ireland

Jan Nicolas Weskamp Fraunhofer IOSB, IOSB-INA Lemgo, Fraunhofer Institute of Optronics, System Technologies and Image Exploitation, Lemgo, Germany

Christian Wolff ATB Bremen, Bremen, Germany

Achille Zappa Insight SFI Research Centre for Data Analytics, University of Galway, Galway, Ireland

Wojciech Zietak CAPVIDIA, Houston, TX, USA

Data Spaces: Design, Deployment, and Future Directions

Edward Curry, Simon Scerri, and Tuomo Tuikka

Abstract Digital transformation, data ecosystems, and Data Spaces are inevitable parts of our future. The book aims to educate the reader on data sharing and exchange techniques using Data Spaces. It will address and explore the cutting-edge theory, technologies, methodologies, and best practices for Data Spaces for both industrial and personal data. The book provides the reader with a basis for understanding the scientific foundation of Data Spaces, how they can be designed and deployed, and future directions.

Keywords Data Spaces · Data ecosystem · Big Data value · Data innovation

1 Introduction

Digital transformation creates a data ecosystem with data on every aspect of our world. The rapidly increasing volumes of diverse data from distributed sources create significant opportunities for extracting valuable knowledge. Data ecosystems can create the conditions for a marketplace competition among participants or enable collaboration among diverse, interconnected participants that depend on each other for their mutual benefit. A data space can provide a clear framework to support data sharing within a data ecosystem. For example, industrial Data Spaces can support the trusted and secure sharing and trading of commercial data assets with automated and robust controls on legal compliance and remuneration of data

E. Curry (✉)
Insight SFI Research Centre for Data Analytics, University of Galway, Galway, Ireland
e-mail: edward.curry@insight-centre.org

S. Scerri
metaphacts, Walldorf, Germany

T. Tuikka
VTT Technical Research Centre of Finland, Oulu, Finland
e-mail: tuomo.tuikka@vtt.fi

© The Author(s) 2022
E. Curry et al. (eds.), *Data Spaces*, https://doi.org/10.1007/978-3-030-98636-0_1

1

owners. Personal Data Spaces enforce legislation and allow data subjects and data owners to control their data and its subsequent use.

Many fundamental technical, organizational, legal, and commercial challenges exist in developing and deploying Data Spaces to support data ecosystems. For example, how do we create trusted and secure Data Spaces and privacy-aware analytics methods for secure sharing of personal data and industrial data? How can small- and medium-sized enterprises get access to Data Spaces and technology? How can we support the utility trade-offs between data analysis and privacy? What are user-friendly privacy metrics for end-users? What are the standardization challenges for Data Spaces, including interoperability? How do Data Spaces ensure secure and controlled sharing of proprietary or personal data? What are the necessary technical, organizational, legal, and commercial best practices for data sharing, brokerage, and trading?

The book aims to educate the reader on data sharing and exchange techniques using Data Spaces. The book will address and explore the cutting-edge theory, technologies, methodologies, and best practices for Data Spaces for both industrial and personal data. In addition, the book provides the reader with a basis for understanding the scientific foundation of Data Spaces, how they can be designed and deployed, and future directions.

The chapter is structured as follows: Sect. 2 defines the notion of data ecosystems. Section 3 introduces the concepts of Data Spaces and their role as a platform and their role for sharing industrial and personal data. Section 4 discusses common European Data Spaces and outlines how their foundations have been established by the Big Data Value Public-Private Partnership (PPP) with the data platform projects. Section 5 details the book's structure in the three key areas of design, deployment, and future directions, together with an analysis of the contribution of the chapter's Data Spaces in terms of value, data, technology, organization people, governance, and trust. Finally, Sect. 6 provides a summary.

2 Data Ecosystems

A data ecosystem is a sociotechnical system enabling value to be extracted from data value chains supported by interacting organizations and individuals [1]. Data value chains are oriented to business and societal purposes within an ecosystem. The ecosystem can create the conditions for a marketplace competition among participants or enable collaboration among diverse, interconnected participants that depend on each other for their mutual benefit.

Digital transformation is creating a data ecosystem with data on every aspect of our world, spread across a range of intelligent systems, with structured and unstructured data (e.g., images, video, audio, and text) that can be exploited by data-driven intelligent systems to deliver value.

There is a need to bring together data from multiple sources within the data ecosystem. For example, smart cities show how different systems within the city (e.g., energy and transport) can collaborate to maximize the potential to optimize overall city operations. At the level of an individual, digital services can deliver a personalized and seamless user experience by bringing together relevant user data from multiple systems [2]. This requires a system of systems (SoS) approach to connect systems that cross organizational boundaries, come from various domains (e.g., finance, manufacturing, facilities, IT, water, traffic, and waste), and operate at different levels (e.g., region, district, neighborhood, building, business function, individual).

Data ecosystems present new challenges to the design of data sharing that require a rethink in how we should deal with the needs of large-scale data-rich environments with multiple participants. There is a clear need to support knowledge sharing among participants within data ecosystems. Meeting these challenges is critical to maximizing the potential of data-intensive intelligent systems [3].

3 Data Spaces

The term "dataspace" or "data space" can now be seen as an umbrella term categorizing several closely related concepts. First introduced by Franklin, Halvey, and Maier in 2005 [4] within the data management community, a data space can contain all the data sources for an organization regardless of its format, location, or model. Each data source (e.g., database, CSV, web service) in the data space is known as a participant. The Franklin et al. data space can model the relations (or associations) between data in different participants. In its purest form, a data space is a set of participants and the inter-relations between them [4]. The modeling of the data space can capture different types of relations among participants, from mapping the schemas between two participants to capturing that Participant A is a replica of Participant B.

The data space concept has gained traction with several groups exploring its usefulness for managing data from different domains and regions within a global data ecosystem. These works have provided many definitions for a data space, as captured in Table 1. For example, the Big Data Value Association (BDVA) view of Data Spaces is any ecosystem of data models, datasets, ontologies, data sharing contracts, and specialized management services (i.e., as often provided by data centers, stores, repositories, individually, or within "data lakes"), together with soft competencies around it (i.e., governance, social interactions, business processes) [16]. These competencies follow a data engineering approach to optimize data storage and exchange mechanisms, preserving, generating, and sharing new knowledge.

Table 1 Definitions of a "dataspace" from literature (Adapted from Curry [5])

Definition	Source
"Dataspaces are not a data integration approach; rather, they are more of a data co-existence approach. The goal of dataspace support is to provide base functionality over all data sources, regardless of how integrated they are."	Halevy et al. [6]
"A data space is defined as a decentralised infrastructure for trustworthy data sharing and exchange in data ecosystems based in commonly agreed principles."	Nagel [7]
"A dataspace system manages the large-scale heterogeneous collection of data distributed over various data sources in different formats. It addresses the structured, semi-structured, and unstructured data in coordinated manner without presuming the semantic integration among them."	Singh [8]
"to provide various of the benefits of classical data integration, but with reduced up-front costs, combined with opportunities for incremental refinement, enabling a "pay-as-you-go" approach."	Hedeler et al. [9]
"enable agile data integration with much lower upfront and maintenance costs."	Hedeler et al. [10]
"A dataspace system processes data, with various formats, accessible through many systems with different interfaces, such as relational, sequential, XML, RDF, etc. Unlike data integration over DBMS, a dataspace system does not have full control on its data, and gradually integrates data as necessary."	Wang et al. [11]
"Dataspace Support Platforms envision data integration systems where the amount of upfront effort is much smaller. The system should be able to bootstrap itself and provide some useful services with no human intervention. Over time, through user feedback or as sources are added and the data management needs become clearer, the system evolves in a pay-as-you-go fashion."	Das Sarma et al. [12]
"Dataspace is defined as a set of participants and a set of relationships among them."	Singh and Jain [13]
"Real-time Linked Dataspace combines the pay-as-you-go paradigm of dataspaces with Linked Data, Knowledge Graphs, and real-time stream and event processing capabilities to support the large-scale distributed heterogeneous collection of streams, events, and data sources."	Curry [14], Curry et al. [15]
"any ecosystem of data models, datasets, ontologies, data sharing contracts and specialised management services (i.e., as often provided by data centres, stores, repositories, individually or within 'data lakes'), together with soft competencies around it (i.e., governance, social interactions, business processes)."	Scerri et al. [16]

3.1 Data Spaces: A Platform for Data Sharing

Data-driven Artificial Intelligence is revolutionizing many industries, including transportation and logistics, security, manufacturing, energy, healthcare, and agriculture, by providing intelligence to improve efficiency, quality, and flexibility. Data sharing is a critical enabler for competitive AI solutions. Data for AI is recognized as an innovation ecosystem in the European AI, data, and robotics framework [17]. In addition, data sharing and trading are enablers in the data economy, although closed and personal data present particular challenges for the free flow of data.

Platform approaches have proved successful in many areas of technology [18], from supporting transactions among buyers and sellers in marketplaces (e.g., Amazon), innovation platforms that provide a foundation on top of which to develop complementary products or services (e.g., Windows), to integrated platforms which are a combined transaction and innovation platform (e.g., Android and the Play Store).

The idea of large-scale "data" platforms has been touted as a possible next step to support data ecosystems [3]. An ecosystem data platform would have to support continuous, coordinated data flows, seamlessly moving data among intelligent systems. The design of infrastructure to support data sharing and reuse is still an active area of research [19]. The following two conceptual solutions—Industrial Data Spaces (IDS) and Personal Data Spaces (PDS)—introduce new approaches to addressing this particular need to regulate closed proprietary and personal data.

3.1.1 Industrial Data Spaces (IDS)

IDS has increasingly been touted as potential catalysts for advancing the European data economy as solutions for emerging data markets, focusing on the need to offer secure and trusted data sharing to interested parties, primarily from the private sector (industrial implementations). The IDS conceptual solution is oriented toward proprietary (or closed) data. Its realization should guarantee a trusted, secure environment where participants can safely and legally monetize and exchange their data assets within a clear legal framework. A functional realization of a continent-wide IDS promises to significantly reduce the existing barriers to a free flow of data within an advanced European data economy. Furthermore, the establishment of a trusted data sharing environment will have a substantial impact on the data economy by incentivizing the marketing and sharing of proprietary data assets (currently widely considered by the private sector as out of bounds) through guarantees for fair and safe financial compensations set out in black-and-white legal terms and obligations for both data owners and users. The "opening up" of previously guarded private data can thus vastly increase its value by several orders of magnitude, boosting the data economy and enabling cross-sectoral applications that were previously unattainable or only possible following one-off bilateral agreements between parties over specific data assets.

Notable advances in IDS include the highly relevant white paper and the reference architecture[1] provided by the International Data Spaces Association (IDSA). In addition, the layered databus, introduced by the Industrial Internet Consortium,[2] and the MindSphere Open Industrial Cloud Platform[3] are all examples of the need for data-centric information-sharing technology that enables data market players to exchange data within a virtual and global data space.

The implementation of Data Spaces needs to be approached on a European level, and existing and planned EU-wide, national, and regional platform development activities could contribute to these efforts as recognized by the European data strategy (*Communication: A European Strategy for Data*, 2020).

3.1.2   Personal Data Spaces (PDS)

So far, consumers have trusted, including companies like Google, Amazon, Facebook, Apple, and Microsoft, to aggregate and use their personal data in return for free services. While EU legislation, through directives such as the Data Protection Directive (1995) and the ePrivacy Directive (1998), has ensured that personal data can only be processed lawfully and for legitimate use, the limited user control offered by such companies and their abuse of a lack of transparency have undermined the consumer's trust. In particular, consumers experience everyday leakage of their data, traded by giant aggregators in the marketing networks for value only returned to consumers in the form of often unwanted digital advertisements. This has recently led to a growth in the number of consumers adopting adblockers to protect their digital life. At the same time, they are becoming more conscious of and suspicious about their personal data trail.

In order to address this growing distrust, the concept of personal Data Spaces (PDS) has emerged as a possible solution that could allow data subjects and data owners to remain in control of their data and its subsequent use.[4] PDS leverages "the concept of user-controlled cloud-based technologies for storage and use of personal data." However, consumers have only been able to store and control access to a limited set of personal data, mainly by connecting their social media

---

[1] Reference Architecture Model for the Industrial Data Space, April 2017, https://www.fraunhofer. de/content/dam/zv/de/Forschungsfelder/industrial-data-space/Industrial-Data-Space_Reference-Architecture-Model-2017.pdf

[2] The Industrial Internet of Things, Volume G1: Reference Architecture, January 2017, https:// www.iiconsortium.org/IIC_PUB_G1_V1.80_2017-01-31.pdf

[3] *MindSphere: The cloud-based, open IoT operating system for digital transformation*, Siemens, 2017, https://www.plm.automation.siemens.com/media/global/en/Siemens_MindSphere_ Whitepaper_tcm27-9395.pdf

[4] See a Commission paper on "Personal information management services—current state of service offers and challenges" analyzing feedback from public consultation: https://ec.europa. eu/digital-single-market/en/news/emerging-offer-personal-information-management-services-current-state-service-offers-and

profiles to various emerging Personal Information Management Systems (PIMS). More successful (but limited in number) uses of PDS have involved the support of large organizations in agreeing to their customers accumulating data in their own self-controlled spaces. The expectation here is the reduction of their liability in securing such data and the opportunity to access and combine them with other data that individuals will import and accumulate from other aggregators. However, a degree of friction and the lack of a successful business model are still hindering the potential of the PDS approach.

A recent driver behind such a self-managed personal data economy is the General Data Protection Regulation (GDPR), which constitutes the single pan-European law on data protection, which requires companies dealing with European consumers to (1) increase transparency, (2) provide users with granular control for data access and sharing, and (3) guarantee consumers a set of fundamental individual digital rights (including the right to rectification, erasure, and data portability and to restrict processing). This creates new opportunities for PDS to emerge. Furthermore, the rise of PDS and the creation of more decentralized personal datasets will also open up new opportunities for SMEs that might benefit from and investigate new secondary uses of such data by gaining access to them from user-controlled personal data stores – a privilege so far available only to large data aggregators. However, further debate is required to understand the best business models (for demand and supply) to develop a marketplace for personal data donors and the mechanisms required to demonstrate transparency and distribute rewards to personal data donors. Finally, questions around data portability and interoperability also have to be addressed.

4   Common European Data Spaces

The European strategy for data aims at creating a single market for data that will ensure Europe's global competitiveness and data sovereignty. The strategy aims to ensure:

- Data can flow within the EU and across sectors.
- Availability of high-quality data to create and innovate.
- European rules and values are fully respected.
- Rules for access and use of data are fair, practical, and clear and precise Data Governance mechanisms are in place.

Common European Data Spaces will ensure that more data becomes available in the economy and society while keeping companies and individuals who generate the data in control (*Communication: A European Strategy for Data,* 2020). Furthermore, as illustrated in Fig. 1, common European Data Spaces will be central to enabling AI techniques and supporting the marketplace for cloud and edge-based services.

As the first concrete steps toward implementing common European Data Spaces, a set of research and innovation actions for data platforms have been funded as part

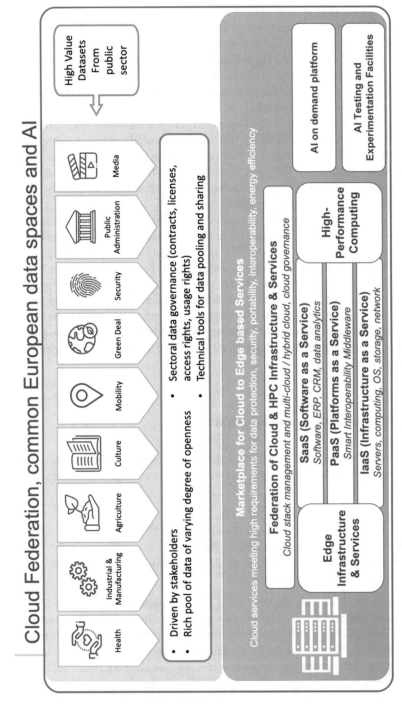

Fig. 1 Overview of cloud federation, common European Data Spaces, and AI (*Communication: A European Strategy for Data*, 2020)

of the Big Data Value PPP. Data platforms refer to architectures and repositories of interoperable hardware/software components, which follow a software engineering approach to enable the creation, transformation, evolution, curation, and exploitation of static and dynamic data in Data Spaces. In the remainder of this section, we describe the Big Data Value PPP, the Big Data Value Association, and the data platform project portfolio of the PPP.

4.1 The Big Data Value PPP (BDV PPP)

The **European contractual Public-Private Partnership on Big Data Value (BDV PPP)** commenced in 2015. It was operationalized with the Leadership in Enabling and Industrial Technologies (LEIT) work program of Horizon 2020. The BDV PPP activities addressed the development of technology and applications, business model discovery, ecosystem validation, skills profiling, regulatory and IPR environments, and many social aspects.

With an initial indicative budget from the European Union of €534M by 2020, the BDV PPP had projects covering a spectrum of data-driven innovations in sectors including advanced manufacturing, transport and logistics, health, and bioeconomy [20]. These projects have advanced state of the art in key enabling technologies for Big Data value and non-technological aspects such as providing solutions, platforms, tools, frameworks, best practices, and invaluable general innovations, setting up firm foundations for a data-driven economy and the future European competitiveness in data and AI [21].

4.2 Big Data Value Association

The Big Data Value Association (BDVA) is an industry-driven international not-for-profit organization that grew over the years to over 220 members all over Europe, with a well-balanced composition of large-, small-, and medium-sized industries as well as research and user organizations. BDVA has over 25 working groups organized in Task Forces and subgroups, tackling all the technical and nontechnical challenges of Big Data value.

BDVA served as the private counterpart to the European Commission to implement the Big Data Value PPP program. BDVA and the Big Data Value PPP pursued a common shared vision of positioning Europe as the world leader in creating Big Data value. BDVA is also a private member of the EuroHPC Joint Undertaking and one of the leading promoters and driving forces of the AI, Data, and Robotics Partnership planned for the next framework program MFF 2021–2027.

The mission of the BDVA was "to develop the Innovation Ecosystem that will enable the data-driven digital transformation in Europe delivering maximum economic and societal benefit, and, to achieve and to sustain Europe's leadership on Big Data Value creation and Artificial Intelligence." BDVA enabled existing regional multi-partner cooperation to collaborate at the European level by providing tools and knowledge to support the co-creation, development, and experimentation of pan-European data-driven applications and services and knowledge exchange. The BDVA developed a joint Strategic Research and Innovation Agenda (SRIA) on Big Data Value [22]. Initially, it was fed by a collection of technical papers and roadmaps [23] and extended with a public consultation that included hundreds of additional stakeholders representing the supply and demand sides. The BDV SRIA defined the overall goals, main technical and non-technical priorities, and a research and innovation roadmap for the BDV PPP. In addition, the SRIA set out the strategic importance of Big Data; described the data value chain and the central role of ecosystems; detailed a vision for Big Data value in Europe in 2020; analyzed the associated strengths, weaknesses, opportunities, and threats; and set out the objectives and goals to be accomplished by the BDV PPP within the European research and innovation landscape of Horizon 2020 and at national and regional levels.

4.3   Data Platform Project Portfolio

The data platform projects running under the Big Data Value PPP umbrella develop integrated technology solutions for data collection, sharing, integration, and exploitation to create such a European data market and economy [22]. The Big Data Value PPP portfolio covers the data platform projects shown in Table 2. This table gives an overview of these projects, the type of data platform they develop and the domain, respectively, and the use cases they address. These projects are briefly summarized below based on open data from https://cordis.europa.eu/.

Table 2 Portfolio of the Big Data Value PPP covering data platforms

Chapter	Project name	Type	Domains/use cases
	BD4NRG: Big Data for Next Generation Energy	Industrial	Energy
	BD4OPEM: Big Data for OPen innovation Energy Marketplace	Industrial and personal	Energy
Ch 3	DataPorts: A Data Platform for the Cognitive Ports of the Future	Industrial	Transport and logistics
Ch 3	DataVaults: Persistent Personal Data Vaults Empowering a Secure and Privacy-Preserving Data Storage, Analysis, Sharing and Monetisation Platform	Personal	Sports Mobility Healthcare Smart home Tourism
Ch 3, Ch 7	i3-Market: Intelligent, Interoperable, Integrative and deployable open-source marketplace with trusted and secure software tools for incentivising the industry data economy	Industrial	Automotive Manufacturing Healthcare
Ch 3, Ch 6	KRAKEN: Brokerage and market platform for personal data	Personal	Education Health
	MOSAICROWN: Multi-Owner data Sharing for Analytics and Integration respecting Confidentiality and Owner control	Personal	Connected vehicles, finance, marketing
Ch 5	MUSKETEER: Machine learning to augment shared knowledge in federated privacy-preserving scenarios	Industrial and personal	Smart manufacturing Healthcare
Ch 3	OpertusMundi: A Single Digital Market for Industrial Geospatial Data Assets	Industrial	Geospatial
Ch 3	PIMCITY: Building the next generation personal data platforms	Personal	Generic
	PLATOON: Digital PLAtform and analytic TOOls for eNergy	Industrial and personal	Energy
	Safe-DEED: Safe Data-Enabled Economic Development	Industrial and personal	Marketing Manufacturing
Ch 3	SmashHit: Smart dispatcher for secure and controlled sharing of distributed personal and industrial data	Industrial and personal	Connected cars Smart cities
Ch 12	SYNERGY: Big Energy Data Value Creation within SYNergetic enERGY-as-a-service Applications through trusted multi-party data sharing over an AI Big Data analytics marketplace.	Industrial	Energy
Ch 3, Ch 10	TheFSM: The Food Safety Market: an SME-powered industrial data platform to boost the competitiveness of European food certification	Industrial	Food supply chain
Ch 3	TRUSTS: Trusted Secure Data Sharing Space	Industrial and personal	Finance Telecom

5   Book Overview

This book captures the early lessons and experience in creating Data Spaces. The book arranges these contributions into three parts (see Fig. 2) covering Part I) design, Part II) deployment, and Part III) future directions.

- The first part of the book explores the design space of Data Spaces. Then, the chapters detail organizational design for Data Spaces, data platforms, Data Governance federated learning, personal data sharing, data marketplaces, and hybrid AI for Data Spaces.
- The second part of the book explores the use of Data Spaces within real-world deployments. The chapters include case studies of Data Spaces in sectors including Industry 4.0, food safety, FinTech, health care, and energy.
- The third and final part of the book details future directions for Data Spaces, including challenges and opportunities for common European Data Spaces and privacy-preserving techniques for trustworthy data sharing.

5.1   Chapter Analysis

As depicted in Fig. 3, the success of widespread data sharing activities revolves around the central key concept of trust: in the validity of the data itself and the algorithms operating on it, in the entities governing the data space, in its enabling technologies, as well as in and among its wide variety of users (organizations and private individuals as data producers, consumers, or intermediaries). To achieve the required levels of trust, each of the following five pillars must meet some of the necessary conditions:

- Organizations—More organizations (including business, research, and governmental) need to rethink their strategy to fully embrace a data culture that places data at the center of their value proposition, exploring new data-driven business models and exploiting new data value flows.
- Data—As a touted fifth European fundamental freedom, free movement of data relies on organizational data strategies that embed methodologies for data sharing by-design (e.g., interoperability) and clear standard guidelines that help determine the market value of data assets.
- Technology—Safer experimentation environments are needed to catalyze the maturation of relevant technology behind trustworthy data, data access, and algorithms (privacy, interoperability, security, and quality). In addition, standardization activities need to adjust for faster reaction times to emerging standards and the identification of new ones.
- People—Data sharing needs to guarantee individual privacy and offer fair value or compensation of shared personal data. For Europe to drive data sharing

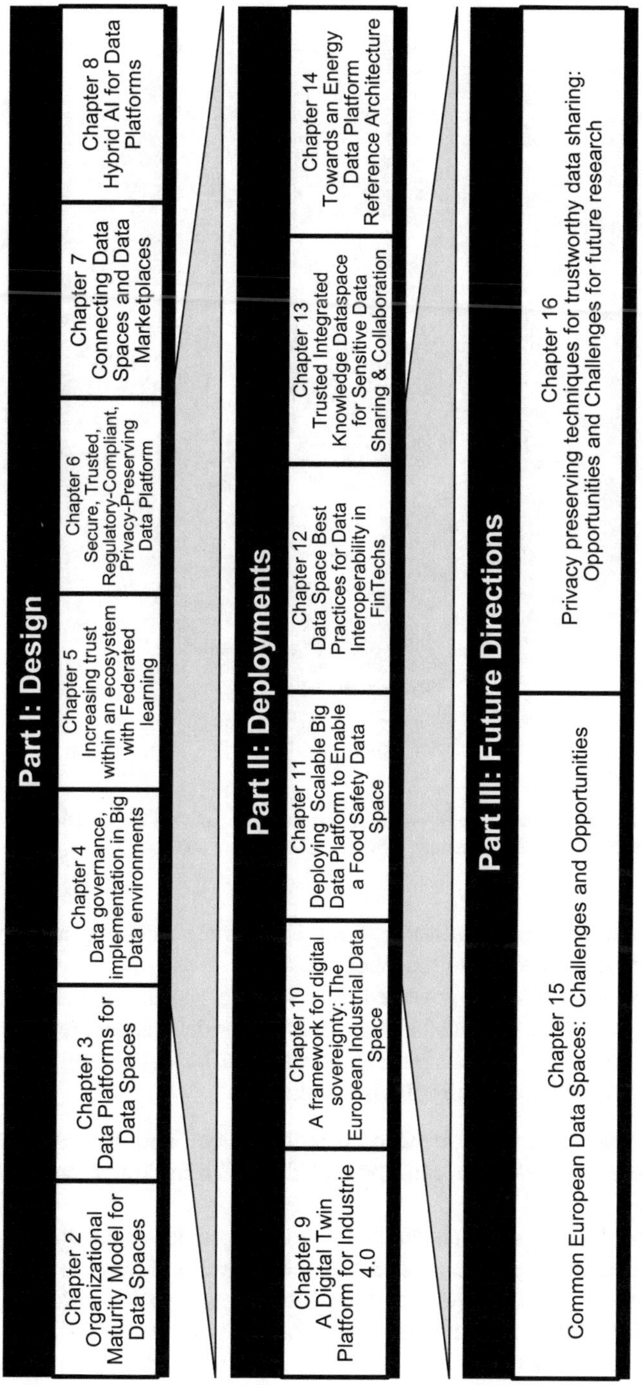

Part I: Design

| Chapter 2 Organizational Maturity Model for Data Spaces | Chapter 3 Data Platforms for Data Spaces | Chapter 4 Data governance, implementation in Big Data environments | Chapter 5 Increasing trust within an ecosystem with Federated learning | Chapter 6 Secure, Trusted, Regulatory-Compliant, Privacy-Preserving Data Platform | Chapter 7 Connecting Data Spaces and Data Marketplaces | Chapter 8 Hybrid AI for Data Platforms |

Part II: Deployments

| Chapter 9 A Digital Twin Platform for Industrie 4.0 | Chapter 10 A framework for digital sovereignty: The European Industrial Data Space | Chapter 11 Deploying Scalable Big Data Platform to Enable a Food Safety Data Space | Chapter 12 Data Space Best Practices for Data Interoperability in FinTechs | Chapter 13 Trusted Integrated Knowledge Dataspace for Sensitive Data Sharing & Collaboration | Chapter 14 Towards an Energy Data Platform Reference Architecture |

Part III: Future Directions

| Chapter 15 Common European Data Spaces: Challenges and Opportunities | Chapter 16 Privacy preserving techniques for trustworthy data sharing: Opportunities and Challenges for future research |

Fig. 2 Structure of the book

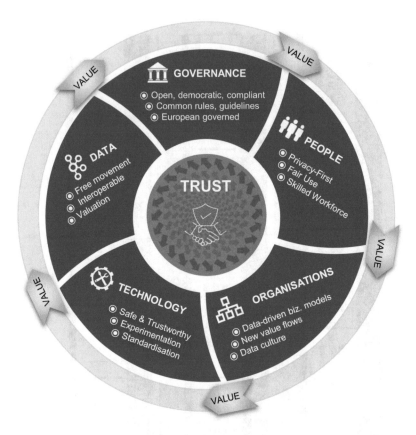

Fig. 3 The data sharing value "wheel"—core pillars and principles of the envisioned European-governed data sharing space that generate value for all sectors of society

activities, the European workforce needs appropriate reskilling and upskilling to meet the evolving needs of the labor market.

- Governance—A European-governed data sharing space can inspire trust by adhering to the more advanced European rules, guidelines, and regulations and promoting European values. Participation should be equally open to all and subject to transparent and fair rules of conduct.

Table 3 gives an overview to which extent the contributions described in the different chapters of this book contribute to the different dimensions of the data sharing wheel.

As this table indicates, the chapters in this book provide broad coverage of the pillars of the data sharing wheel, reinforcing the relevance of these concerns.

The majority of the chapters cover the value, data, and technology pillars of the wheel which illustrate the data-driven focus of the works in Data Spaces. Governance and trust are also well covered, highlighting the importance of these pillars to the deployment and operation of Data Spaces. While organization aspects

Table 3 Coverage of the pillars of the data sharing wheel by the book chapters

Chapter	Value	Data	Technology	Organization	People	Governance	Trust
Part I: Design							
Ch 2	X	X	X	X	X	X	X
Ch 3		X	X	X	X	X	X
Ch 4	X	X	X	X		X	
Ch 5	X	X	X			X	X
Ch 6		X	X	X		X	X
Ch 7	X	X	X				X
Ch 8	X	X	X	X		X	
Part II: Deployment							
Ch 9	X	X	X	X			
Ch 10	X	X		X		X	X
Ch 11	X	X	X				X
Ch 12	X	X	X				
Ch 13		X	X			X	X
Ch 14	X	X	X	X			X
Part III: Future directions							
Ch 15		X	X	X		X	X
Ch 16	X	X	X	X	X	X	

are well covered with an understanding of how organizations can leverage the benefits of Data Spaces to transform their business models and operations, there is a paucity of work in the area of the people pillar. Skills, reskilling, and upskilling to meet the emerging needs of Data Spaces and society.

6 Summary

We are now seeing digital transformation toward global digital markets and Data Spaces. However, this will not be a fast transition and may take a decade before we understand the methods and the means of mature Data Spaces. In comparison, the World Wide Web as a mature platform for trade took from the mid-1990s to well beyond 2000 before it became an everyday tool to search for information and order weekly groceries.

As this development is systemic, it requires scientific, technical, and social foundations. This book addresses and crystallizes the developments of many efforts to establish Data Spaces and learnings from the efforts. Data Spaces are feature-rich technical constructs within a social and regulation framework that support data ecosystems with fair and trusted approaches to share data. It is an ambitious goal, and therefore data ecosystems present new challenges to the design of data sharing. We need to rethink how we should deal with the needs of these large-scale data-rich environments with multiple participants. This chapter gave the foundations of the concepts, but obviously, many challenges exist. A data space can provide a clear framework to support data sharing within a data ecosystem. This book is a step toward such a framework by delineating real experiences from pilots and experiments. The book addresses and explores the cutting-edge theory, technologies, methodologies, and best practices for Data Spaces for both industrial and personal data. It provides the reader with a basis for understanding the scientific foundation of Data Spaces, how they can be designed and deployed, and future directions.

The development of data space technology is societal. For example, the development of electricity networks required agreements on a common approach for the electricity grid. In the same manner, common agreements are needed from large and small industry, policymakers, educators, researchers, and society at large to create the basis of the data economy and common European Data Spaces. The BDV PPP has advanced the value of Big Data and AI, laying the basis for new combinations of technologies that will go beyond the digital market toward a new and productive Digital Society.

References

1. Curry, E. (2016). The big data value chain: Definitions, concepts, and theoretical approaches. In J. M. Cavanillas, E. Curry, & W. Wahlster (Eds.), *New horizons for a data-driven economy* (pp. 29–37). Springer International Publishing. https://doi.org/10.1007/978-3-319-21569-3_3

2. Curry, E., Hasan, S., Kouroupetroglou, C., Fabritius, W., ul Hassan, U., & Derguech, W. (2018). Internet of Things enhanced user experience for smart water and energy management. *IEEE Internet Computing, 22*(1), 18–28. https://doi.org/10.1109/MIC.2018.011581514

3. Curry, E., & Sheth, A. (2018). Next-generation smart environments: from system of systems to data ecosystems. *IEEE Intelligent Systems, 33*(3), 69–76. https://doi.org/10.1109/MIS.2018.033001418

4. Franklin, M., Halevy, A., & Maier, D. (2005). From databases to dataspaces: a new abstraction for information management. *ACM SIGMOD Record, 34*(4), 27–33. https://doi.org/10.1145/1107499.1107502

5. Curry, E. (2020). Fundamentals of real-time linked dataspaces. In *Real-time linked dataspaces* (pp. 63–80). Springer International Publishing. https://doi.org/10.1007/978-3-030-29665-0_4

6. Halevy, A., Franklin, M., & Maier, D. (2006). Principles of dataspace systems. In *25th ACM SIGMOD-SIGACT-SIGART symposium on Principles of database systems - PODS '06* (pp. 1–9). ACM Press. https://doi.org/10.1145/1142351.1142352

7. Nagel, L. (2021). *Design principles for data spaces.*

8. Singh, M. (2013). A framework for data modeling and querying dataspace systems. In *Seventh International Conference on Data Mining and Warehousing (ICDMW).*

9. Hedeler, C., Belhajjame, K., Paton, N. W., Campi, A., Fernandes, A. A. A., & Embury, S. M. (2010). Dataspaces. In S. Ceri & M. Brambilla (Eds.), *Search computing* (pp. 114–134). Springer. https://doi.org/10.1007/978-3-642-12310-8_7

10. Hedeler, C., Belhajjame, K., Paton, N. W., Fernandes, A. A. A., Embury, S. M., Mao, L., & Guo, C. (2011). Pay-as-you-go mapping selection in dataspaces. In *Proceedings of the 2011 international conference on Management of data - SIGMOD '11* (pp. 1279–1282). ACM Press. doi:https://doi.org/10.1145/1989323.1989476.

11. Wang, Y., Song, S., & Chen, L. (2016). A survey on accessing dataspaces. *ACM SIGMOD Record, 45*(2), 33–44. https://doi.org/10.1145/3003665.3003672

12. Das Sarma, A., Dong, X., & Halevy, A. Y. (2009). Data modelling in dataspace support platforms. In *Lecture Notes in Computer Science (including subseries Lecture Notes in Artificial Intelligence and Lecture Notes in Bioinformatics)* (Vol. 5600 LNCS, pp. 122–138). doi:https://doi.org/10.1007/978-3-642-02463-4_8.
13. Singh, M., & Jain, S. K. (2011). A survey on dataspace. In *Communications in Computer and Information Science* (pp. 608–621). doi:https://doi.org/10.1007/978-3-642-22540-6_59.
14. Curry, E. (2020). *Real-time linked dataspaces*. Springer International Publishing. https://doi.org/10.1007/978-3-030-29665-0
15. Curry, E., Derguech, W., Hasan, S., Kouroupetroglou, C., & ul Hassan, U. (2019). A real-time linked dataspace for the Internet of Things: Enabling "pay-as-you-go" data management in smart environments. *Future Generation Computer Systems, 90*, 405–422. https://doi.org/10.1016/j.future.2018.07.019
16. Scerri, S., Tuikka, T., & Lopez de Vallejoan, I. (Eds.). (2020). Towards a European data sharing space.
17. Zillner, S., Bisset, D., Milano, M., Curry, E., Hahn, T., Lafrenz, R., et al. (2020). *Strategic research, innovation and deployment agenda - AI, data and robotics partnership. Third Release* (Third). Brussels: BDVA, euRobotics, ELLIS, EurAI and CLAIRE.
18. Gawer, A., & Cusumano, M. A. (2014). Industry platforms and ecosystem innovation. *Journal of Product Innovation Management, 31*(3), 417–433. https://doi.org/10.1111/jpim.12105
19. Curry, E., & Ojo, A. (2020). Enabling knowledge flows in an intelligent systems data ecosystem. In *Real-time linked dataspaces* (pp. 15–43). : Springer International Publishing. doi:https://doi.org/10.1007/978-3-030-29665-0_2.
20. Curry, E., Metzger, A., Zillner, S., Pazzaglia, J.-C., & García Robles, A. (2021). *The elements of big data value*. Springer International Publishing. https://doi.org/10.1007/978-3-030-68176-0
21. Zillner, S., Gomez, J. A., García Robles, A., Hahn, T., Le Bars, L., Petkovic, M., & Curry, E. (2021). Data economy 2.0: From big data value to AI value and a european data space. In E. Curry, A. Metzger, S. Zillner, J.-C. Pazzaglia, & A. García Robles (Eds.), *The elements of big data value* (pp. 379–399). Springer International Publishing. https://doi.org/10.1007/978-3-030-68176-0_16
22. Zillner, S., Curry, E., Metzger, A., Auer, S., & Seidl, R. (Eds.). (2017). *European big data value strategic research & innovation agenda*. Big Data Value Association. Retrieved from http://www.edwardcurry.org/publications/BDVA_SRIA_v4_Ed1.1.pdf
23. Cavanillas, J. M., Curry, E., & Wahlster, W. (2016). *New horizons for a data-driven economy: A roadmap for usage and exploitation of big data in Europe*. Springer Nature PP. https://doi.org/10.1007/978-3-319-21569-3

Part I
Design

An Organizational Maturity Model for Data Spaces: A Data Sharing Wheel Approach

Edward Curry and Tuomo Tuikka

Abstract This chapter presents a maturity model for Data Spaces, which provides a management system with associated improvement roadmaps that guide strategies to continuously improve, develop, and manage the data space capability within their organization. It highlights the challenges with data sharing and motivates the benefit of maturity models. This chapter describes the Maturity Model for Data Spaces (MM4DS) and its use to determine an organization's data space capability maturity. The MM4DS takes an organization's user-centric/demand-side perspective utilizing a data space. The development process for the MM4DS is discussed, along with the role of design science in the model development process. Finally, the chapter details an illustrative case using the model to benchmark data space capabilities in five fictitious organizations. The MM4DS can be applied within organizations to better manage their data space capabilities, with assessment, providing insights into what they are doing well and where they need to improve.

Keywords Data space · Maturity model · Data ecosystem · Big Data value · Data innovation

1 Introduction

To leverage the benefits of data sharing, many organizations are now looking at developing data space capabilities to create new value and business opportunities. A data space capability goes beyond technology to encompass other factors such as alignment with organization strategy, project planning, developing expertise,

E. Curry (✉)
Insight SFI Research Centre for Data Analytics, University of Galway, Galway, Ireland
e-mail: edward.curry@insight-centre.org

T. Tuikka
VTT Technical Research Centre of Finland, Oulu, Finland
e-mail: tuomo.tuikka@vtt.fi

© The Author(s) 2022
E. Curry et al. (eds.), *Data Spaces*, https://doi.org/10.1007/978-3-030-98636-0_2

culture, and governance. Unfortunately, because the field is new and evolving, few guidelines and best practices are available, resulting in many organizations not fully exploiting data sharing potential. As a result, organizations face many challenges in developing and driving their overall data strategies and programs. The point of departure for this work is the call for the community to engage substantively with the topic of Data Spaces [1]. The chapter contributes to theory by discussing organizational capabilities for Data Spaces. We have developed a model for systematically assessing and improving data space capabilities. We have used an open-innovation collaboration model, engaging academia and industry in creating the Maturity Model for Data Spaces (MM4DS), and especially when developing BDVA Data Sharing Value Wheel which is used as a conceptual basis for MM4DS. The core of this maturity model for Data Spaces provides a management system with associated improvement roadmaps that guide strategies to continuously improve, develop, and manage the data space capability. The maturity model can be applied within an organization to better manage its data space capabilities. The assessment provides insights into what they are doing well and where they need to improve.

The chapter highlights the opportunities of data ecosystems and Data Spaces and motivates the need for maturity models to develop and manage organizational capabilities. First, the chapter describes the MM4DS and its use to determine the maturity of data space capability. Next, the development process for the MM4DS is discussed, detailing the role of design science and the model development process. Finally, the chapter details an illustrative use of the model to benchmark organizations.

2   Background and Context

The European data strategy identifies data as an essential resource for economic growth, competitiveness, innovation, job creation, and societal progress. IDC forecasts worldwide investments in Big Data and analytics to reach 294 B€ by 2025, of which 16%, corresponding to 47 B€, was generated in the EU27. A key enabler for AI and data-driven business opportunities is the growth in data, with more than 175 zettabytes of data available by 2025. In parallel, we are witnessing a shift of data to the edge and cloud environments. While, in 2020, 80% of processing and analysis takes place within data centers, the transition is onto more data being processed at the edge of the network in smart connected devices and machines. IDC predicts that 46% of the world's stored data in 2025 will be in the public cloud. This creates new opportunities for Europe to lead edge data processing and maintain control of their data [2]. As EU Commissioner Thierry Breton stated, "the goal is to prepare ourselves so the data produced by Europeans will be used for Europeans, and with our European values."

2.1 Data Ecosystems

A data ecosystem is a socio-technical system that enables value to be extracted from data value chains that interact with organizations and individuals. Data value chains can be oriented to business and societal purposes within an ecosystem. The ecosystem can create a marketplace competition between participants or enable collaboration among diverse, interconnected participants who depend on each other for mutual benefit. Data ecosystems can be formed in different ways around an organization, community technology platforms, or within or across sectors [3]. A well-functioning working data ecosystem must bring together the key stakeholders with a clear benefit for all. The key actors in a data ecosystem include data suppliers and consumers, technology and infrastructure providers, data end-users, marketplaces, regulators, and standardization bodies.

There is a need to bring together data from multiple participants within a data ecosystem [4]. For example, smart cities show how different systems within the city (e.g., energy and transport) can collaborate to maximize the potential to optimize overall city operations. At the level of an individual, digital services can deliver a personalized and seamless user experience by bringing together relevant user data from multiple systems [5] that cross organizational boundaries, come from various domains (e.g., finance, manufacturing, facilities, IT, water, traffic, and waste), and operate at different levels (e.g., region, district, neighborhood, building, business function, individual).

Data ecosystems present new challenges to data sharing. How can we support data sharing within a data ecosystem? What are the technical and nontechnical barriers to data sharing within the ecosystem [4]?

2.2 Data Value Chains and Data-Driven AI

Data enables AI innovation, and AI makes data actionable. Data flows link the emerging value chains improved or disrupted by new AI services and tools, where new skills, business models, and infrastructures are needed [3]. The Data Governance models and issues such as data access, data sovereignty, and data protection are essential factors in developing sustainable AI- and data-driven value chains respecting all stakeholder interests, particularly SMEs. The latter is currently lagging in AI adoption. AI and data innovation can generate value not only for business but also for society and individuals. There is increasing potential to use AI and data for social good by contributing solutions to the UN Social Development Goals (SDGs) and the goals of the EU New Green Deal. Enterprises are developing sustainability programs in the context of their corporate social responsibility strategies, leveraging data and AI to reduce their ecological footprint, cutting costs, and contributing to social welfare at the same time. Public authorities are also looking into unlocking private data for general purposes. Business and social value can be pursued simultaneously, encouraging the reuse and sharing of data collected and processed for AI and data innovation (sharing private data for the public good, B2G, and not only B2B). Expertise is needed to increase awareness about the potential value for society and people and the business of data-driven innovation combined with AI [6].

2.3 High-Level Europe Opportunity and Challenges

For the European data economy to develop further and meet expectations, large volumes of cross-sectoral, unbiased, high-quality, and trustworthy data must be made available [7]. There are, however, significant business, organizational, and legal constraints that can block this scenario, such as the lack of motivation to share data due to ownership concerns, loss of control, lack of trust, the lack of foresight in not understanding the value of data or its sharing potential, the lack of data valuation standards in marketplaces, the legal blocks to the free flow of data, and the uncertainty around data policies [8]. Therefore, the exploration of ethical, secure, and trustworthy legal, regulatory, and governance frameworks is needed. European values, e.g., democracy, privacy safeguards, and equal opportunities, can become the trademark of European data economy technologies, products, and practices. Rather than be seen as restrictive, legislation enforcing these values should be considered a unique competitive advantage in the global data marketplace.

3 Data Spaces and Organizational Capabilities

Data Spaces, platforms, and marketplaces are enablers, the key to unleashing the potential of data. Significant technical challenges such as interoperability, data verification and provenance support, quality and accuracy, decentralized data sharing and processing architectures, maturity, and uptake of privacy-preserving technologies for Big Data directly impact the data available for sharing [1]. Aligning and integrating established data sharing technologies and solutions and further developments in architectures and governance models to unlock data silos would enable data analytics across a European data sharing ecosystem. This will allow AI-enhanced digital services to make analyses and predictions on European-wide data, thereby combining data and service economies. New business models will help exploit the value of those data assets by implementing AI among participating stakeholders, including industry, local, national, and European authorities and institutions, research entities, and even private individuals. The European data strategy sets out a vision for the EU to become a role model for a data-driven society and create a single data market to ensure Europe's global competitiveness and data sovereignty. As highlighted by Breton, "to be ahead of the curve, we need to develop suitable European infrastructures allowing the storage, the use, and the creation of data-based applications or Artificial Intelligence services. I consider this as a major issue of Europe's digital sovereignty."

3.1   BDVA Data Sharing Value Wheel

The Big Data Value Association has used an open-innovation model of collaboration, engaging academia and industry in creating the Data Sharing Value Wheel. In the Wheel, as depicted in Fig. 1 and introduced in Scerri et al. [1], the success of widespread data sharing activities revolves around the central key concept of trust: in the validity of the data itself and the algorithms operating on it, in the entities governing the data space; in its enabling technologies, as well as in and among its wide variety of users (organizations and private individuals as data producers, consumers, or intermediaries). To achieve the required levels of trust, each of the following five pillars must meet some of the necessary conditions:

- Organizations—More organizations (including business, research, and governmental) need to rethink their strategy to fully embrace a data culture that places

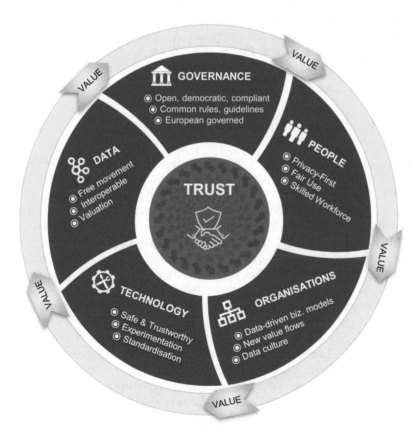

Fig. 1 The Data Sharing Value "Wheel"—core pillars and principles of the envisioned European-governed data sharing space that generate value for all sectors of society [1]

data at the center of their value proposition, exploring new data-driven business models and exploiting new data value flows.

- Data—As a touted 5th European fundamental freedom, free movement of data relies on organizational data strategies that embed methodologies for data sharing by-design (e.g., interoperability) and clear standard guidelines that help determine the market value of data assets.
- Technology—Safer experimentation environments are needed to catalyze the maturation of relevant technology behind trustworthy data, data access, and algorithms (privacy, interoperability, security, and quality). In addition, standardization activities need to adjust for faster reaction times to emerging standards and the identification of new ones.
- People—Data sharing needs to guarantee individual privacy and offer fair value or compensation of shared personal data. For Europe to drive data sharing activities, the European workforce needs appropriate reskilling and upskilling to meet the evolving needs of the labor market.
- Governance—A European-governed data sharing space can inspire trust by adhering to the more advanced European rules, guidelines, and regulations and promoting European values. Participation should be equally open to all and subject to transparent and fair rules of conduct.

3.2   Organizational Capabilities

The resource-based view (RBV) is one of the significant firm-theoretical perspectives with solid tradition within the business research community [9]. Within the RBV, an organization is conceptualized as a collection of resources, where a resource is "anything which could be thought of as a strength or weakness of a given firm" [10]. According to Wade and Hulland [9], resources comprise (a) capabilities and (b) assets. The term capability refers to the ability of an organization to perform a coordinated set of tasks to achieve a particular result [11]. Assets are defined as anything tangible or intangible that can be used in the firm's processes [9]. Capabilities can be viewed as repeatable patterns of actions [9] or coordinated set of tasks [11] that utilize the firm's assets as input [11]. IT capabilities enable the firm to acquire, deploy, combine, and reconfigure IT resources to support and enhance business strategies and processes [12]. Bharadwaj [13] describes IT capabilities as the "firm's ability to mobilise and deploy IT-based resources in combination or co-present with other resources and capabilities."

Teece et al. [14] differentiate between different types of capabilities which exist in the firm. Operational capabilities are the firm's ability "to perform the basic functional activities of the firm, such as plant layout, distribution logistics, and marketing campaigns, more efficiently than competitors" [15]. These capabilities are targeted toward the operational functioning of the firm [16]. On the other hand, dynamic capabilities are "the firm's ability to integrate, build, and reconfigure internal and external competencies to address rapidly changing environments" [14].

Dynamic capabilities do not directly affect the firm's output but indirectly contribute to the firm's output through an impact on operational capabilities [11]. In turbulent settings, IT dynamic capabilities become even more critical. These processes and routines facilitate learning and transform firm asset/resource positions [17].

The research reported here aims to explore the capabilities needed for Data Spaces within organizations. The study aims to identify the critical foundations needed within the organization that permit firms to build the capabilities that can deliver value from Data Spaces. Focusing on foundations enables researchers to build a detailed conceptual foundation for data space capability and devise strategies for implementation by a firm's management.

3.3 Maturity Models

Maturity models are conceptual models that outline anticipated, typical, logical, and desired evolution paths toward maturity [18], where maturity is a measure to evaluate the capabilities of an organization concerning a particular discipline [18]. Maturity models are tools that have been used to improve many capabilities within organizations, from business process management (BPM) [18] and project management [19] to software engineering [20]. In addition, several maturity frameworks have recently been developed related to information technology (IT) management and IT/business alignment [21].

Maturity models contain two aspects, one capturing the assessment of the current status and another one guiding organizations toward higher maturity levels. They can have multiple uses within an organization, from helping them find a place to start, providing a foundation to build a common language and shared vision, to helping organizations prioritize actions and define roadmaps [22]. If a community of organizations defines the model, it can capture the collective knowledge of the community's prior experiences. A maturity model could also be used as an assessment tool and benchmark for comparative assessments of the capabilities of different organizations. Furthermore, the model can help transform organizations toward higher maturity levels by suggesting how these capabilities are developed.

4 A Maturity Model for Data Spaces

This chapter presents the Maturity Model for Data Spaces (MM4DS), which provides a management system with associated improvement roadmaps and strategies to continuously improve, develop, and manage the data space capability within an organization. The MM4DS takes an organization's user-centric/demand-side perspective utilizing a data space to gain business value. The MM4DS has been designed following the high-level dimensions of the BDVA Data Sharing Wheel and is used to determine an organization's data space capability maturity. The MM4DS offers a comprehensive value-based model for organizing, evaluating, planning, and managing data space capabilities.

The initial model was developed by a subgroup of Data Space Task Force of the Big Data Value Association (BDVA), which is comprised of university-based academic researchers and industry-based practitioner-researchers drawn from over 200 organizations across Europe using "engaged scholarship" [23] and "open-innovation" principles [24]. The initial version of the model presented in this chapter will be developed further by the task force to refine it and validate it within real-world Data Spaces.

The section details the design methodology, describes its capabilities, associated maturity curves, and outlines the assessment approach for the MM4DS.

4.1 Model Design Methodology

The design science paradigm seeks to extend the boundaries of human and organizational capabilities by creating new and innovative artifacts, including constructs, models, methods, and instantiations [25]. Maturity models in design-oriented research are located between models and methods in the form of state descriptions (e.g., the maturity levels) and guidelines [26]. In order to transform organizations from one maturity level to another, the method component is usually described by "maturity curves" or "maturity profiles." Thus, a maturity model represents both model elements in the form of assessments and method components in the form of improvement guidelines. In this regard, "method engineering" is central to our approach and can be seen as elements of design science-oriented information systems research [25, 27].

The MM4DS follows design science principles within a rigorous design process that facilitates scholars' engagement and ensures consistency by providing a meta-model for structuring the maturity model. The design science approach used in the MM4DS is closely aligned with the three design science research cycles (relevance cycle, rigor cycle, and design cycle) proposed by Hevner [28]. A group was established to develop the model, including a mix of subject matter experts (SMEs) and key opinion leaders (KOLs), including academic researchers and industry-based practitioners. The objective was to capture the collective learnings and experiences of the group within a maturity model for data.

4.2 Capabilities

The MM4DS model consists of 15 capabilities (see Table 1) across the following 7 pillars of the data sharing wheel.

Table 1 Organizational capabilities for Data Spaces

Pillars	Capability	Description
(O) Organization *Definition and execution of data space strategy to influence and align with the organization's business goals*	(O1) Strategy and planning	*Definition and agreement of the strategy and scope of objectives for the data space initiative*
	(O2) Business alignment	*Influencing and aligning with the organization's business goals*
	(O3) Performance monitoring	*Monitoring progress against specific data space objectives within the organization and the ecosystem*
(V) Value *Sensing and capture of business value opportunities*	(V1) Sensing	*Value sensing for the business strategy via constant monitoring of data space business opportunities*
	(V2) Capture	*Value capture via constant improvement of core business activities and new business opportunities*
(D) Data *Facilitating data sharing, management, and stewardship in the organization*	(D1) Life cycle	*Provision of data sharing in the data product and services' data management life cycle*
	(D2) Management and stewardship	*Processes for the management and stewardship of data assets for the data space*
(T) Technology *Sourcing and operation of technical infrastructure and support services for Data Spaces*	(T1) Infrastructure	*Sourcing and operation of technical infrastructure to deliver data space objectives*
	(T2) Support services	*Provision of support services that facilitate data space usage and application development*
(P) People *Develop data space skills and culture. Drive adoption of Data Spaces*	(P1) Skills and culture	*Establish a structured approach to data space skills and development and promote a data space culture*
	(P2) Adoption and communication	*Embed data space principles and communicate a common understanding across the organization*
(G) Governance *Establish clear policies, compliance, and accountability for Data Spaces*	(G1) Policies	*Establish common and consistent policies to support data space strategy to meet current and future objectives*
	(G2) Compliance	*Enablement and demonstration of compliance with data legislation, regulation, and directives*
	(G3) Accountability	*Clear accountability for data space roles and decision making within the organization and the ecosystem*
(T) Trust *Level of trust for data owners*	(T1) Assurance	*Level of assurance provided to data owners (organizations and individuals) on their data*

- Organization (O) includes data space strategy and planning and its alignment and reporting with the organization's overall business strategy, objectives, and goals.
- Value (V) develops the sensing of data space business opportunities and value capture.
- Data (D) includes the provision of data sharing within the life cycle and the management and stewardship of data in the data space.
- Technology (T) includes the operation of infrastructure and support services that facilitate data space usage.
- People (P), which develops skills and the organization culture together with communication and adoption activities to help embed data space principles across the organization and the broader ecosystem.
- Governance develops common and consistent policies and requires accountability and compliance with relevant regulations and legislation.
- Trust, which needs to provide assurances to data owners and users.

4.3 Maturity Curve

A maturity curve serves two important purposes. First, it is the basis of an assessment process that helps determine the current maturity level. Second, it provides a view of the growth path by identifying the next set of capabilities an organization should develop to drive business value from Data Spaces. A contrast of low- and high-level capability maturity for Data Spaces is offered in Fig. 2 to illustrate the comprehensiveness and range of data space maturity; such comparisons can facilitate understanding the concept of process maturity [20]. Humphrey [29] emphasizes that there is no ultimate state of process maturity, but that maturity implies a firm foundation established from where continuous improvement initiatives can be

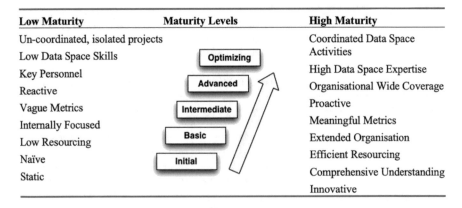

Low Maturity	Maturity Levels	High Maturity
Un-coordinated, isolated projects		Coordinated Data Space Activities
Low Data Space Skills	Optimizing	
Key Personnel		High Data Space Expertise
Reactive	Advanced	Organisational Wide Coverage
Vague Metrics	Intermediate	Proactive
Internally Focused		Meaningful Metrics
Low Resourcing	Basic	Extended Organisation
Naïve	Initial	Efficient Resourcing
Static		Comprehensive Understanding
		Innovative

Fig. 2 Comparison of low and high maturity of Data Spaces (adapted from Rosemann and de Bruin [18])

launched. The model defines a five-level maturity curve, as detailed in Table 2, for identifying and developing data space capabilities:

- **Initial:** Data space capabilities are ad hoc; there is little understanding of the subject and few or no related policies. Data space activities are not defined and are not considered in the organizational processes.
- **Basic:** There is a limited data space strategy with associated execution plans. It is mainly reactive and lacks consistency. There is an increasing awareness of the subject, but accountability is not clearly established. Some policies may exist but with inconsistent adoption.
- **Intermediate:** A data space strategy exists with associated plans and priorities. The organization has developed capabilities and skills and encourages individuals to contribute to data space programs. The organization includes Data Spaces across its processes and tracks targets and metrics.
- **Advanced:** Data Spaces are a core component of the data and business planning life cycles. Cross-functional teams jointly drive programs and progress. The organization recognizes Data Spaces as a significant contributor to its business strategy. It aligns business and data space metrics to achieve success across the organization. It also designs policies to enable the achievement of best practices.
- **Optimizing:** The industry recognizes the organization as a Data Space leader and uses its data space practices as an example to set industry standards and best practices. In addition, the organization recognizes Data Spaces as a key factor in driving data-driven innovation as a competitive advantage.

4.4 Assessment Approach

The MM4DS assessment determines how data space capabilities contribute to the organization's overall data innovation goals and objectives. This gap analysis between what the business wants and their current capabilities is delivering positions the MM4DS as a management tool for aligning and developing data space capabilities to meet business objectives. The model focuses on the execution of four key actions for increasing data space value:

- Define the scope and goal of data space.
- Understand the current data space capability maturity level.
- Systematically develop and manage the data space capability.
- Assess and manage data space capability progress over time.

Here we outline these actions in more detail and discuss their implementation.

Table 2 Maturity curve for each data space capability

Pillars	Capability	Initial	Basic	Intermediate	Advanced	Optimizing
(O) Organization	(O1) Strategy and planning	Any data space objectives that have been defined are limited and inconsistent	A minimum set of data space objectives are available and benchmarked	Data space objectives are part of an improvement roadmap for the medium term covering all aspects of the data space	The data space strategy is managed at the senior executive level and is executed as an integrated part of the organization's overall business strategy	Strategic planning for the data space extends outside the organization to include stakeholders from the broader ecosystem
	(O2) Business alignment	Any data space alignment that takes place is informal and inconsistent	The data space group reviews objectives that can be aligned to business goals	A complete set of short- and medium-term objectives for Data Spaces are agreed with the business	Longer-term data space objectives are agreed upon and integrated into business goals	Data space objectives are reviewed and set as part of board-level reviews of the organizational goals and aligned with the broader ecosystem value chain
	(O3) Performance monitoring	Performance measurement or reporting is ad hoc	Some data space performance metrics may exist, but reporting occurs at the project level	Data space performance is aggregated across all Data Spaces. Thus, there is the beginning of an alignment with corporate objectives	Data space performance is aggregated and reported across Data Spaces and aligned with business metrics	Data space performance is aggregated and reported across the organization and aligned with business metrics for the broader ecosystem
(V) Value	(V1) Sensing	Limited value opportunity sensing	Occasional business improvements are identified	A dedicated data space team identifies value opportunities for the organization	Cross-functional capabilities to identify new data space business opportunities	Data space value opportunities drive business strategy and collaborations with ecosystem partners
	(V2) Capture	Value undefined and not managed	Value capture is driven by individual efforts with limited recognition of value	A dedicated data space team advises on value capture to the business groups	Cross-functional capabilities to capture repeatable impact to the business	Data space value capture drives business revenue in cooperation with the ecosystem value network

(D) Data	(D1) Life cycle	Data sharing criteria included in data life cycles are inconsistent and ad hoc	Basic data sharing criteria are implemented for the data life cycle of a limited number of data products and services	Data sharing criteria and policies are regularly implemented within the life cycle of data products and services. Data sharing follows open standards	Data sharing criteria and policies are consistently implemented within the life cycle of data products and services	Industry-leading implementation of data sharing life cycle for Data Spaces. The organization influences and leads industrial best practices and standardization for data sharing
	(D2) Management and stewardship	Ad hoc management and stewardship processes	Processes are basic and project-based	Limited dedicated resources are dedicated to management and stewardship	Cross-functional capabilities for data management and stewardship. Processes are integrated into the workflow of business groups and are aligned with the data sharing life cycle	Industry-leading implementation of processes for data management and stewardship for Data Spaces. Influences and leads industrial best practices and standardization activities
(T) Technology	(T1) Infrastructure	Infrastructure is implemented using ad hoc technology choices	Basic infrastructure architecture guidelines and reference models are in place. Infrastructure interoperability complies with open standards	New data/IT systems deployed with defined data space interfaces	Roadmaps guide infrastructure technology choices and interoperability	The organization leads data space infrastructure research and driving industry best practices and standardization
	(T2) Support services	Limited or no data space support services	Services provide a minimal level of support with basic functionality over data assets (e.g., browsing)	Services are available to provide essential functionality at the data-item/entity level within data assets (e.g., search) and basic identification/naming of items	Data sources are integrated with most support service features (e.g., queries) with support for federation. Application development is supported with a toolbox to simplify development	Full semantic integration of data assets into the support services with a consistent global view of the data space. The organization leads industry best practices in data space support services

(continued)

Table 2 (continued)

Pillars	Capability	Initial	Basic	Intermediate	Advanced	Optimizing
(P) People	(P1) Skills and culture	Ad hoc informal training and culture at the project level	An informal mentoring network exists for Data Spaces. Dissemination of Data Spaces best practices	Data space competencies are integrated into HR strategies for managing technical and business personnel, including recruitment, training, and career development	Data space goals are an integral part of the company's core values and mission statement. There is a common culture focused on Data Spaces across the organization	Data space culture is engrained across the enterprise such that everyone feels empowered to develop innovative data-driven solutions using Data Spaces. Staff create best practices and drive industry thought leadership
	(P2) Adoption and communication	The adoption, language, and communication of Data Spaces are at best ad hoc and at the project level	Communication of basic data space principles, including basic terminology, to technical and business groups. Team events, informal training, and other communication channels encourage informal knowledge sharing and adoption	Data space adoption targets are defined for business groups, and regular reviews are undertaken to assess progress. Regular communication initiatives occur across the organization to improve awareness of key data space concepts and practices	Embedding of data space principles across the organization. Incentivization of data space membership and active participation. Establishment of ecosystem stakeholder engagement process. Dissemination of data space success stories	The organization is recognized internationally as a thought leader in Data Spaces and evangelizes best practices. Data Spaces are a key part of board level thinking and are included in internal and external communications
(G) Governance	(G1) Policies	No individual or team has overall responsibility for data space policies	A person or team is responsible for data space policies	A joint cross-functional team has responsibility for compliance with relevant regulations and standards applicable to Data Spaces	Performance is regularly reviewed against data space standards and policies. Changes are made in line with internal organization targets and external regulatory requirements	The organization is part of international bodies which define best practices and relevant regulations and standards applicable to Data Spaces

	(G2) Compliance	Compliance with external requirements or standards is at best ad hoc	Common access point for assessing relevant data space standards and regulations. Inconsistent application across the data space	A dedicated individual or team advises on incorporating data space standards and regulations into relevant business activities with a focus on minimum regulatory compliance	Regular audits and compliance reviews with internal targets and external data space standards and regulations	The organization participates in industry-wide peer reviews of compliance with data space standards/regulatory requirements. New developments in standards and regulations are implemented at an early stage
	(G3) Accountability	No formal accountability roles and processes are in place	A person/team is responsible for data space standards and regulations. The relevance of regulation to the organization is understood	A dedicated team is given responsibility for compliance with standards and regulations for the data space. The organization can demonstrate regulatory compliance	Regular audits and compliance reviews with internal targets and external data space standards and regulations. The structure and roles in the team changes in line with changes to regulatory requirements	The team head reports to the Chief Information Officer (CDO), Chief Data Officer (CDO), or directly to the Board
(T) Trust	(T1) Assurance	Limited or ad hoc control on a project basis	Coarse-grained (data asset level) access control	Fine-grained (entity-level/record-level) access control and data provenance	On-demand data anonymization and basic usage control	Full usage control across internal and external Data Spaces

4.4.1 Defining the Scope and Goal

First, the organization must define the scope of its data space effort. As a prerequisite, the organization should identify how it views data sharing and its aspirations. Typically, organizational goals involve one or more of the following:

- Develop significant capabilities and a reputation for leadership in Data Spaces.
- Keep pace with competitors or stakeholder expectations.
- Meet minimum compliance requirements and reap readily available benefits.

Second, the organization must define the goals of its data space effort. It is essential to be clear on the organization's business objectives and the role of the data space in enabling those objectives. A transparent agreement between business and technical stakeholders can tangibly help achieve those objectives. Significant benefits can be gained by simply understanding the relationship between business and data strategy goals.

Over time the goals and scope of a data space can evolve and change. As a data space grows, it may develop many subgoals or shared goals with other Data Spaces. The design and development of goals is a continuous interactive process to manage this systematically. Agreeing on the desired business goals for data innovation will significantly impact business and thus data strategy goals and priorities. After deciding to improve data space, organizations are often keen to aim for a consistent and widespread approach across the organization. Developing appropriate and effective capabilities is an iterative process and requires investment from both business and technical groups to learn from experience and deliver the desired benefits. This is because data innovation goes beyond technology. It is also about helping the whole business leverage data-driven innovation to meet its targets.

Once the scope and goals of data space capability are clear, the organization must identify its current capability maturity level by examining its data space capabilities.

4.4.2 Assessment Data Collection and Analysis

The first step is to assess the organization's status for the 15 capabilities within the MM4DS model. The assessment begins with the survey to understand their assessments of the maturity and importance of their data space capabilities. The survey consisted of 45 questions. The survey structure is aligned with the assessment approached and divided into three sections:

- **Current maturity (15 questions):** Participants are invited to score the organization's current maturity for data space capabilities. Each question describes the characteristics of a maturity level that follow maturity level logic across five stages: initial, basic, intermediate, advanced, and optimized.
- **Desired maturity (15 questions):** Participants are invited to score the organization's future desired maturity for data space capabilities. Each question describes

the characteristics of a maturity level that follow maturity level logic across five stages: initial, basic, intermediate, advanced, and optimized.

- **Importance of capability (15 questions):** Participants are asked to value each data space capability by grading them on a 1 to 5 scale, with 1 being not important and 5 being very important.

4.4.3 Using the Assessment Results to Develop and Manage Capabilities

With the assessment complete, organizations will have a clear view of current capability and key areas for action and improvement. However, to further develop data space capabilities, the organization should assess and manage progress over time by using the assessment results to:

- Develop a roadmap and action plan
- Add a yearly follow-up assessment to measure progress and the value of data space adoption over time

Agreeing on stakeholder ownership for each priority area is critical to developing short-term and long-term action plans for improvement. The assessment results can be used to prioritize the opportunities for quick wins. Those capabilities have smaller gaps between current and desired maturity and those recognized as more important but might have a more significant gap to bridge.

5 Illustrative Benchmarking Example

In this section, we use five fictitious organizations to illustrate the usage of the MM4DS. In addition, this section details the assessment process and the analysis which can be performed to benchmark capabilities across the organization.

5.1 Benchmark Results

The survey should be taken by a range of stakeholders from different parts of the organization to get a holistic view. The results of the surveys are then averaged to determine the overall level of maturity for the organization. The results for the MM4DS of the example organizations are presented in Table 3. From the benchmark, we can understand the state of maturity of data space capabilities within each of the benchmarked organizations.

Table 3 MM4DS assessment results for data space capability maturity of five organizations (average from survey responses)

Pillars	Capability	Org1	Org2	Org3	Org4	Org5
(O) Organization	(O1) Strategy and planning	2.7	2.5	2.4	3.2	2.3
	(O2) Business alignment	2.8	2.3	2.1	2.8	2.1
	(O3) Performance monitoring	2.4	2.3	2.4	2.8	2.4
(V) Value	(V1) Sensing	3	2.5	2.9	2.6	2.5
	(V2) Capture	3	2.5	2.5	2.8	1.2
(D) Data	(D1) Life cycle	2.1	1.9	1.9	2.3	2.0
	(D2) Management and stewardship	2.4	2.0	2.1	2.9	2.2
(T) Technology	(T1) Infrastructure	2.9	2.4	2.1	1.8	1.8
	(T2) Support services	2.9	2.5	2.1	2.5	1.4
(P) People	(P1) Skills and culture	2.7	2.5	2.4	3.2	2.3
	(P2) Adoption and communication	2.8	2.3	2.1	2.8	2.1
(G) Governance	(G1) Policies	2.4	2.3	2.4	2.8	2.4
	(G2) Compliance	3	2.5	2.9	2.6	2.5
	(G3) Accountability	3	2.5	2.5	2.8	1.2
(T) Trust	(T1) Assurance	2.1	1.9	1.9	2.3	2.0

5.1.1 Capability Gap Analysis

Using the benchmark results, we can determine a capability gap analysis by contrasting the current and desired maturity of the organization's data space capabilities. The results of this capability gap analysis are presented in Table 4. Looking at the organizations' current average maturity of capabilities versus the desired capability maturity, we can see a clear gap across all capabilities, as detailed in Table 4.

5.1.2 Capability Importance

As detailed in Table 5 the assessment provides valuable insight into the importance of individual capabilities. Understanding the current maturity levels and importance of a capability enables an organization to identify an action plan for improvement. Analyzing the maturity gaps between the current and desired state can identify where the organizations prioritize their actions. Where the importance of a capability is correlated with its current maturity, we can derive a prioritized ranking of capability improvements.

Table 4 Capability gap analysis

Pillars	Capability	Current			Desired			Gap
		Avg.	Low	High	Avg.	Low	High	Avg.
(O) Organization	(O1) Strategy and planning	2.6	2.4	3.2	4.0	3.8	4.5	1.4
	(O2) Business alignment	2.4	2.1	2.8	3.8	3.4	4.1	1.4
	(O3) Performance monitoring	2.5	2.3	2.8	3.7	3.5	3.9	1.2
(V) Value	(V1) Sensing	2.7	2.5	3.0	3.9	3.7	4.3	1.3
	(V2) Capture	2.4	1.2	3.0	3.9	3.7	4.2	1.5
(D) Data	(D1) Life cycle	2.0	1.9	2.3	3.6	3.3	4.0	1.6
	(D2) Management and stewardship	2.3	2.0	2.9	3.7	3.2	4.3	1.4
(T) Technology	(T1) Infrastructure	2.2	1.8	2.9	3.5	3.1	4.1	1.3
	(T2) Support services	2.3	1.4	2.9	3.7	2.9	4.0	1.4
(P) People	(P1) Skills and culture	2.6	2.4	3.2	4.0	3.8	4.5	1.4
	(P2) Adoption and communication	2.4	2.1	2.8	3.8	3.4	4.1	1.4
(G) Governance	(G1) Policies	2.5	2.3	2.8	3.7	3.5	3.9	1.2
	(G2) Compliance	2.7	2.5	3.0	3.9	3.7	4.3	1.3
	(G3) Accountability	2.4	1.2	3.0	3.9	3.7	4.2	1.5
(T) Trust	(T1) Assurance	2.0	1.9	2.3	3.6	3.3	4.0	1.6

Table 5 Capability importance analysis

Pillars	Capability	Importance		
		Avg.	Low	High
(O) Organization	(O1) Strategy and planning	4.2	4.0	4.6
	(O2) Business alignment	4.4	4.2	4.8
	(O3) Performance monitoring	4.0	3.7	4.3
(V) Value	(V1) Sensing	3.8	3.6	4.3
	(V2) Capture	4.3	4.1	4.7
(D) Data	(D1) Life cycle	4.0	3.5	4.5
	(D2) Management and stewardship	3.4	3.2	4.0
(T) Technology	(T1) Infrastructure	3.7	3.3	4.2
	(T2) Support services	4.3	4.0	4.6
(P) People	(P1) Skills and culture	4.2	4.0	4.6
	(P2) Adoption and communication	4.4	4.2	4.8
(G) Governance	(G1) Policies	4.0	3.7	4.3
	(G2) Compliance	3.8	3.6	4.3
	(G3) Accountability	4.3	4.1	4.7
(T) Trust	(T1) Assurance	4.0	3.5	4.5

6 Conclusion

The MM4DS gives user-centric/demand-side organizations a vital tool to manage their data space capability to gain business value. The model provides a comprehensive value-based model for organizing, evaluating, planning, and managing data space capabilities. Using the model, organizations can assess the maturity of their data space capability and systematically improve capabilities to meet the business objectives. The model was developed using an open-innovation collaboration model, engaging academia and industry in scholarly work following a design science research approach. In addition, an illustrative benchmark of the data space capabilities of five organizations using the model was undertaken. The initial version of the model presented in this chapter will be developed further by the task force to refine it and validate it within real-world Data Spaces.

Acknowledgments We would like to thank the contributors from the BDVA Task Force on Data Sharing Spaces. We would also like to acknowledge the comments received from BDVA members and external communities.

References

1. Scerri, S., Tuikka, T., & Lopez de Vallejo, I. (Eds.). (2020). *Towards a European data sharing space*.
2. Curry, E., Metzger, A., Zillner, S., Pazzaglia, J.-C., & García Robles, A. (2021). *The elements of big data value*. Springer International Publishing. https://doi.org/10.1007/978-3-030-68176-0
3. Curry, E. (2016). The big data value chain: Definitions, concepts, and theoretical approaches. In J. M. Cavanillas, E. Curry, & W. Wahlster (Eds.), *New horizons for a data-driven economy* (pp. 29–37). Springer International Publishing. https://doi.org/10.1007/978-3-319-21569-3_3
4. Curry, E. (2020). *Real-time linked dataspaces. Springer International Publishing.*https://doi.org/10.1007/978-3-030-29665-0
5. Curry, E., & Sheth, A. (2018). Next-generation smart environments: From system of systems to data ecosystems. *IEEE Intelligent Systems, 33*(3), 69–76. https://doi.org/10.1109/MIS.2018.033001418
6. Zillner, S., Bisset, D., Milano, M., Curry, E., Hahn, T., Lafrenz, R., Liepert, B., Robles, A. G., Smeulders, A., & O'Sullivan, B. (2020). *Strategic research, innovation and deployment agenda - AI, data and robotics partnership. Third Release* (Third). BDVA, euRobotics, ELLIS, EurAI and CLAIRE.
7. Zillner, S., Gomez, J. A., García Robles, A., Hahn, T., Le Bars, L., Petkovic, M., & Curry, E. (2021). Data economy 2.0: From big data value to AI value and a European data space. In E. Curry, A. Metzger, S. Zillner, J.-C. Pazzaglia, & A. García Robles (Eds.), *The elements of big data value* (pp. 379–399). Springer International Publishing. https://doi.org/10.1007/978-3-030-68176-0_16
8. Zillner, S., Curry, E., Metzger, A., Auer, S., & Seidl, R. (Eds.). (2017). *European big data value strategic research & innovation agenda*. Big Data Value Association. http://www.edwardcurry.org/publications/BDVA_SRIA_v4_Ed1.1.pdf
9. Wade, M., & Hulland, J. (2004). Review: The resource-based view and information systems research: Review, extension, and suggestions for future research. *MIS Quarterly, 28*(1), 107–142. https://doi.org/10.2307/25148626
10. Wernerfelt, B. (1984). A resource-based view of the firm. *Strategic Management Journal, 5*, 171–180. https://doi.org/10.1002/smj.4250050207
11. Helfat, C. E., & Peteraf, M. A. (2003). The dynamic resource-based view: Capability life-cycles. *Strategic Management Journal, 24*(10), 997–1010. https://doi.org/10.1002/smj.332

12. Sambamurthy, V., & Zmud, R. W. (1997). At the heart of success: organisationwide management competencies. In C. Sauer & P. Yetton (Eds.), *Steps to the future: Fresh thinking on the management of IT-based organisational transformation* (pp. 143–163). Jossey-Bass.
13. Bharadwaj, A. S. (2000). A resource-based perspective on information technology capability and firm performance: An Empirical investigation. *MIS Quarterly, 24*(1), 169. https://doi.org/10.2307/3250983
14. Teece, D. J., Pisano, G., & Shuen, A. M. Y. (1997). Dynamic capabilities and strategic management. *Strategic Management Journal, 18*(March), 509–533. https://doi.org/10.1002/(SICI)1097-0266(199708)18:7%3C509::AID-SMJ882%3E3.0.CO;2-Z
15. Collis, D. J. (1994). How valuable are organisational capabilities. *Strategic Management Journal, 15*, 143–152. https://doi.org/10.2307/2486815
16. Zollo, M., & Winter, S. G. (2002). Deliberate learning and the evolution of dynamic capabilities. *Organization Science, 13*(3), 339–351. https://doi.org/10.1287/orsc.13.3.339.2780
17. Butler, T., & Murphy, C. (2008). An exploratory study on IS capabilities and assets in a small-to-medium software enterprise. *Journal of Information Technology, 23*(4), 330–344. https://doi.org/10.1057/jit.2008.19
18. Rosemann, M., & de Bruin, T. (2005). Application of a holistic model for determining BPM maturity. *BPTrends, February*, 1–21. http://scholar.google.com/scholar?hl=en&btnG=Search&q=intitle:Application+of+a+Holistic+Model+for+Determining+BPM+Maturity#0
19. Crawford, J. K. (2006). The project management maturity model. *Information Systems Management, 23*(4), 50–58. https://doi.org/10.1201/1078.10580530/46352.23.4.20060901/95113.7
20. Paulk, M. C., Curtis, B., Chrissis, M. B., Weber, C., & v. (1993). The capability maturity model for software. *Software Engineering Project Management, 10*(CMU/SEI-93-TR-24 ESC-TR-93-177), 1–26.
21. Luftman, J. (2003). Assessing it/business alignment. *Information Systems Management, 20*(4), 9–15. https://doi.org/10.1201/1078/43647.20.4.20030901/77287.2
22. Curry, E., & Donnellan, B. (2012). Understanding the maturity of sustainable ICT. In J. vom Brocke, S. Seidel, & J. Recker (Eds.), *Green business process management - Towards the sustainable enterprise* (pp. 203–216). Springer. Retrieved from http://www.edwardcurry.org/publications/Curry_SICTMaturity_PrePrint.pdf
23. van de Ven, A. H. (2007). Engaged scholarship: A guide for organizational and social research. In A. H. van de Ven (Ed.), *Process studies* (Vol. 33, 4). Oxford University Press. https://doi.org/10.2307/20159463
24. Chesbrough, H. W. (2003). Open innovation: The new imperative for creating and profiting from technology. In H. W. Chesbrough, W. Vanhaverbeke, & J. West (Eds.), *Perspectives academy of management* (Vol. 20(2). Harvard Business School Press. https://doi.org/10.1016/j.jengtecman.2004.05.003
25. Hevner, A. R., March, S. T., Park, J., & Ram, S. (2004). Design science in information systems research. *MIS Quarterly, 28*(1), 75–105. https://doi.org/10.2307/249422
26. Mettler, T., & Rohner, P. (2009). Situational maturity models as instrumental artifacts for organisational design. In V. Vaishanvi & S. Purao (Eds.), *Proceedings of the 4th International Conference on Design Science Research in Information Systems and Technology* (pp. 1–9). ACM Press. https://doi.org/10.1145/1555619.1555649
27. Braun, C., Wortmann, F., Hafner, M., & Winter, R. (2005). Method construction - A core approach to organisational engineering. *Proceedings of the 2005 ACM Symposium on Applied Computing SAC 05, 2*(c), 1295–1299. https://doi.org/10.1145/1066677.1066971
28. Hevner, A. R. (2007). A three cycle view of design science research. *Scandinavian Journal of Information Systems, 19*(2), 87–92. http://aisel.aisnet.org/sjis/vol19/iss2/4
29. Humphrey, W. S. (1989). *Managing the software process*. Addison-Wesley Longman Publishing Co., Inc.

Data Platforms for Data Spaces

Amin Anjomshoaa, Santiago Cáceres Elvira, Christian Wolff,
Juan Carlos Pérez Baún, Manos Karvounis, Marco Mellia, Spiros Athanasiou,
Asterios Katsifodimos, Alexandra Garatzogianni, Andreas Trügler,
Martin Serrano, Achille Zappa, Yury Glikman, Tuomo Tuikka,
and Edward Curry

A. Anjomshoaa (✉)
Maynooth University, Maynooth, Ireland
e-mail: amin.anjomshoaa@mu.ie

S. C. Elvira
ITI – Instituto Tecnológico de Informática, Valencia, Spain

C. Wolff
ATB Bremen, Bremen, Germany

J. C. Pérez Baún
Atos Spain S.A., Madrid, Spain

M. Karvounis
Agroknow, Athens, Greece

M. Mellia
Politecnico di Torino, Torino, Italy

S. Athanasiou
Athena Research Center, Athens, Greece

A. Katsifodimos
TU Delft, Delft, Netherlands

A. Garatzogianni
Leibniz University Hannover, Hannover, Germany

A. Trügler
Know-Center GmbH, Graz, Austria

Graz University of Technology, Graz, Austria

M. Serrano · A. Zappa · E. Curry
Insight SFI Research Centre for Data Analytics, University of Galway, Galway, Ireland

Y. Glikman
Fraunhofer FOKUS, Berlin, Germany

T. Tuikka
VTT Technical Research Centre of Finland, Oulu, Finland

E. Curry et al. (eds.), *Data Spaces*, https://doi.org/10.1007/978-3-030-98636-0_3

43

Abstract In our societies, there is a growing demand for the production and use of more data. Data is reaching the point that is driving all the social and economic activities in every industry sector. Technology is not going to be a barrier anymore; however, where there is large deployment of technology, the production of data creates a growing demand for better data-driven services, and at the same time the benefits of the production of the data are at large an impulse for a global data economy, Data has become the business's most valuable asset. In order to achieve its full value and help data-driven organizations to gain competitive advantages, we need effective and reliable ecosystems that support the cross-border flow of data. To this end, data ecosystems are the key enablers of data sharing and reuse within or across organizations. Data ecosystems need to tackle the various fundamental challenges of data management, including technical and nontechnical aspects (e.g., legal and ethical concerns). This chapter explores the Big Data value ecosystems and provides a detailed overview of several data platform implementations as best-effort approaches for sharing and trading industrial and personal data. We also introduce several key enabling technologies for implementing data platforms. The chapter concludes with common challenges encountered by data platform projects and details best practices to address these challenges.

Keywords Data platforms · Data Spaces · Data ecosystem · Design

1 Introduction

Many industries and enterprises have recognized the real potential of Big Data value for exploring new opportunities and making disruptive changes to their business models. However, to realize the vision of Big Data value systems and create strong and sustaining Big Data ecosystems, several concerns and issues must be addressed. This includes [3] availability of high-quality data and data resources, availability of rightly skilled data experts, addressing legal issues, advancing technical aspects of data systems, developing and validating market-ready applications, developing appropriate business models, and addressing the societal aspects.

To foster, strengthen, and support the development and wide adoption of Big Data value technologies within an increasingly complex landscape requires an interdisciplinary approach that addresses the multiple elements of Big Data value. To this end, the introduction of the Big Data Value Reference Model (BDV-RM) [3] is an effort to address the common challenges and concerns of the Big Data value chain and create a data-driven ecosystem for Big Data. The BDVA Reference model is structured into core data processing concerns (horizontal) and cross-cutting concerns (vertical) as depicted in Fig. 1. The horizontal concerns include specific aspects along the data processing chain, starting with data collection and ingestion and extending to data visualization. On the other hand, vertical concerns address cross-cutting issues, which may affect all the horizontal concerns and involve nontechnical aspects.

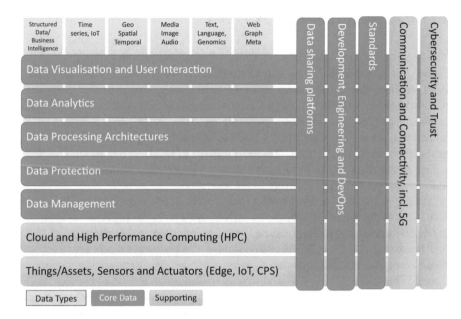

Fig. 1 Big Data Value Reference Model

This book chapter first explores the Big Data value ecosystems. It introduces state-of-the-art data management systems that follow the BDV Reference Model to realize data value chains and data flows within ecosystems of intelligent systems. Then, we provide a detailed overview of several data platform implementations as best-effort approaches for sharing and trading industrial and personal data. We also compare the data management and Data Governance services of the data platform projects. Finally, the key enabling technologies for implementing data platforms will be introduced. We conclude this book chapter by providing an overview of common challenges encountered by data platform projects and best practices to address these challenges.

2 Big Data Value Ecosystems

A data ecosystem is a sociotechnical system that enables value to be extracted from data value chains supported by interacting organizations and individuals. Data value chains can be oriented to business and societal purposes within an ecosystem. The ecosystem can create the conditions for a marketplace competition between participants or enable collaboration among diverse, interconnected participants that depend on each other for their mutual benefit. Data ecosystems can be formed in different ways around an organization or community technology platforms or within or across sectors. This section introduces some best practices and proposed architectures to realize the Big Data value ecosystems.

2.1 Data Spaces and Data Platforms

The Big Data Value Association (BDVA)—that is, the private counterpart of the European Commission in the Big Data Value Public-Private-Partnership (BDV PPP)—defines data space as an umbrella term corresponding to any ecosystem of data models, datasets, ontologies, data sharing contracts, and specialized management services (i.e., as often provided by data centers, stores, repositories, individually, or within "data lake'"), together with soft competencies around it (i.e., governance, social interactions, business processes) [1]. These competencies follow a data engineering approach to optimize data storage and exchange mechanisms, preserving, generating, and sharing new knowledge.

In comparison, data platforms refer to architectures and repositories of interoperable hardware/software components, which follow a software engineering approach to enable the creation, transformation, evolution, curation, and exploitation of static and dynamic data in Data Spaces. To this end, a data platform would have to support continuous, coordinated data flows, seamlessly moving data among intelligent systems [2].

Although distinct, the evolution of the data space and data platform concepts goes hand in hand and needs to be jointly considered, and both can be considered the two faces of the same data economy coin. However, their complementary nature means that commercial solutions often do not distinguish between the two concepts. Furthermore, due to the particular requirements for the preservation of individual privacy, a distinction between technology and infrastructures that store and/or handle personal and other data has emerged. As a result, the evolution of industrial data platforms (considered key enablers of overall industrial digitization) and personal data platforms (services that use personal data, subject to privacy preservation, for value creation) has continued to follow different paths.

2.2 Gaia-X Ecosystem

Gaia-X[1] is a project to develop an efficient and competitive, secure, and trustworthy federation of data infrastructure and service providers for Europe, supported by representatives of business, science, and administration from European countries. Gaia-X follows the principles of openness and transparency of standards, interoperability, federation (i.e., decentralized distribution), and authenticity and trust.

The Gaia-X ecosystem is structured into a data ecosystem and the infrastructure ecosystem as depicted in Fig. 2. The data ecosystem enables Data Spaces as envisioned by the European data strategy, where data is exchanged, and advanced smart services are provided. The infrastructure ecosystem comprises building blocks from hardware nodes to application containers, where data is stored and services are executed, as well as networks for transmission of data between nodes and applications. addition, the infrastructure itself may be provided as a service.

---

[1] https://www.gaia-x.eu/

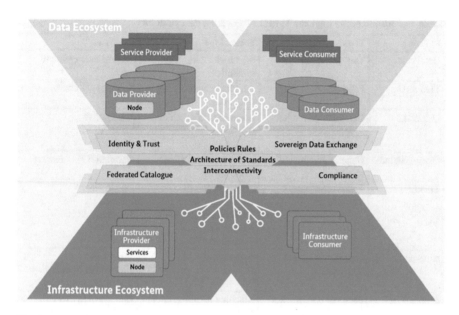

Fig. 2 Gaia-X architecture

3 Data Platform Project Portfolio

The data platform projects running under the umbrella of the Big Data Value Public-Private Partnership (BDV PPP) develop integrated technology solutions for data collection, sharing, integration, and exploitation to facilitate the creation of such a European data market and economy [3]. The portfolio of the Big Data value covers the data platform projects shown in Table 1. This table gives an overview of these projects, the type of data platform they develop, and the domain, the core enabling technologies, and the use cases they address. Each of these projects is briefly summarized in this section.

3.1 DataPorts Project

The DataPorts project[2] is devoted to creating a secure data platform that allows sharing the information between seaport agents in a reliable and trustworthy manner, with access permits and contracts to allow data sharing and the exploration of new Artificial Intelligence and cognitive services. It provides seaports with a secure and privacy-aware environment where the stakeholders can share data from different

---

[2] https://dataports-project.eu/

Table 1 Portfolio of the Big Data Value PPP covering data platforms

Project	Type	Technology	Use cases
DataPorts	Transportation	AI, blockchain, semantics	Seaport management
TheFSM	Food	AI, blockchain, semantics	Food supply chain
i3-Market	Generic data market support tools	Semantics, blockchain, OpenID	Automotive Manufacturing Wellbeing
OpertusMundi	Generic geodata market	Microservices, BPMN workflows	Geospatial data market
Trusts	Personal/industrial data market	Data encryption Blockchain KAN-based open data repositories Semantics	Data market Finance Telecom
smashHit	Personal/industrial data market	Semantics	Insurance, automotive industry, insurance, smart city
PimCity	Personal data market	Machine learning, data provenance, privacy-preserving	Generic data market
Kraken	Personal data market	Blockchain, privacy-preserving, self-sovereign identity, data encryption	Education health
DataVaults	Personal data market	Machine learning, blockchain	Sports Mobility Healthcare Tourism Smart home Smart city

sources to get real value, providing a set of novel AI and cognitive tools to the port community.

The platform takes advantage of huge data provided by stakeholders for improving existing processes and enabling new business models. To this end, the project offers several common analytics services such as auto model training and machine learning pipelines that seaports agents can reuse.

The Data Governance components of the project benefit from Semantic Web technologies to enable interoperability between stakeholders and blockchain technology that realizes the business rules via smart contracts. Figure 3 provides an overview of data services offered by DataPorts data platform.

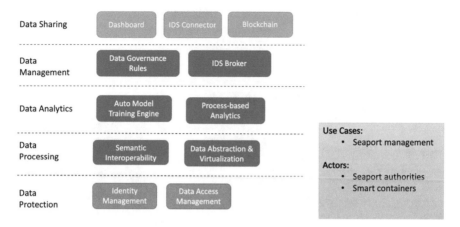

Fig. 3 System overview and data services of DataPorts data platform

3.2 TheFSM Project

TheFSM platform[3] aspires to ensure transparent and safe food production by digitizing food certification processes that assess safety via audits. More specifically, during the past 5 years, we have witnessed major changes in the food sector, with tremendous emphasis being put on food safety. A series of food safety scandals and health incidents have led to the international alignment of food safety standards through the Global Food Safety Initiative (GFSI). Governments also apply stricter policies and legislation, such as the integrated food safety policy of the European Commission and the US Food Safety Modernization Act (FSMA). There is increased pressure for the agri-food and grocery sector to ensure that their suppliers comply with food safety standards recognized by the GFSI. This translates into more pressure for all stakeholders in the supply chain to exchange data critical to food safety assessment and assurance in a timely, trusted, and secure manner. Finally, the global COVID-19 pandemic has further emphasized the need for supporting digital and remote auditing and certification processes.

The Food Safety Market (TheFSM) aims to deliver an industrial data platform that will significantly boost food certification in Europe. To achieve this goal, and as the food certification market is multifaceted, there is the need for all the actors in the food supply chain to share food safety data in a well-defined and automated way. Therefore, the platform aims to establish remote auditing in the European food market and serves as a data marketplace that enables all actors in the food chain to monitor, trace, and predict food safety risks in the food supply chain, to allow food safety inspectors and auditors to manage inspection/certification workflow

---

[3] https://foodsafetymarket.eu/

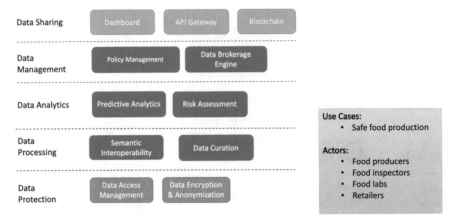

Fig. 4 System overview and data services of TheFSM data platform

digitally, and to allow farmers and food producers to manage their resources and their certification data.

The platform provides data curation and semantic enrichment services to create and manage a Knowledge Graph of domain objects. Furthermore, the platform benefits from blockchain technology to provide a collaborative hub for connecting organizations aiming to work together and solve complex supply chain challenges.

Eventually, TheFSM aspires to catalyze the digital evolution of global food certification's traditional but very data-intensive business ecosystem. Figure 4 provides an overview of data services offered by TheFSM data platform.

3.3   i3-MARKET Project

It has been largely discussed that there is a growing demand for a global data economy, where the different data stakeholders can participate in the distribution of the benefits from selling/trading data assets. The i3-MARKET project[4] addresses this growing demand from the perspective of a single European Data Market Economy by innovating marketplace platforms, enabling them with software artifacts that allow the deployment of data-related services, and demonstrating that data economy growth is possible with industrial implementations. The i3-MARKET solution(s) aims at providing technologies for trustworthy (secure and reliable), data-driven collaboration and federation of existing and new future marketplace platforms, with particular attention on industrial data. Furthermore, the i3-MARKET architecture is

---

[4] https://www.i3-market.eu/

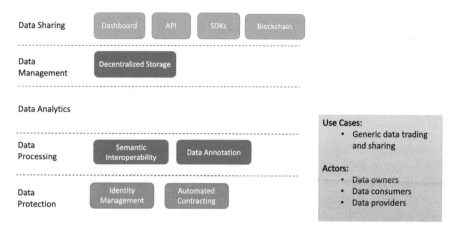

Fig. 5 System overview and data services of i3-Market data platform

designed to enable secure and privacy-preserving data sharing across Data Spaces and marketplaces by deploying a backplane across operational data marketplaces.

The i3-MARKET project does not try to create another new marketplace, involving the multiple data marketplace characteristics and functionalities; rather, it implements a backplane solution that other data marketplaces and Data Spaces can use to expand their market data offering capacities; facilitate the registration and discovery of data assets; facilitate the trading and sharing of data assets among providers, consumers, and owners; and provide tools to add functionalities they lack for a better data sharing and trading processes across domains. By bringing together data providers (supply side) and data consumers (demand side), i3-MARKET acts as an enabler for data monetization, realizing promising business ideas based on data trading, and trustworthy and data-driven collaboration. This way, i3-MARKET is the missing link acting as reference implementation that will allow all the interested parties to connect while offering incentives to data owners, data providers, and data consumers to engage in data trading. It may also serve as best practices for enabling data-driven economy and pave the way to the European data sharing economy in a safer, secured, and fair manner. The i3-MARKET project targets the possibility to interconnect data assets in a distributed manner enabling a federated query system that facilitates increasing the data offerings without the need to collect and host data locally. In this form any data marketplace that registers to the i3-MARKET ecosystem is able to provide access to cross-domain description and use the smart contract approach to be able to allow the access to the data asset remotely. Figure 5 provides an overview of data services offered by i3-Market data platform.

3.4   OpertusMundi

The OpertusMundi project[5] aims to deliver a trusted, secure, and scalable pan-European industrial geospatial data market, Topio,[6] acting as a single point for the streamlined and trusted discovery, sharing, trading, remuneration, and use of geospatial data assets, guaranteeing low cost and flexibility to accommodate current and emerging needs of data economy stakeholders regardless of size, domain, and expertise.

Topio empowers *geospatial data suppliers* to trade their assets under (a) *homogenized, configurable, digital,* and *automated* contracting facilities enforceable across EU; (b) multiple *standardized pricing models* and *tiers* suited to the type of their assets and business models; (c) full autonomy in *publishing, vetting,* and *monitoring* the sales and use of their assets via rich integrated analytics and IPR protection schemes; (d) novel *monetization schemes* by *automatically* exposing *data* as *services* created and operationalized by Topio in *a revenue-sharing* scheme; and (e) unrestricted *opt-in/out* of its services. From the *consumer's* perspective, Topio enables them to *fairly* purchase assets that are *fit for purpose* via (a) rich automated metadata for traded assets (i.e., *data profiling*) independently provided by Topio to support *informed purchasing decisions*, (b) clear and transparent *terms, conditions*, and *pricing* for assets *before purchase*, (c) automated digital *contracting* and *payments* with dispute resolution provided by Topio, and (d) *streamlined, low-effort,* and *direct use* of purchased and private assets via a plethora of web services. For all types of users, including those not actively trading or purchasing data, Topio enables them to *use* and *extract value* from geospatial data through a plethora of low-cost and intuitive *value-added services*, ranging from cloud storage and custom maps to Jupyter notebooks and analysis bundles for select thematic domains.

Topio, as a sustainable commercial endeavor and business entity, is designed and built on the principles of *trust, fairness,* and adherence to *law*. On a technical level, it is *fully auditable* via automated BPMN workflows, with all transactions taking place by KYB/KYC-validated entities under anti-money laundering (AML) safeguards. On a legal level, it is in full conformance with the EU's current and emerging legal framework on Data Spaces and markets, e-marketplaces, consumer rights, competition, and especially the Data Governance Act.[7] Finally, on a business level, Topio's business model is founded on our vision to *grow* and *serve* the emerging data economy in the EU, estimated[8] to reach 550b€ in size by 2025,[9] with 99b€ in data supplier revenues. Considering that ~80% of data are anecdotally considered as *geospatial*, and with 583K EU data users and 173K suppliers by 2025, Topio's

---

[5] https://www.opertusmundi.eu/

[6] https://topio.market/

[7] Data Governance Act, COM/2020/767 final.

[8] https://ec.europa.eu/digital-single-market/en/news/european-data-market-study-update

[9] 829b€ according to https://ec.europa.eu/info/strategy/priorities-2019-2024/europe-fit-digital-age/european-data-strategy

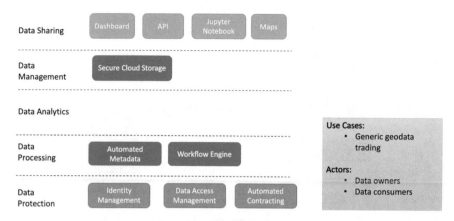

Fig. 6 System overview and data services of OpertusMundi data platform

impact in materializing this vision can be substantial. For this reason, Topio's business model and service offerings *do not, and will not, impose fees* on data trading. Instead, Topio generates its profits *solely* from operationalizing supplier-provided *data as services* and *subscriptions* from its value-added services. Figure 6 provides an overview of data services offered by OpertusMundi data platform.

3.5 TRUSTS Project

The TRUSTS project[10] aims to ensure the sustainable business uptake of secure data markets by enabling a fully operational and GDPR-compliant European data marketplace for personal and industrial data in the finance and telecom sectors, while allowing the integration and adoption of future platforms. To this end, the platform provides services to identify and overcome legal, ethical, and technical challenges of cross-border data markets. The platform follows the reference architecture designed by the International Data Spaces (IDS) Association which uses Semantic Web technologies for describing data schemas to configure connectors and interpret the data shared through these connectors. Furthermore, the proposed approach aims to create trust between participants through certified security functions, to allow secure collaboration over private data, and to establish governance rules for data usage and data flows. The IDS architecture ensures data sovereignty for those who make data available in data ecosystems. Figure 7 provides an overview of data services offered by TRUSTS data platform.

---

[10] https://www.trusts-data.eu/

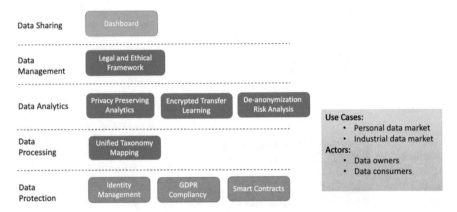

Fig. 7 System overview and data services of TRUSTS data platform

3.6   smashHit Project

The objective of smashHit[11] is to assure trusted and secure sharing of data streams from both personal and industrial platforms, needed to build sectorial and cross-sectorial services. The project establishes a framework for processing data owner consent and legal rules and effective contracting, as well as joint security and privacy-preserving mechanisms. The vision of smashHit is to overcome obstacles in the rapidly growing data economy, which is characterized by heterogeneous technical designs and proprietary implementations, lacking business opportunities due to the inconsistent consent and legal rules among different data sharing platforms, actors, and operators. By using the smashHit project solutions, it is expected to achieve improved citizen trust (by providing data owners awareness of their given consent), improved OEM and data customer trust (due to fingerprinted data to ensure traceability/unchangeability along the value chain as well as due to consent tracing), simplified consent process (by providing certification of consent and a single point of consent management), and support in consent/contract generation (facilitating the generation of legally binding contracts, taking into account relevant legislation/legal rules). Figure 8 provides an overview of data services offered by smashHit data platform.

3.7   PimCity Project

The PimCity project[12] aims to increase transparency in online data markets by giving users control over their data, ensuring that citizens, companies, and

---

[11] https://smashhit.eu/

[12] https://www.pimcity-h2020.eu

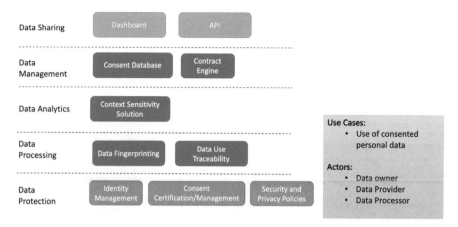

Fig. 8 System overview and data services of smashHit data platform

organizations are informed and can make respectful and ethical use of personal data. The project follows a human-centric paradigm aimed at a fair, sustainable, and prosperous Digital Society. The sharing of personal data is based on trust and a balanced and fair relationship between individuals, businesses, and organizations.

The project provides a PIMS Development Kit (PDK) that allows developers to engineer and experiment with new solutions. It allows them to integrate new data sources and connect them to new services. The PDK focuses on interoperability, which is at the same time the most significant challenge because it requires a process of standardization of consent mechanisms, formats, and semantics. All platform components offer Web APIs that are documented using the Open API specifications to allow seamless integration. This enables communications and interactions among components in the PDK, easing integration with existing PIMS, and the design and development of new ones. Figure 9 provides an overview of data services offered by PimCity data platform.

3.8 KRAKEN Project

KRAKEN[13] (brokerage and market platform for personal data) aims to develop a trusted and secure personal data platform with state-of-the-art privacy-aware analytics methods that return the control of personal data to users. The project also aims to enable the sharing, brokerage, and trading of potentially sensitive personal data by returning the control of this data to citizens throughout the entire data lifecycle. The project combines, interoperates, and extends the best results from two

---

[13] https://www.krakenh2020.eu/

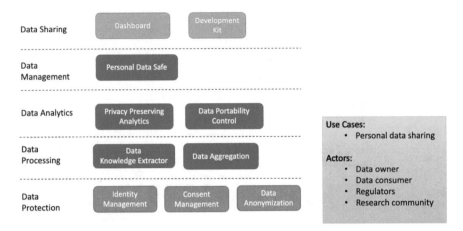

Fig. 9 System overview and data services of PimCity data platform

existing mature computing platforms developed within two H2020 actions, namely, CREDENTIAL[14] and MyHealthMyData.[15]

The project addresses the challenge of removing obstacles that prevent citizens from controlling and widely sharing their personal data. With this objective, KRAKEN is investigating data processing mechanisms working in the encrypted domain to increase security, privacy, functionality, and scalability for boosting trust. In this sense, KRAKEN will provide a highly trusted, secure, scalable, and efficient personal data sharing and analysis platform adopting cutting-edge technologies. The KRAKEN project is based on three main pillars:

- The self-sovereign identity (SSI) paradigm provides user-centric access control to data. The data owner controls their data by using an SSI mobile app where the verifiable credentials and key material are stored.
- The cryptographic techniques support the other two pillars, such as functional encryption (FE) and secure multi-party computation (SMPC). These tools enable building a data-analytics-as-a-service platform integrated with the marketplace. They also ensure end-to-end secure data sharing in terms of confidentiality and authenticity.
- Data marketplace brings together the other two pillars, demonstrating in two high-impact pilots health and education the applicability of the KRAKEN solution. The marketplace acts as an open and decentralized exchange system connecting data providers and data consumers, leveraging a blockchain network facilitating the business and legal logic related to the data transactions.

---

[14] https://credential.eu/

[15] http://www.myhealthmydata.eu/

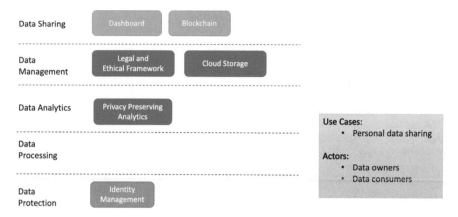

Fig. 10 System overview and data services of Kraken data platform

To follow regulatory frameworks and be GDPR and eIDAS compliant, an ethical and legal framework is implemented, affecting both the design and the implementation aspects. Figure 10 provides an overview of data services offered by Kraken data platform.

3.9 DataVaults Project

DataVaults[16] aims to deliver a framework and a platform that has personal data, coming from diverse sources in its center and that defines secure, trusted, and privacy-preserving mechanisms allowing individuals to take ownership and control of their data and share them at will, through flexible data sharing and fair compensation schemes with other entities (companies or not). Furthermore, data queries and analysis in DataVaults will allow the linking and merging of data from various sources and combining those with personal data, based on the DataVaults core data model. These activities, which rely on the semantic annotation of data and the curation of those to make them linkable, will raise the economic value of both personal and other kinds of data, as more detailed and interesting insights will be generated.

The overall approach will rejuvenate the personal data value chain, which could from now on be seen as a multi-sided and multi-tier ecosystem governed and regulated by smart contracts which safeguard personal data ownership, privacy, and usage and attributes value to the ones who produce it. Addressing the concerns on privacy, ethics and IPR ownership over the DataVaults value chain is one of the cornerstones of the project. It aims to set, sustain, and mobilize an ever-growing

---

[16] https://www.datavaults.eu/

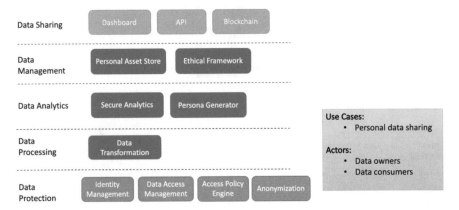

Fig. 11 System overview and data services of DataVaults data platform

ecosystem for personal data and insights sharing and for enhanced collaboration between stakeholders (data owners and data seekers) on the basis of DataVaults personal data platform's extra functionalities and methods for retaining data owner-ship, safeguarding security and privacy, notifying individuals of their risk exposure, as well as on securing value flow based on smart contract. Figure 11 provides an overview of data services offered by DataVaults data platform.

4 Comparison of Data Management Services

The projects presented in the previous section will be explored based on their data management and Data Governance features. The comparison is based on the following primary requirements for a catalogue and entity management service (EMS) [4], which are needed to support the incremental data management approach of Data Spaces:

- Data Source Registry and Metadata: The requirement to provide a registry for static and dynamic data sources and their descriptions.
- Entity Registry and Metadata: The requirement to provide a registry for entities and their descriptions.
- Machine-Readable Metadata: The requirement to store and provide metadata about data sources and entities in machine-readable formats using open standards such as JavaScript Object Notation (JSON) and Resource Description Frame-work (RDF).
- Schema Mappings: The capability to define mappings between schema elements.
- Entity Mappings: The capability to define mappings between entities.
- Semantic Linkage: The capability to define semantic relationships and linkages among schema elements and entities.

- Search and Browse Interface: The requirement to provide a user interface over the catalogue and EMS, which allows searching and browsing all the stored elements.
- Authentication and Authorization: The requirement to verify the credentials of users and applications accessing the catalogue and EMS, which can limit access to sources/entities based on access policies or rules.
- Data Protection and Licensing: The requirement to fulfill the privacy and confidentiality requirements of data owners and provide licensing information on the use of data.
- Provenance Tracking: The requirement of tracking the lineage of changes made to the catalogue and EMS by users and applications.

	DP	FSM	I3M	OM	T	SH	PC	K	DV
Data Source Registry and Metadata	+	+	+	+	+	+	+	+	+
Entity Registry and Metadata	+	+	+	+	+	+	+	+	+
Machine-Readable Metadata	+	+	+	+	+		+	+	+
Schema Mappings			+	+	+		+		+
Entity Mappings		+			+				+
Semantic Linkage	+	+	+	+	+				+
Search and Browse Interface	+	+	+	+	+	+	+	+	+
Authentication and Authorization	+	+	+	+	+	+	+	+	+
Data Protection and Licensing		+		+	+	+	+	+	+
Provenance Tracking	+	+	+	+			+	+	

5 Key Enabling Technologies

Data platform implementations address a set of common and known problems and challenges which can be tackled through existing guidelines, appropriate technologies, and best practices. In this section we provide an overview of some key enabling technologies that have broad application in the implementation of data platforms.

5.1 Semantics

Semantics is becoming increasingly important within Data Spaces and potentially becomes an area of competitive advantage in European data space and data market projects where semantics plays the role of federator and ecosystem facilitator. In order to facilitate the use of data platforms, they should be able to semantically annotate and enhance the data without imposing extra effort on data owners and data producers. The semantically enhanced data will improve various data processes and unlock data silos using interoperability standards and efficient technologies of the

Semantic Web domain. For instance, data integration, one of the hardest challenges in computer science, benefits significantly from semantics. The challenging issue in data integration is that people have different conceptualizations of the world. As a result, we would need a computational semantics layer to automate the process of data integration. The realization of the semantic layer could range from implementing taxonomies and metadata schema to more profound knowledge engineering approaches, including ontologies and semantic deduction. Semantic technologies and related solutions are also used for indexing and discovery of data services and creating service repositories in the context of data platform ecosystems.

5.2   Blockchain and Smart Contracts

As data and its corresponding services move beyond the organization's borders, we need mechanisms that support data integrity and traceability. In decentralized data platforms where peers need to share and consume data resources, trustworthiness and transparency of data processes are of great importance. Such platforms should provide features such as data provenance, workflow transparency, and authenticity of data providers and data consumers. To this end, blockchain offers disruptive methods to view the Data Governance, open data, and data ownership problems from a new perspective.

Recent advances in blockchain technology have shown the potential of building trustworthy data sharing solutions while maintaining a sufficient level of transparency in decentralized ecosystems. To this end, several data platforms have employed smart contracts as the conceptual basis for capturing and realization of business requirements. Furthermore, the smart contract and blockchain-driven approaches also incentivize user engagement and monetizes data platform solutions.

5.3   AI and Machine Learning

The success of AI and machine learning approaches is determined by the availability of high-quality data, which allows running various statistical learning methods to build efficient models and processes. To this end, data platforms are one of the key enablers to AI and machine learning processes. The value-added services of data platforms make high-quality and trustworthy data available to data consumers who use the data to create innovative solutions and turn data into products and services. Furthermore, data platforms apply machine learning and AI processes to improve data quality and enrich the data with the required information during the data preparation pipelines. Due to the increasing amount of data and the growing need for data and data services, applying AI and machine learning methods in data processes is inevitable. Several data platforms are already equipped with such processes, which range from data analysis (e.g., data cleaning and data integration) and data enrichment (e.g., data annotation and clustering) to more advanced processes such as natural language processing (NLP) and automatic enforcement of legal and ethical requirements (e.g., identifying sensitive data and data anonymization).

6 Common Challenges and Lessons Learned

The design and implementation of data platforms pose significant technical challenges such as data integration, interoperability, privacy-related issues, and non-technical challenges such as legal and technical issues, engaging end-users, and organizational challenges. This section provides an overview of such challenges encountered by data platform projects and discusses some best practices to address these challenges.

6.1 AI and Machine Learning Challenges

The recent advances in AI and machine learning have led to widespread use in data-driven use cases and industries [5]. Although the results are promising, and in many cases comparable to human performance, there are several situations where the outcomes are unreliable and require human intervention to address irregular and unpredictable situations. For instance, the lack of comprehensive datasets in some cases may lead to algorithmic discrimination against minorities or misleading outcomes.

In the AI and machine learning domains, high-quality data plays a significant role and greatly determines the quality of outcomes. As such, data platforms need to integrate best practices for data quality into their architecture in order to provide trustworthy and reliable data sources.

One particular challenge ahead is the automation of AI methods in data platforms, which depends on several configurations such as choice of algorithm, configuring relevant parameters of selected algorithms, and identification of features from available datasets. So, including metadata and machine-readable description of AI and machine learning components and including them in transfer learning processes seem to be the key to addressing these issues.

Another challenge in machine learning services of data platforms is to supplement the outcomes with human-understandable explanations. Explainable AI is an active field of research in the AI domain, and the data platform community needs to consider the explainability feature in data platform processes when appropriate.

6.2   Legal Challenges

Data platforms require to implement and realize legal requirements. In the case of domain-specific data platforms that deal with known vendors and predetermined business requirements, the policy and regulation are assessed and included in the architecture of the corresponding data platform. However, in the case of generic data platforms such as data markets, the uncertainty of policy and regulations makes this task very difficult, if not impossible. The legal challenges span many policies and regulations, including privacy and data protection law, national/international data protection legislation (e.g., GDPR), human rights law (e.g., EU Convention on Human Rights), and business regulations.

In addition to identifying the relevant regulations for different datasets and use cases, the regulations need to be captured and articulated in a machine-readable way and be integrated into data platform processes.

6.3   Ethical Challenges

Moral and ethical guidelines are challenging parts of data platform implementation. Similar to legal challenges, there is a handful of relevant policies and guidelines for justified use of data in data platforms. However, there is no one-size-fits-all scenario, and we need to define the ethical regulations based on the specific requirements of the target domain.

The ethical challenges are weighted more importantly when dealing with the storage and sharing of personal information. Therefore, we would need privacy-aware analytics methods to realize the moral and ethical requirements. Furthermore, in the case of personal information, we would also need fine-grained consent management processes that clearly define and include user preferences and the conditions of granting and revoking data usage permissions [6].

6.4   Sustainability Challenges

In data platforms, sustainability includes two dimensions: sustainability of software architecture and sustainability of data services. The development of sustainable architecture is a well-studied domain. As a result, there are a handful of guidelines and solutions for implementing a sustainable software architecture that can be applied to the specific case of data platform systems. In contrast, the sustainability of data services in generic data platforms is not a straightforward task. The reason is that data services are usually built on top of a complex web of data, policies, and regulations that might change over time. For instance, if the international e-privacy regulation is changed, data platforms need to evaluate the new requirements and make sure the data platform complies with new requirements, which is a challenging task. Also, the sustainability of data services is even more complicated if the policy changes need to be applied to legacy data processes and existing data processes. For instance, if the data sharing process uses blockchain technology, reconsidering shared data might be infeasible due to the immutable nature of public blockchains.

6.5 User Engagement Challenges

Active participation of users from the very early stages of a data platform project is a fundamental element for the successful implementation and realization of data platforms. User participation guarantees continued support and sustainable development of data platform systems in the future. Therefore, adopting a user-oriented approach, while analyzing business requirements and exploring legal and ethical aspects through user stories, is a key factor here. However, despite the fundamental role of user engagement in data platform projects, a handful of challenges make user involvement difficult. For instance, lack of trust for sharing personal and industrial data, before realizing a data platform and envisioning benefits and gained values for each stakeholder, is one of the main challenges.

7 Conclusion

Big Data ecosystems offer enormous potential to support cross-organization data sharing and trading in a trusted, secure, and transparent manner. This chapter presented several data platform projects and highlighted their common features and services. These platforms provide various services and facilitate reliable and transparent data flow between systems. However, to achieve maximum economic and societal benefits, several challenges in AI and machine learning, and ethical, legal, and sustainability domains need further investigation.

References

1. Big Data Value Association. (2019). *Towards a European Data Sharing Space: Enabling data exchange and unlocking AI potential*. BDVA. http://www.bdva.eu/node/1277
2. Curry, E., & Ojo, A. (2020). Enabling knowledge flows in an intelligent systems data ecosystem. In *Real-time linked dataspaces* (pp. 15–43). Springer.
3. Zillner, S., Curry, E., Metzger, A., & Auer, S. (2017). European big data value partnership strategic research and innovation. *Agenda, 2017*.
4. ul Hassan, U., Ojo, A., & Curry, E. (2020). Catalog and entity management service for internet of things-based smart environments. In *Real-time Linked Dataspaces* (pp. 89–103). Springer.

5. Zillner, S., Bisset, D., Milano, M., Curry, E., Södergård, C., & Tuikka, T. (2020). *Strategic research, innovation and deployment agenda: AI, data and robotics partnership.*
6. EC. (2020). *Communication: A European strategy for data.* Retrieved from https://ec.europa.eu/info/sites/info/files/communication-european-strategy-data-19feb2020_en.pdf

Technological Perspective of Data Governance in Data Space Ecosystems

Ana I. Torre-Bastida, Guillermo Gil, Raúl Miñón, and Josu Díaz-de-Arcaya

Abstract Data has been identified as a valuable input to boost enterprises. Nowadays, with the vast quantity of data available, a favorable scenario is established to exploit it, but crucial challenges must be addressed, highlighting its sharing and governance. In this context, the data space ecosystem is the cornerstone which enables companies to share and use valuable data assets. However, appropriate Data Governance techniques must be established to benefit from such opportunity considering two levels: internal to the organization and at the level of sharing between organizations. At a technological level, to reach this scenario, companies need to design and provision adequate data platforms to deal with Data Governance in order to cover the data life-cycle. In this chapter, we will address questions such as: How to share data and extract value while maintaining sovereignty over data, confidentiality, and fulfilling the applicable policies and regulations? How does the Big Data paradigm and its analytical approach affect correct Data Governance? What are the key characteristics of the data platforms to be covered to ensure the correct management of data without losing value? This chapter explores these challenges providing an overview of state-of-the-art techniques.

Keywords Data Spaces · Big Data Governance · Big Data platform · Data sharing · Big Data life-cycle · DataOps

1 Introduction

Today we find ourselves in a completely digital world where content generated by devices and people increases every second. People are witnessing an explosion in the types, volume, and availability requirements of the different data sources. To get an idea of the magnitude of the problem and its growth, the data generated during 2

A. I. Torre-Bastida (✉) · G. Gil · R. Miñón · J. Díaz-de-Arcaya
TECNALIA, Basque Research & Technology Alliance (BRTA), Bilbao, Spain
e-mail: isabel.torre@tecnalia.com; guillermo.gil@tecnalia.com; raul.minon@tecnalia.com; josu.diazdearcaya@tecnalia.com

© The Author(s) 2022
E. Curry et al. (eds.), *Data Spaces*, https://doi.org/10.1007/978-3-030-98636-0_4

days in 2011 is larger than the accumulated from the origin of the civilization to the beginning of 2003. Therefore, we can ensure without any doubt that we live in the era of information. Consequently, a wide range of technologies associated with this term have emerged.

Nevertheless, this new concept not only has this technological vertical and information volume, it represents what has come to be called the power of data. Big Data is more than just a technological revolution and its use raises a radical transformation of mentality and business model in companies. Its purpose is to take advantage of the incalculable value of the data collected from different internal and external sources, such as customers, products, and operations to successfully optimize main business process performance or even to generate new business opportunities and models. Due to this phenomenon, organizations must face new challenges to go beyond what traditional tools are reporting with their information when analyzing, discovering, and understanding their data. Besides a matter of size, it is a change in mentality, a vision to boost business opportunities driven by data. In fact, companies are already leveraging the power of data processing to better understand their customers' profiles, needs, and feelings when interacting with their products and services. Now, it is time to explore the next level and identify new business opportunities and initiatives based on the study, exploitation, and sharing of data.

Big Data represents a new era in the use of data, and together with new paradigms such as Internet of Things (IoT) or Artificial Intelligence (AI), the relevance and value of data have been redefined. Organizations are becoming more and more aware of the notion of "data as an asset," but, on the other hand, they are also starting to be conscious of the need to report a "single version of the truth" [46] to take advantage of the opportunity of deploying business around data. But besides possible benefits, in this new scenario a number of additional challenges must be addressed to extract trustworthy and high-quality data.

The most frequent barriers to the successful exploitation of data are the diversity of sources and types of data, the tremendous volume of data, the disparate approaches and use cases, and the different processing and storage flows with which they can be implemented, in addition to the high volatility and the lack of unified data quality standards. Moreover, these problems are multiplied when considering open scenarios where different data agents or organizations interact, and beyond that, at macro level, when national or international strategies, policies, and regulations must be taken into consideration. In this context, the concept of Data Governance [48], which has existed for decades in the ICT field, has come to the fore in this realm. Data Governance should be understood as all those supporting mechanisms for decision making and responsibilities for processing related with information. Previous work must be accomplished to clearly identify which models will be established and who can take the actions, what data is going to be used, when it is going to be taken, and with what methods will be described. Data Governance concerns any individual or group that has any interest in how data is created and how it is collected, processed, manipulated, stored, and made available for use/exploitation during the whole life-cycle.

In the last decades, many efforts [49] have been dedicated to the specification of frameworks and standards in which the methodology and indicators for adequate government of traditional data processes are defined. Nowadays, these traditional approaches are insufficient due to the complex scenarios that arise where sharing, control of use, or ethical aspects become relevant. For instance, in [2] Alhassan et al. discover more than 110 new activities related with Data Governance that need to be acted upon in current data ecosystems. In contrast, there are certain aspects of Data Governance that remain unchanged, such as the establishment of clear objectives and the management of two essential levels of definition. The objectives are the following:

- Ensure that data meets the needs of the organization as well as other applicable requirements (i.e., legal issues).
- Protect and manage data as a valuable asset.
- Reduce data management costs.

Derived from these objectives, considerations which are grouped into the following two levels must be implemented:

1. Organizational level: The change of mentality of the organizations, involving three key concepts: data, participants, and processes, and the necessary legal, regulatory, administrative, or economic areas of action to establish an adequate Data Governance methodology.
2. Technological level: The technological platform that must be deployed to make Data Governance viable. In this chapter, the reference key architectures and the general data platform concepts are detailed in later sections.

In this chapter, new prospects for Data Governance will be defined: sovereignty, quality, trust and security, economics and ethics, as well as the objectives, guiding principles, methodologies, and agents that are necessary to contribute in an environment as complex as Big Data and data ecosystems.

In the described scenario, three key conditioning factors must be highlighted: (1) the availability of large and heterogeneous volumes of internal and external data, (2) the interest of companies in exploiting them through new approaches, and (3) the study of new methodologies and areas of action to ensure its correct governance. These three points together shape what has been called Big Data ecosystems, a new wave of large-scale data-rich smart environments. These Data Spaces or ecosystems present new challenges and opportunities in the design of the architecture necessary to implement them, as stated in the work presented by Curry et al. [23]. Therefore, in this chapter, the possibilities and challenges of these new ecosystems are introduced taking into consideration aspects such as the challenges that sovereignty presents in shared spaces and how it is a pressing need to enable the "wealth of data" through governing policies or agreements of use.

Finally, a crucial aspect that all Data Governance approaches must cover cannot be forgotten, the data life-cycle. Thanks to the life-cycle and its correct definition, data processes can be better understood considering the nature and the phases they require to become a valuable asset. In recent years, this life-cycle concept has been perfected and specialized to address large volumes of data, and, as a result, emerging new phases have appeared to embrace the five Vs of Big Data ecosystems (volume, velocity, variety, veracity, and value). At the same time, new life-cycle schemes more oriented to analysis using AI techniques have appeared. However, ensuring the life-cycle through Data Governance methodologies is not a simple task, and, unfortunately, in multiple organizations it has resulted in failure due to its large technological and application load. In recent times, the study of new ways of implementing Data Governance in a more practical way has proliferated, trying to reconstruct it as an engineering discipline to promote its integration in product development and adapt it to the data and software life-cycle. It is the set of good practices for Data Governance, known as DataOps. In this chapter, a section to deal with this new phenomenon is dedicated. Concretely, this section will expose, in a practical way, software engineering and analytical techniques that implement the life-cycle with correct Data Governance.

The chapter relates to the technical priorities in the implementation of Data Governance into Big Data platforms of the European Big Data Value Strategic Research and Innovation Agenda [77]. It addresses the horizontal concern Big Data Governance of the BDV Technical Reference Model. The chapter deals with the implementation of governance, as a cornerstone of Big Data platforms and Data Spaces, enablers of the AI, Data and Robotics Strategic Research, Innovation, and Deployment Agenda [78]. The specific contributions of this chapter can be summarized as follows:

1. Specification of the concept of Data Governance in the scenario of large volumes of data in an external-internal organizational context, from two aspects: conceptual and technological
2. Introduction of new practical paradigms for the implementation of data governance as a software engineering discipline: DataOps
3. Review of existing technological stacks to implement tasks related with Big Data Governance

The rest of the manuscript is structured as follows: Sect. 2 and subsections therein provide the aforementioned definition of Data Governance in the context of the Big Data problems, detailing the new emerging environment of Data Spaces. Next, Sect. 3 delves into the phases of the Big Data life-cycle and its connotations for Data Governance, introducing new paradigms such as DataOps. Section 4 presents crucial architectural patterns for data sharing, storage, or processing that should be considered when dealing with the concept of Big Data Governance. The chapter ends with Sect. 5, where a set of conclusions and open challenges are summarized regarding Data Governance action areas.

2 Data Governance: General Concepts

Data Governance, in general terms, is understood as the correct management and maintenance of data assets and related aspects, such as data rights, data privacy, and data security, among others. It has different meanings depending on the considered level: micro, meso, and macro levels. At the micro level, or intra-organizational, it is understood as the internal managing focus inside an individual organization with the final goal of maximizing the value impact of its data assets. At the meso level, or inter-organizational, it can be understood as the common principles and rules that a group of participating organizations agree in a trusted data community or space for accessing or using data of any of them. At the macro level, it should be understood as a set of measures to support national or international policies, strategies, and regulations regarding data or some types of data. Even if Data Governance goals and scopes vary at micro, meso, or macro levels, in most practical cases the goals should be aligned and connected for the whole data life-cycle, being part of a governance continuum. As an example, the General Data Protection Regulation (GDPR) [33] is a governance measure at the macro level for European member states regarding personal data and the free movement of such data that requires its translation internally in organizations at the micro level to cope with it.

The culture of data and information management into an organization is a very widespread area of study [46, 48], and for years there have been multiple frameworks or standards [15] in which the methodology and indicators of a correct governance of traditional data are defined, such as:

- DAMA-DMBOK [18]—Data Management Book of Knowledge. Management and use of data.
- TOGAF [44]—The Open Group Architecture Framework. Data architecture as part of the general architecture of a company.
- COBIT [25]—Control Objectives for Information and Related Technology. General governance of the information technology area.
- DGI Data Governance Framework [71]. It is a simple frame of reference to generate Data Governance.

In environments as complex and heterogeneous as it is in the case of Big Data ecosystems, the traditional version of governance is not enough and this concept of "government and sovereignty of data" needs an adaptation and redefinition, as it is studied in the works presented in [1, 54, 69, 70]. All these works present common guidelines, which are the general definition of the term "Big Data Governance," the identification of the main concepts/elements to be considered, and the definition of certain principles or disciplines of action which pivot or focus on the methodology. From the detailed analysis of these studies, we can therefore establish the appropriate background to understand the governance of data on large volumes, starting by selecting the following definition:

> Big Data Governance is part of a broader Data Governance program that formulates policy relating to the optimization, privacy, and monetization of Big Data by aligning the objectives of multiple functions. **Sunil Soares, 2013**

An emerging aspect that is complementing all these works of Data Governance is being developed when several participants are sharing information in common spaces, or Data Spaces, and the business models that can arise from these meso scenarios.

The governance in a data space sets the management and guides the process to achieve the vision of the community, that is, to create value for the data space community by facilitating the finding, access, interoperability, and reuse of data irrespective of its physical location, in a trusted and secure environment. For each participant of the community, this implies that the authority and decisions that affect its own data assets should be orchestrated also with the rights and interests of other collaborating participants of the data space ecosystem.

The governance in this meso level should provide the common principles, rules, and requirements for orchestrating data access and management for participants according to the different roles they may play within the data space ecosystem. It is implemented with a set of overlapping legal, administrative, organizational, business, and technical measures and procedures that define the roles, functions, rights, obligations, attributions, and responsibilities of each participant.

Some of the principles that are usually considered for the governance of a data space are:

- General conditions for participation and data sharing within the data space in the manner that it is intended for this space, including aspects such as confidentiality, transparency, fair competence, non-discrimination, data commercialization and monetization, (the exercise of) data property rights provision and protection, etc. These conditions could be administrative, regulatory, technical, organizational, etc.
- Conditions for the protection and assurance of the rights and interests of data owners, including data sovereignty to grant and, eventually, withdraw consent on data access, use, share, or control.
- Compliance of the rules and requirements of the data space with the applicable law and prevention of unlawful access and use of data; for instance, regarding personal data and highly sensitive data, or the requirements of public policies, such as law enforcement related to public security, defense or criminal law, as well as other policies, such as market competition law and other general regulations.
- Specific provisions that may be made for certain types of participants and their roles in the data space, such as the power to represent another participant, the power to intermediate, the obligation to be registered, etc.
- Specific provisions that may be made for certain types and categories of data in the data space, like the anonymization and other privacy-proof techniques of personal data, the grant to reuse public owned data under certain conditions, the grant to be processed under certain administration, etc.

Some of the types of rules and requirements that are used in the governance of a data space are:

- Specific technical, legal, or organizational requirements that participants may comply with, depending on their specific role in the ecosystem: the publication of associated metadata to data assets with common vocabularies, to follow specific policies, to adopt some specific technical architectures, etc.
- Conditions, rules, or specific arrangements for data sharing among participants of the data space, as well as specific provisions that may be applicable to the processing environment or to certain types of data. Note that data sharing does not necessarily imply the transfer of data from one participating entity of the data space to another.
- Access or use policies or restrictions that may be applied to certain types of data during the full life-cycle.
- The inclusion of compliance and certification procedures at different levels: participants, technologies, data, etc.
- Conditions, rules, or specific arrangements for, eventually, data sharing with third parties. For instance, one of the participants may need to share some data under the governance of the data space with a third party which is not subject to data space governance.

In addition to these micro and meso governance levels, there would be also a macro level where public administration bodies, from local to international ones, may set specific governing measures and mechanism to support their strategies, in many cases as an alternative to the de facto models of the private sector and, in particular, of big tech companies. The previously mentioned GDPR or the directive on open data and the re-use of public sector information, also known as the "Open Data Directive," are examples of the measures of the specific legislative framework that the European Union is putting in place to become a leading role model for a society empowered by data to make better decisions—in business and the public sector. As these perspectives of Data Governance are being developed independently, there is a potential break in continuity among them. In order to close the gap and create a governance continuum, it is necessary to support its coverage from, ideally, both the organizational and technical perspectives.

In Fig. 1, we synthesize our vision of the components that should be considered in an open, complex Data Governance continuum, defining two levels: organizational and technological.

In the organizational section we consider two important aspects, the governance dimensions and perspectives. Dimensions refer to the scope of governance, either internal or within the organization, or external affecting ecosystems formed by multiple participants, such as Data Spaces. Perspectives refer to the areas of action, which are those tasks or fields that the governance measures and mechanisms must cover.

In the work of Soares et al. [69], a Big Data Governance framework was introduced, with the following disciplines: organization, metadata, privacy, data quality, business process integration, master data integration, and information life-cycle

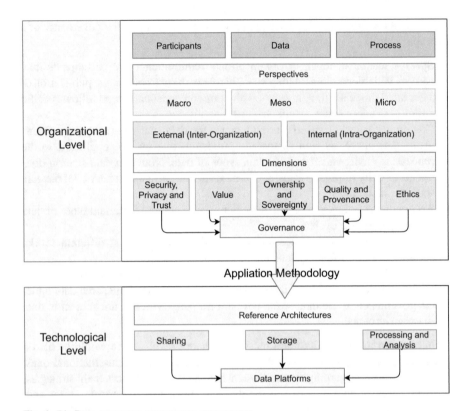

Fig. 1 Big Data ecosystem governance components

management. Based on these disciplines, the proposed governance perspectives of Fig. 1 are grouped into the following themes:

1. Ownership and sovereignty [40]. The data ownership is an important aspect when the intention is offering data and negotiating contracts in digital business ecosystems. And the associated term "data sovereignty" indicates the rights, duties, and responsibilities of that owner.
2. Trust, privacy, and security [64]. The data throughout its entire life-cycle must be safe and come to add value without being compromised at any point. For this, it is important that throughout the cycle all participants are trusted.
3. Value [28]. New digital economic models based on data as an asset of organizations are required. The concept of monetization of data, which seeks using data to increase revenue, is essential in this perspective.
4. Quality and provenance [75]. Data quality refers to the processes, techniques, algorithms, and operations aimed at improving the quality of existing data in companies and organizations, and associated with this comes the provenance, which indicates the traceability or path that the data travels through the organization.
5. Ethics [63]. Refers to systematizing, defending, and recommending concepts of correct and incorrect conduct in relation to data, in particular, personal data.

This list extends Soares's work with security aspects, and life-cycle management has been considered independently as a crosscutting element, and it is covered in Sect. 3 of this chapter. Finally, it is important to reconcile the methodology defined at the conceptual level with the technical implementation, including its components and modules in adequate reference architecture. In this chapter, three reference architecture approaches have been considered, according to their main mission: sharing, storage, or processing of data. An explanation with detailed examples of each theme is introduced in Sect. 4.1. The materialization of the architecture or possible combinations of architectures, with a selection of effective technologies that meet the requirements of data and information governance, have been identified as a data platform.

3 Big Data Life-Cycle

As mentioned in the previous section, Data Governance should improve data quality, encourage efficient sharing of information, protect sensitive data, and manage the data set throughout its life-cycle. For this reason, a main concept to introduce within the Data Governance is the life-cycle. This is not a new concept, but it must be adapted with the five Vs of Big Data characteristics [67]. The Big Data life-cycle is required in order to transform the data into valuable information, due to its permits that data can be better understood, as well as, a better analysis of its nature and characteristics.

In Fig. 2 you can see a life-cycle scheme, composed of five phases: collection, integration, persistence, analysis, and visualization and how each of these phases

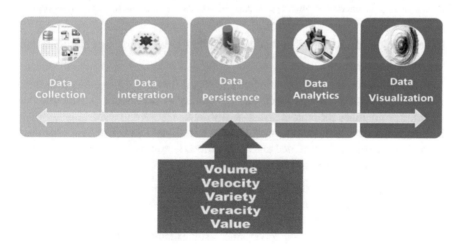

Fig. 2 Big Data life-cycle

in the case of a Big Data ecosystem should be adapted to offer a response to the requirements imposed by the five Vs [59]: volume, velocity, variety, veracity, and value.

Data life-cycle management is a process that helps organizations to manage the flow of data throughout its life-cycle—from initial creation through to destruction. Having a clearly defined and documented data life-cycle management process is key to ensuring Data Governance can be conducted effectively within an organization. In the Big Data ecosystem, there are already approaches such as that of Arass et al. [11] that try to provide frameworks to carry out this management, but there is still some work to be done. In the next sections, a series of new software engineering disciplines that try to pave this way are introduced, such as DataOps.

3.1 DataOps

Big Data security warranty is associated with a correct establishment of privacy and "good Data Governance" policies and methodologies. However, the technology dimension and stack of Big Data tools are enormous, and most of them have arisen without taking into account the requirements and necessary components to implement an adequate Data Governance policy.

In this context, DataOps was born as a variant of the software engineering discipline, which tries to redirect the strict guidelines established in the typical Data Governance methodologies into a set of good practices that are easy to implement technologically. In the work [26] presented by Julian Ereth, a definition is proposed:

> DataOps is a set of practices, processes and technologies that combines an integrated and process-oriented perspective on data with automation and methods from agile software engineering to improve quality, speed, and collaboration and promote a culture of continuous improvement. Julian Ereth, 2018

In our opinion, we believe that this definition has to be outlined in two important points:

1. Establish two basic guiding principles [12]: (a) improve cycle times of turning data into a useful data product, and (b) quality is paramount, a principle that requires practices for building trustworthy data.
2. Integrate a new "data component" team to traditional DevOps teams. This team is in charge of all data-related tasks and primarily comprised of the roles of data providers, data preparers, and data consumers.

The main scopes of DataOps in Big Data environments can be seen in Fig. 3. The objective is to interconnect all these specialties with each other and with the traditional tasks of the life-cycle.

With DataOps automation, governance can execute continuously as part of development, deployment, operations, and monitoring workflows. Governance automation is one of the most important parts of the DataOps movement. In practice,

Fig. 3 Fields and subfields
of DataOps discipline

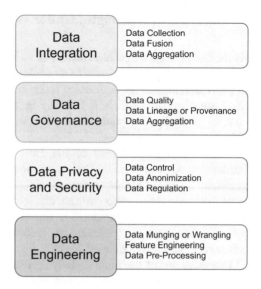

when Data Governance is backed by a technical platform and applied to data analytic, it is called DataOps. To achieve this, components and tools are required to develop or support these automation processes being part of the implementation of the platform. Ideally, many of these modules will use AI techniques to achieve greater intelligence and level of automation.

In this field, it is important to highlight how large and important companies and organizations are working on this type of governance implementation using the DataOps discipline:

1. Uber [72] has a dedicated project called Michelangelo that helps manage DataOps in the same way as DevOps by encouraging iterative development of models and democratizing the access to data and tools across teams.
2. Netflix [13] follows a DataOps approach to manage its historical data and their model versioning in the context of its tool of content recommendation.
3. Airbnb [17] controls the features engineering and parameters selection process over the models that they generated to improve search rankings and relevance, using a DataOps approach.

In conclusion, DataOps [55] aims to increase speed, quality, and reliability of the data and the analytic processes around it by improving the coordination between data science, analytic, data engineering, and operations.

4 Big Data Technologies Under Governance Perspective

As the Big Data paradigm has taken hold, so have the possible technological options to implement it. This gives rise to an endless number of systems and tools that try to provide solutions for addressing tasks in the data life-cycle; it is the so-called Big Data ecosystem [36, 53]. The way these solutions are bundled together into complete and effective product suites is commonly referred as Big Data platforms [30].

The technological origins of Big Data platforms are found in the Hadoop framework. Apache Hadoop originated from a paper on the Google file system published in 2003 [31]. Nowadays, Hadoop distributions continue to evolve, and from their base emerge increasingly efficient and comprehensive data or analytics platforms, as it will be explained in the next sections in further detail. The main reason is that as the this paradigm has matured, the diversity of use cases and their complexity have also increased.

However, one of the main underlying problems is that security technologies in these environments have not been developed at the same pace, and, consequently, this area has become a challenge. Concretely, governance has been a lately addressed issue, and currently it still poses many unresolved challenges. Therefore, in this section we focus on Big Data technologies, since they are currently under development and evolution to solve associated risks that do not occur in normal volumes of data, for example, the associated technical, reputational, and economic risks posited by Paul Tallon in special issue [52]. In the next sections, the current state of technology (see Sect. 4.2) is examined, and at the end of the chapter, a series of challenges are presented which even nowadays remain without a clear solution.

4.1 Architectures and Paradigms

In this section, Big Data platforms will be analyzed from different perspectives. First, architecture paradigms for data sharing in data ecosystems followed by storing of Big Data are analyzed, and then, computing paradigms are examined in detail.

4.1.1 Architectures for Data Sharing

There are different attempts in the literature to define reference architectures around the concept of data sharing in distributed environments. The approaches vary significantly according to their characteristics such as security conditions, purpose of sharing, or technological focus and especially depending on the domain of application to which they are addressed. Below, the most important ones trying to cover different domains are named:

- Industry 4.0. There are several initiatives at the international level where attempts are being made to organize the paradigm in an easily understandable and widely adopted reference architecture. This is the case of the International Data Spaces (IDS) architecture promoted by the German-based IDS Association (IDSA) [41], the Reference Architecture Model Industry 4.0 (RAMI4.0) [66] developed by the German Electrical Manufacturer's Association, and Electronic and Industrial Internet Reference Architecture (IIRA) of the Industrial Internet Consortium (IIC) [50]. Among all of them, the leading reference architecture in Europe, IDSA, stands out as optimal in terms of its capabilities to represent data security and sovereignty needs.
- Health domain. Data privacy is the central axis of most of the studies. An example is that of article [76], where a system is presented addressing the problem of the exchange of medical data between medical repositories in an environment without trust.
- Smart cities. Regarding this new paradigm of data sharing approaches, article [20] proposes a trust model to share data in smart cities.
- Social and cultural domains. In this domain, digital content sharing environments are considered, such as social networks, e-governance, or associations around research/cultural data. Article [39] provides a systematic mechanism to identify and resolve conflicts in collaborative social network data.
- It is worth mentioning the European project Gaia-X [19] is devoted to promote a secure, open, and sovereign use of data. It promotes the portability, interoperability, and interconnectivity within a federated infrastructure of data while also preserving the European values of openness, transparency, trust, and sovereignty, among others. Gaia-X's vision is to enable companies to share their data in ways they control. And to achieve an enabling infrastructure for this objective, Industrial Data Space (IDS) appear. For this reason, the IDS standard, which enables open, transparent, and self-determined data exchange, is a central element of the Gaia-X architecture.

4.1.2 Architectures for Data Storage

Traditionally to drive complex analysis on different data sources, there exist a great number of architectural patterns to store data in a centralized fashion. But the requirements and needs of today's large volumes of data make new distributed approaches necessary. Below, the three main architectural patterns that have been deployed in recent years by large corporations are presented: data warehouse, which follows a centralized approach in which data normally resides in a single repository, and data hub and data lake, where data can be stored in a decentralized manner, with physically distributed repositories sharing a common access point.

The data warehouse concept was originated in 1988 with the work of IBM researchers, later William H. Inmon [43], describing it as *A collection of data oriented to a specific, integrated, time-varying and non-volatile subject that supports the decision-making process.* As a general rule, this system is used mainly for reporting and data analysis. The repository can be physical or logical and emphasizes capturing data from various sources primarily for access and analytical purposes. The typical components of any data warehouse are as follows:

- ETL process: component that allows preprocessing operations on the data so that they can be persisted using a correct structure
- Central repository of the data warehouse, a central area where the data is stored
- Views or DataMarts: logical views of the data

Traditionally, corporate servers are used to allocate the data warehouse, but in the last years, with the boom of the cloud providers, enterprises are starting to deploy them in the cloud. As examples of data warehouse in the cloud, Amazon Redshift [3], Microsoft Azure Synapse Analytics [14], Google BigQuery [34], and Snowflake [68] can be highlighted.

The second approach is the data hub [16]. In this case, it does not mean that the data is centralized under the same structure, but that there is a central point of access to them. The difference is significant since the data is not what is centralized, but the information about it is the metadata. The main objective is to be able to integrate all the information about the data in the same point taking into account the different business needs that a company may have. In this case, data can be physically moved and have the ability to be ordered again in a different system, since its metainformation continues to be kept in a single point.

Finally, the datalake [27] architecture is presented, a clearly analytical approach, which is also the most modern, referring to a repository of data stored in its natural raw format, generally objects or files. Datalake is named for the metaphor that the data is in "water," which is a transparent and clear substance, so they are preserved in a natural state and not modified. What it is meant by this is that the data is original and there is a storage where a large amount of information of all kinds is together, from structured to unstructured data. It is an architectural design pattern that favors the use of data in analytical processes.

Datalakes contain a large amount of raw data kept there until needed. Unlike a hierarchical data warehouse that stores data in files or folders, a datalake uses a flat architecture to store data. The term is being accepted as a way to describe any large data set in which the schema and data requirements are not defined until the data is queried which is called as schema on reading.

Each element in a datalake is assigned a unique identifier and is tagged with a set of extended metadata tags. In this way, when a business issue needs to be resolved, the datalake can be queried for identifying data related to that issue. Once obtained, the data acquired can be analyzed to help organizations obtain a valuable answer.

A datalake is made up of two clear components: the repository of raw data and the transformations that allow adapting the data to the structure necessary for further processing.

In Table 1 a comparative of these three approaches is provided. In the case of choosing an analysis-oriented approach for specific use cases, the most appropriate option to consider would be the data warehouse. By contrast, if the analysis has not been defined and should be as open as possible to future use cases, the chosen

Table 1 Storage paradigm comparison

Feature	Data warehouse	Data hub	Data lake
Data	Structured	Semi-structured	Structured, semi-structured, non-structured
Processing	Schema on write	Mixed	Schema on read
Storage	Low scalability	Medium scalability	High scalability
Flexibility	Rigid	Adaptable	Flexible
Security	Very mature	Mature	Immature

option should be the datalake pattern, and finally, if the possible data sources or the purpose of using them is not clear, the best option would be data hub. In article [62], the benefits and disadvantages of each of the three architectural patterns are clearly exposed.

In order to boost Data Governance good practices, when utilizing these three storage paradigms, we propose some considerations:

- When a new data source is integrated, access control policies should be defined identifying who should have access, with which privileges, and to which content.
- For data lakes, all data sources should be defined in a centralized data catalog to avoid silos and to facilitate its access and discovery. In addition, relevant metainformation of each data source should be included to allow adequate filters and to improve classification mechanisms.
- A usage control policy should be defined for each data source to clearly identify aspects such as privacy fields that require anonymization, expiration of the data, and additional required operations that should be considered when treating the data source.
- An intelligent mechanism for detecting duplicated data and for analyzing data quality would be really useful to provide better insights.
- Encryption techniques should be considered when storing data.

The implementation of the aforementioned architectures can be achieved by using different technologies, some of which may be distributed such as file systems or NOSQL databases. An example of the former would be SAN systems, which offer distributed block storage between different servers. And some other feasible examples would be GlusterFS [32] for POSIX-compliant storage or the well-known Hadoop HDFS [35] which is able to leverage data locality for different types of workloads and tools.

4.1.3 Architectures for Data Processing

In the last decade, there has been a tendency for acquiring more and more data. In this context, traditional technologies, like relational databases, started to experience performance issues not being able to process and manage big quantities of data, with low latency and diverse structures. As a consequence, certain paradigms such as MapReduce [22] faced the problem by dividing the data in several machines, then processing data isolated in each machine (when possible), and finally merging the results. This way, distributed computing is enabled. In addition, in this paradigm data can be replicated among a set of machines to get fault tolerance and high availability. This approach paves the way for Big Data technologies enabling batch processing architectures.

As defined firstly in 2011 by [38] and in [29], the Lambda architecture is a generic, scalable, and fault-tolerant real-time processing architecture. Its general purpose consists of applying a function over a dataset and extracting a result with low latency. However, in some scenarios, the computation of raw data tends to derive in heavy processes; as a consequence, it is proposed to perform the specific queries over already preprocessed data to minimize the latency and get a real-time experience. For this objective, a speed layer can be supported from a batch layer to provide such preprocessed data, and finally, a serving layer is proposed to query over data merged from both batch and speed layers.

Later, in 2014, Kreps [47] questioned the disadvantages of the Lambda architecture and proposed the Kappa architecture, a simplified perspective where the batch layer was eliminated from the equation. One of the main advantages of this architecture is that there is no need for implementing two heterogeneous components for batch and speed layer. As a consequence, the development phases of implementation, debugging, testing, and maintenance are significantly simplified. By contrast, not having the batch layer hinders (or makes impossible) the execution of heavy processes since the data preprocessing conducted in that layer for the speed layer in the Lambda architecture cannot be done. Consequently, in the Kappa architecture the speed layer might execute heavier processes making it more inefficient.

The conclusion is that the processing architecture should be selected taking the type of problem to resolve into consideration. For executing heavy workloads without requiring real-time decisions, the batch architecture could be enough. On the contrary, when only real-time data is required and there is no need to provide historical data and, consequently, data processed can be discarded, Kappa architecture is adequate. Finally, when both heavy processes and data in real time are required, Lambda architecture should be selected.

Besides the governance considerations exposed in the previous subsection, when processing data, it is highly recommended to be able to track the operations performed over each data source in order to provide an adequate data lineage and to supply a robust audit mechanism.

4.2 Current Tool Ecosystem for Big Data Governance

Currently, there are a number of useful tools that cover Data Governance requirements for Big Data platforms to some extent. This subsection examines the more successful ones.

Kerberos [45] is an authentication protocol that allows two computers to securely prove their identity to each other. It is implemented on a client server architecture and works on the basis of tickets that serve to prove the identity of the users.

Apache Ranger [9] improves the authorization support for Hadoop ecosystems. It provides a centralized platform to define, administer, and manage security policies uniformly across many Hadoop components. In this way, security administrators can manage access control editing policies with different access scopes such as files, folders, databases, tables, or columns. Additionally, Ranger Key Management Service (Ranger KMS) offers scalable encryption in HDFS. Moreover, Ranger also gives the possibility of tracking access requests.

Apache Knox [7] Gateway offers perimeter security to allow data platforms to be accessed externally while satisfying policy requirements. It enables HTTP access using RESTful APIs to a number of technologies of the stack. For this purpose, Knox acts as an HTTP interceptor to provide authentication, authorization, auditing, URL rewriting, web vulnerability removal, and other security services through a series of extensible interceptor processes.

Apache Atlas [4] offers a set of governance utilities for helping companies in fulfilling their compliance requirements in the Hadoop data ecosystem. It allows managing the metadata to better catalog company data enabling classification and collaboration support for different profiles of a team.

As stated in [61], Cloudera Navigator [21] provides a set of functionalities related with Data Governance in the context of Big Data platforms. Concretely, it enables metadata management, data classification, data lineage, auditing, definition of policies, and data encryption. In addition, it can be integrated with Informatica [42] to extend the lineage support outside a data lake.

Additional technologies were going to be integrated in this section but, unfortunately, some of them, like Apache Sentry, Apache Eagle and Apache Metron, have been recently moved to the Apache Attic project, which means that they have arrived to its end of life.

Regarding technologies related to the task of data exchange and sharing, we can highlight that almost all solutions and approaches are in initial versions. Then we name as examples (a) certain platforms that have begun to investigate this field, Snowflake, (b) solutions related to data sharing in the cloud, and (c) finally technologies related to data usage control.

Snowflake [68] is yet another tool that offers Data Governance capabilities. In particular, one of its features is to securely and instantly share governed data across the business ecosystem. Sharing data within the same organization has also evolved over the years, and cloud-based technologies (e.g., ownCloud [58], OneDrive [56]) have replaced shared folders as the preferred method for sharing documentation.

Today, there are multiple technologies related to secure access control, such as Discretionary Access Control (DAC), Mandatory Access Control (MAC), Role-Based Access Control (RBAC), or Attribute-Based Access Control (ABAC); a comparative analysis can be found in the survey [73]. But their scope is limited to access and therefore they do not solve challenges encountered in the task of data sharing. For this task, distributed use control technologies are necessary. To our

Table 2 Big Data Governance ecosystem tools

Tools	Access control	Audit lineage	Metadata management data catalog	Encryption	Sovereignty	Quality	Sharing use control
Kerberos	✓	–	–	–	–	–	–
Ranger	✓	✓	–	–	–	–	–
Knox	✓	✓	–	–	–	–	–
Atlas	–	✓	✓	✓	–	–	–
Cloudera Navigator	–	✓	✓	✓	–	–	-
Snowflake	✓	✓	✓	✓	–	✓	✓
Lucon	✓	–	–	✓	–	–	✓

knowledge, there are not many research works in this field and the existing ones have failed to provide mature technological solutions to the industry [60]. For example, within the context of the IDS architecture [57], new solutions are being considered and in some cases developed that allow the control of data use in distributed storage platforms. Specifically, the technologies currently under development are LUCON [65], MYDATA-IND2UCE [37], and Degree [51], but all three are in the early stages of maturity.

As stated in this subsection and summarized in Table 2, besides the cloud proprietary solutions, there exist certain Big Data tools supporting governance aspects. However, there is still some missing governance support such as editing control of use policies for boosting the sovereignty, identifying duplicated data, checking the quality and veracity, or anonymizing sensible data when shared.

5   Conclusion and Future Challenges

In an environment as complex as Big Data, it is not easy to establish adequate Data Governance policies, and this difficulty is increased when the tools that contribute or help in this task have not followed the same level of growth as the Big Data technological stack. In this chapter, these facts have been presented and an attempt has been made to provide some light to pave the way for those companies or organizations that wish to establish a Big Data Governance methodology. In order to accomplish this, the main contributions have been focused on providing an adequate definition of the Big Data environment, introducing new disciplines such as DataOps that materialize Data Governance from a perspective close to software engineering. Moreover, finally, a number of significant architectures and solutions useful for the technological implementation have been examined.

In addition, throughout the whole chapter, new challenges have been introduced that currently are not adequately addressed. In this section, a compilation of the most important ones is proposed, highlighting the possible future works that would arise in their field. The following list is an enumeration of these challenges to consider:

- Usage control to warranty data sovereignty during data sharing process. The problem is how to technologically enable sharing while keeping sovereignty and the use of data safe and controlled. In article [74] this dilemma of finding the balance between correct regulation and technological disruption is posed. One of the main problems is what happens to the data once it leaves the repository or platform of the proprietary provider, such as controlling, monitoring, or tracing its use to facilitate future profitability. Until now, the concern was to protect access to data, but its use by potential consumers was not controlled, and, therefore, this is the greatest current challenge, especially at the technological level.

- Data modeling is cool again. Data modeling can help generate more efficient processing techniques for the data collected. But nowadays the problems are on (a) how to balance modeling between not losing data quality and not generating unnecessary metadata that takes up extra storage and (b) how to directly extract semantic metadata from moving data streams.

- New AI approaches and life-cycles. The new life-cycles associated with analytical processes developed with AI techniques have many challenges associated with the use of data, among which the following stands out: (a) heterogeneity in development environments and lack of uniformity or standards; with this we refer to the multiple techniques, libraries, existing frameworks, etc. (b) The high cost of infrastructure and the necessary storage scalability, for example, deep learning techniques require a very high sample volume, which leads to large repositories and high scalability.

- Ethical Artificial Intelligence (AI). Explainability, defined as why models make given predictions or classifications, is a hot topic in this area. However, equally important and often overlooked are the related topics of model auditability and governance, understanding and managing access to models, data, and related assets.

- Data literacy: data-driven culture company-wide. Understanding as enabling employees to derive meaningful insights from data. The objective is that the people of the organization are able to analyze situations from a critical perspective and be participants in an analytical culture based on the data. A very important point in this area, therefore, is the democratization and flow of data throughout the organization, a technological challenge that today is extremely complex to combine with access and privacy policies.

- Data catalog: The new technological trend. Gartner [24] even refers to them as "the new black in data management and analytic" and now they are recognized as a central technology for data management. It is a central and necessary technology for the treatment of metadata, which allows giving meaning to the data. Gartner has defined a data catalog as a tool that "creates and maintains an inventory of data assets through the discovery, description and organization of distributed datasets." And the real challenge in this technology is found in keeping the information in a data catalog up to date using automated discovery and search techniques.

• Data Governance at the edge. The architectural challenge faces on the use of edge computing is where to store the data: at the edge or at the cloud. And associated with this challenge in architecture, how to implement Data Governance, access and auditing mechanisms adapted and efficient to this complex environment, made up of multiple infrastructure elements, technological components and network levels.

Acknowledgments The work presented in this chapter has been partially supported by the SPRI—Basque Government through their ELKARTEK program (DAEKIN project, ref.KK-2020/00035).

References

1. Al-Badi, A., Tarhini, A., & I. Khan, A. (2018). Exploring big data governance frameworks. *Procedia Computer Science, 141*, 271–277.
2. Alhassan, I., Sammon, D., & Daly, M. (2018). Data governance activities: A comparison between scientific and practice-oriented literature. *Journal of Enterprise Information Management, 31*(2), 300–316.
3. Amazon redshift, 2021. Retrieved February 8, 2021 form https://aws.amazon.com/es/redshif
4. Apache atlas, 2021. Retrieved February 8, 2021 form https://atlas.apache.org/
5. Apache attic, 2021. Retrieved February 8, 2021 form https://attic.apache.org/
6. Apache eagle, 2021. Retrieved February 8, 2021 form https://eagle.apache.org/
7. Apache knox, 2021. Retrieved February 8, 2021 form https://knox.apache.org/
8. Apache metron, 2021. Retrieved February 8, 2021 form https://metron.apache.org/
9. Apache ranger, 2021. Retrieved February 8, 2021 form https://ranger.apache.org/
10. Apache sentry, 2021. Retrieved February 8, 2021 form https://sentry.apache.org/
11. Arass, M. E., Ouazzani-Touhami, K., & Souissi, N. (2019). The system of systems paradigm to reduce the complexity of data lifecycle management. case of the security information and event management. *International Journal of System of Systems Engineering, 9*(4), 331–361.
12. Atwal, H. (2020). Dataops technology. In *Practical DataOps* (pp. 215–247). Springer.
13. Atwal, H. (2020). Organizing for dataops. In *Practical DataOps* (pp. 191–211). Springer.
14. Azure synapse analytics, 2021. Retrieved February 8, 2021 form https://azure.microsoft.com/es-es/services/synapse-analytics/
15. Begg, C., & Caira, T. (2011). Data governance in practice: The SME quandary reflections on the reality of data governance in the small to medium enterprise (SME) sector. In *The European Conference on Information Systems Management* (p. 75). Academic Conferences International Limited.
16. Bhardwaj, A., Bhattacherjee, S., Chavan, A., Deshpande, A., Elmore, A. J. Madden, S., & Parameswaran, A. G. (2014). Data hub: Collaborative data science & dataset version management at scale. arXiv preprint arXiv:1409.0798.
17. Borek, A., & Prill, N. (2020). *Driving digital transformation through data and AI: A practical guide to delivering data science and machine learning products.* Kogan Page Publishers.
18. Brackett, M., Earley, S., & Henderson, D. (2009). *The dama guide to the data management body of knowledge: Dama-dmbok guide.* Estados Unidos: Technics Publications.
19. Braud, A., Fromentoux, G., Radier, B., & Le Grand, O. (2021). The road to European digital sovereignty with Gaia-x and IDSA. *IEEE Network, 35*(2), 4–5.

20. Cao, Q. H., Khan, I., Farahbakhsh, R., Madhusudan, G., Lee, G. M., & Crespi, N. (2016). A trust model for data sharing in smart cities. In *2016 IEEE International Conference on Communications (ICC)* (pp. 1–7). IEEE.
21. Cloudera navigator, 2021. Retrieved February 8, 2021 form https://www.cloudera.com/products/product-components/cloudera-navigator.html
22. Condie, T., Conway, N., Alvaro, P., Hellerstein, J. M., Elmeleegy, K., & Sears, R. (2010). Mapreduce online. In *Nsdi* (Vol. 10, p. 20).
23. Curry, E., & Sheth, A. (2018). Next-generation smart environments: From system of systems to data ecosystems. *IEEE Intelligent Systems, 33*(3), 69–76.
24. Data catalogs are the new black in data management and analytics, 2017. Retrieved March 16, 2021 from https://www.gartner.com/en/documents/3837968/data-catalogs-are-the-new-black-in-data-management-and-a
25. De Haes, S., Van Grembergen, W., & Debreceny, R. S. (2013). Cobit 5 and enterprise governance of information technology: Building blocks and research opportunities. *Journal of Information Systems, 27*(1), 307–324.
26. Ereth, J. (2018). Dataops-towards a definition. *LWDA, 2191*, 104–112.
27. Fang, H. (2015). Managing data lakes in big data era: What's a data lake and why has it became popular in data management ecosystem. In *2015 IEEE International Conference on Cyber Technology in Automation, Control, and Intelligent Systems (CYBER)* (pp. 820–824). IEEE.
28. Faroukhi, A. Z., El Alaoui, I., Gahi, Y., & Amine, A. (2020). Big data monetization throughout big data value chain: A comprehensive review. *Journal of Big Data, 7*(1), 1–22.
29. Feick, M., Kleer, N., & Kohn, M. (2018). Fundamentals of real-time data processing architectures lambda and kappa. *SKILL 2018-Studierendenkonferenz Informatik.*
30. Ferguson, M. (2012). Architecting a big data platform for analytics. *A Whitepaper prepared for IBM*, 30.
31. Ghemawat, S., Gobioff, H., & Leung, S.-T. (2003). The google file system. In *Proceedings of the Nineteenth ACM Symposium on Operating Systems Principles* (pp. 29–43).
32. Glusterfs, 2021. Retrieved May 19, 2021 form https://www.gluster.org/
33. Goddard, M. (2017). The EU general data protection regulation (GDPR): European regulation that has a global impact. *International Journal of Market Research, 59*(6), 703–705.
34. Google big query, 2021. Retrieved February 8, 2021 form https://cloud.google.com/bigquery/
35. Hadoop hdfs, 2021. Retrieved May 19, 2021 form https://hadoop.apache.org/docs/current/hadoop-project-dist/hadoop-hdfs/HdfsDesign.html
36. Hamad, M. M., et al. (2021). Big data management using hadoop. In *Journal of Physics: Conference Series* (Vol. 1804, p. 012109).
37. Hosseinzadeh, A., Eitel, A., & Jung, C. (2020). A systematic approach toward extracting technically enforceable policies from data usage control requirements. In *ICISSP* (pp. 397–405).
38. How to beat the cap theorem, 2011. Retrieved February 8, 2021 form http://nathanmarz.com/blog/how-to-beat-the-cap-theorem.html
39. Hu, H., Ahn, G.-J., & Jorgensen, J. (2011). Detecting and resolving privacy conflicts for collaborative data sharing in online social networks. In *Proceedings of the 27th Annual Computer Security Applications Conference* (pp. 103–112).
40. Hummel, P., Braun, M., Augsberg, S., & Dabrock, P. (2018). Sovereignty and data sharing. *ITU Journal: ICT Discoveries, 25*(8).
41. Ids association, 2021. Retrieved March 14, 2021 form https://internationaldataspaces.org/
42. Informatica, 2021. Retrieved February 8, 2021 form https://www.cloudera.com/solutions/gallery/informatica-customer-360.html
43. Inmon, W. H. (1996). The data warehouse and data mining. *Communications of the ACM, 39*(11), 49–51.
44. Josey, A. (2016). *TOGAF® version 9.1-A pocket guide*. Van Haren.
45. Kerberos, 2021. Retrieved February 8, 2021 form https://www.kerberos.org/
46. Khatri, V., & Brown, C. V. (2010). Designing data governance. *Communications of the ACM, 53*(1), 148–152.

47. Kreps, J. (2014). Questioning the lambda architecture. *Online article, July*, 205.
48. Ladley, J. (2019). *Data governance: How to design, deploy, and sustain an effective data governance program*. Academic Press.
49. Lee, S. U., Zhu, L., & Jeffery, R. (2018). Designing data governance in platform ecosystems. In *Proceedings of the 51st Hawaii International Conference on System Sciences*.
50. Lin, S.-W., Miller, B., Durand, J., Joshi, R., Didier, P., Chigani, A., Torenbeek, R., Duggal, D., Martin, R., Bleakley, G., et al. (2015). Industrial internet reference architecture. *Industrial Internet Consortium (IIC), Tech. Rep.*
51. Lyle, J., Monteleone, S., Faily, S., Patti, D., & Ricciato, F. (2012). Cross-platform access control for mobile web applications. In *2012 IEEE International Symposium on Policies for Distributed Systems and Networks* (pp. 37–44). IEEE.
52. Michael, K., & Miller, K. W. (2013). Big data: New opportunities and new challenges [guest editors' introduction]. *Computer, 46*(6), 22–24.
53. Monteith, J. Y., McGregor, J. D., & Ingram, J. E. (2013). Hadoop and its evolving ecosystem. In *5th International Workshop on Software Ecosystems (IWSECO 2013)* (Vol. 50). Citeseer.
54. Morabito, V. (2015). Big data governance. In *Big data and analytics* (pp. 83–104).
55. Munappy, A. R., Mattos, D. I., Bosch, J., Olsson, H. H., & Dakkak, A. (2020). From ad-hoc data analytics to dataops. In *Proceedings of the International Conference on Software and System Processes* (pp. 165–174).
56. Onedrive, 2021. Retrieved May 19, 2021 form https://www.microsoft.com/en-ww/microsoft-365/onedrive/online-cloud-storage
57. Otto, B., ten Hompel, M., & Wrobel, S. (2018). Industrial data space. In *Digitalisierung* (pp. 113–133). Springer.
58. owncloud, 2021. Retrieved May 19, 2021 form https://owncloud.com/
59. Pouchard, L. (2015). Revisiting the data lifecycle with big data curation.
60. Pretschner, A., & Walter, T. (2008). Negotiation of usage control policies-simply the best? In *2008 Third International Conference on Availability, Reliability and Security* (pp. 1135–1136). IEEE
61. Quinto, B. (2018). *Big data governance and management* (pp. 495–506).
62. Quinto, B. (2018). Big data warehousing. In *Next-generation big data* (pp. 375–406). Springer.
63. Richards, N. M., & King, J. H. (2014). Big data ethics. *Wake Forest Law Review, 49*, 393.
64. Sänger, J., Richthammer, C., Hassan, S., & Pernul, G. (2014). Trust and big data: a roadmap for research. In *2014 25th International Workshop on Database and Expert Systems Applications* (pp. 278–282). IEEE.
65. Schütte, J., & Brost, G. S. (2018). Lucon: Data flow control for message-based IoT systems. In *2018 17th IEEE International Conference On Trust, Security And Privacy In Computing And Communications/12th IEEE International Conference On Big Data Science And Engineering (TrustCom/BigDataSE)* (pp. 289–299). IEEE.
66. Schweichhart, K. (2016). Reference architectural model industrie 4.0 (rami 4.0). *An Introduction*. https://www.plattform-i40.de
67. Sinaeepourfard, A., Garcia, J., Masip-Bruin, X., & Marín-Torder, E. (2016). Towards a comprehensive data lifecycle model for big data environments. In *Proceedings of the 3rd IEEE/ACM International Conference on Big Data Computing, Applications and Technologies* (pp. 100–106).
68. Snowflake, 2021. Retrieved May 19, 2021 form https://www.snowflake.com/workloads/data-sharing/
69. Soares, S. (2012). *Big data governance*. Information Asset, LLC.
70. Tallon, P. P. (2013). Corporate governance of big data: Perspectives on value, risk, and cost. *Computer, 46*(6), 32–38.
71. Thomas, G. (2006). *The dgi data governance framework: The data governance institute*. Orlando: FL.
72. Thusoo, A. (2017). *Creating a data-driven enterprise with DataOps: Insights from Facebook, Uber, LinkedIn, Twitter, and EBay*. O'Reilly Media.

73. Ubale Swapnaja, A., Modani Dattatray, G., & Apte Sulabha, S. (2014). Analysis of dac mac rbac access control based models for security. *International Journal of Computer Applications, 104*(5), 6–13.
74. van den Broek, T., & van Veenstra, A. F. (2018). Governance of big data collaborations: How to balance regulatory compliance and disruptive innovation. *Technological Forecasting and Social Change, 129*, 330–338.
75. Wang, J., Crawl, D., Purawat, S., Nguyen, M., & Altintas, I. (2015). Big data provenance: Challenges, state of the art and opportunities. In *2015 IEEE International Conference on Big Data (Big Data)* (pp. 2509–2516). IEEE.
76. Xia, Q., Sifah, E. B., Asamoah, K. O., Gao, J., Du, X., & Guizani, M. (2017). Medshare: Trust-less medical data sharing among cloud service providers via blockchain. *IEEE Access, 5*, 14757–14767.
77. Zillner, S., Curry, E., Metzger, A., Auer, S., & Seidl, R. (2017). *European big data value strategic research & innovation agenda*. Big Data Value Association.
78. Zillner, S., Bisset, D., Milano, M., Curry, E., Södergård, C., Tuikka, T., et al. (2020). Strategic research, innovation and deployment agenda: Ai, data and robotics partnership.

Increasing Trust for Data Spaces with Federated Learning

Susanna Bonura, Davide dalle Carbonare, Roberto Díaz-Morales, Ángel Navia-Vázquez, Mark Purcell, and Stephanie Rossello

Abstract Despite the need for data in a time of general digitization of organizations, many challenges are still hampering its shared use. Technical, organizational, legal, and commercial issues remain to leverage data satisfactorily, specially when the data is distributed among different locations and confidentiality must be preserved. Data platforms can offer "ad hoc" solutions to tackle specific matters within a data space. MUSKETEER develops an Industrial Data Platform (IDP) including algorithms for federated and privacy-preserving machine learning techniques on a distributed setup, detection and mitigation of adversarial attacks, and a rewarding model capable of monetizing datasets according to the real data value. The platform can offer an adequate response for organizations in demand of high security standards such as industrial companies with sensitive data or hospitals with personal data. From the architectural point of view, trust is enforced in such a way that data has never to leave out its provider's premises, thanks to federated learning. This approach can help to better comply with the European regulation as confirmed

All authors have contributed equally and they are listed in alphabetical order.

S. Bonura · D. dalle Carbonare (✉)
Engineering Ingegneria Informatica SpA - Piazzale dell'Agricoltura, Rome, Italy
e-mail: susanna.bonura@eng.it; davide.dallecarbonare@eng.it

R. Díaz-Morales
Tree Technology - Parque Tecnológico de Asturias, Llanera, Spain
e-mail: roberto.diaz@treetk.com

Á. Navia-Vázquez
University Carlos III of Madrid, Madrid, Spain
e-mail: angel.navia@uc3m.es

M. Purcell
IBM Research Europe, IBM Campus, Damastown Industrial Estate, Dublin, Ireland
e-mail: markpurcell@ie.ibm.com

S. Rossello
KU Leuven Centre for IT & IP Law - imec, Leuven, Belgium
e-mail: stephanie.rossello@kuleuven.be

© The Author(s) 2022
E. Curry et al. (eds.), *Data Spaces*, https://doi.org/10.1007/978-3-030-98636-0_5

from a legal perspective. Besides, MUSKETEER explores several rewarding models based on the availability of objective and quantitative data value estimations, which further increases the trust of the participants in the data space as a whole.

Keywords Industrial Data Platform · Federated learning · Data Spaces · Data value estimation · GDPR · Trust · MUSKETEER

1 Introduction

Thanks to important advances in the recent years, machine learning has led to disruptive innovation in many sectors, for instance, industry, finance, pharmaceutical, healthcare, or self-driving cars, just to name a few. Since companies are facing increasingly complex tasks to solve, there is a huge demand for data in these areas. However, the task can be challenging also because it does not only depend on the companies themselves. For example, the healthcare sector has started to use machine learning to detect illnesses and support treatments. However, the necessity to use appropriate datasets, composed of data from enough patients suffering a given illness and related treatments, can be hindered by the limited number of patients that can be found in the historical medical records of a single hospital. This issue could be solved if people and companies were given an adequate way to share data tackling the numerous concerns and fears of a large part of the population that form barriers preventing the development of the data economy:

- Personal information leakage: The main concern of the population is the fear about possible information leakage. However, companies, in order to run their analysis, need digital information such as images or healthcare records containing very sensitive information.
- Confidentiality: A company can benefit from jointly created predictive models, but the possibility of leaking some business secrets in the process could lead to disadvantage this company vis-à-vis its competitors.
- Legal barriers: Governments, in order to regulate the use of data, have defined legal constraints that impact the location of data storage or processing.
- Ownership fear: Some data could be very valuable. Some companies and people could benefit economically from providing access to these data. But digital information could be easily copied and redistributed.
- Data value: Data owners could provide data with low quality, or even fake data, so that, effectively, there would only be limited value for other partners in using this data. Hence, a key challenge is to provide mechanisms for monetizing the real value of datasets and avoiding a situation where companies acquire a dataset without information about its usefulness.

In order to remove these barriers, several technologies have emerged to improve the trustworthiness of machine learning. Aligned with priorities of the Big Data Value Association Strategic Research, Innovation and Deployment Agenda such as identifying strong and robust privacy-preserving techniques, exploring and engaging a broad range of stakeholder's perspectives, or providing support in directing research efforts to identify a smart mix of technical, legal, ethical, and business best practices and solutions [33], the MUSKETEER project developed an Industrial Data Platform including algorithms for federated and privacy-preserving machine learning techniques, detection and mitigation of adversarial attacks, and a rewarding model capable of monetizing datasets according to the real data value. We will show in this chapter how these challenges are tackled by the platform architecture but also how these techniques improve the compliance with certain principles of the EU regulation and eventually the necessary data value estimation needed to balance the contributions of the platform stakeholders creating incentive models. Ultimately, the contributions from MUSKETEER help to increase the level of trust among participants engaged in federated machine learning.

2 Industrial Data Platform, an Architecture Perspective

The MUSKETEER platform is a client-server architecture, where the client is a software application that in general is installed on-premise and run at every end user site. This software application is named the Client Connector in the MUSKETEER taxonomy. On the server side of MUSKETEER resides the central part of the platform that communicates with all the client connectors and acts as a coordinator for all operations. Users of the MUSKETEER Industrial Data Platform interact with the Client Connector installed on their side, and that client will communicate with the server to perform several actions on the platform. In Fig. 1, we show the topology of a MUSKETEER installation.

Often in client-server architectures, the means of communication between remote modules is direct, i.e., each module has a communications component that essentially presents an outward-facing interface that allows remote modules to connect. This is usually accomplished by publishing details of an IP address and port number. For operations beyond the local area network, this IP address must be Internet-addressable. The actual implementation of the communications can vary: examples are direct socket communications, REST, gRPC, etc.

There are a number of security and privacy challenges to these traditional approaches that the MUSKETEER architecture addresses. Allowing direct connections from the outside world is a potential security risk, from a malicious actor perspective, but it is also susceptible to man-in-the-middle attacks. These attacks often target known vulnerabilities in the host operating system or software stack. It is also possible for these attacks to operate bidirectionally, whereby a benign entity might be attacked, and potentially sensitive data may be at risk. Furthermore, firewall policies in different organizations may not permit Internet-based traffic, further restricting platform use.

In the MUSKETEER architecture, there are no direct connections between participants and aggregators. All interactions occur indirectly through the MUS-

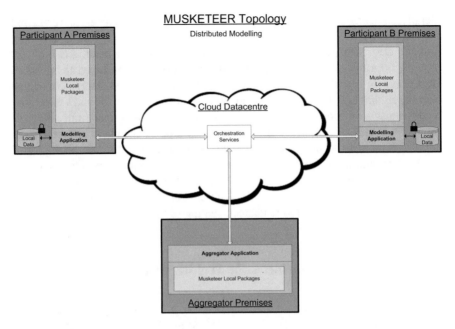

Fig. 1 MUSKETEER topology

KETEER central platform, as depicted by *Orchestration Services* in Fig. 1. The central platform acts as a service broker, orchestrating and routing information between participants and aggregators. In this way, only the connection details for the broker are made available, with all other entities protected from direct attack. Such an architecture slightly differs from current reference models promoted by the International Data Spaces Association (IDSA) and the Gaia-X initiative. Although largely aligned with most of the concepts included in these models (containerization, secured communication, etc.), there is an important difference with the privacy by design dimension included in the MUSKETEER architecture. Both IDSA and Gaia-X models rely on mutual trust between participants in the same ecosystem, while participants in MUSKETEER never have direct interactions.

2.1 Client Connector

The Client Connector is a software component that is installed at the client site, as depicted by *Musketeer Local Packages* in Fig. 1. Within the Client Connector, two types of architectures have been designed: the first one implements a Cluster mode; the second one implements a Desktop mode.

The Cluster Client Connector (Fig. 2) supports the storage and the processing of large datasets before applying the machine learning federation, through horizontal scalability and workload distribution on multiple nodes of the cluster. Within a

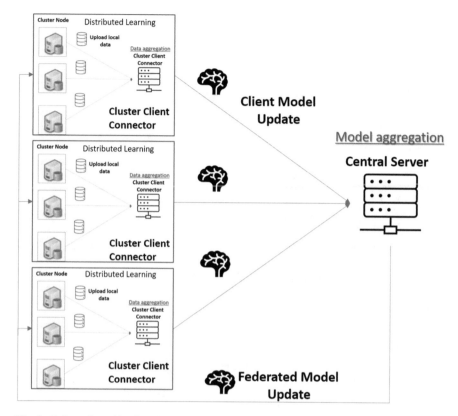

Fig. 2 Federated machine learning through the Cluster Client Connector

Cluster Client Connector, distributed machine learning algorithms have the potential to be efficient with respect to accuracy and computation: data is processed in parallel in a cluster or cloud by adopting any off-the-shelf efficient machine learning algorithm (e.g., Spark's MLlib). In this way, we combine the benefits of distributed machine learning (inside the Client Connector) with the benefits of federated machine learning (outside the Client Connector).

The Desktop Client Connector (Fig. 3) is used when data is collected in a non-centralized way and there is no need to use a cluster to distribute the workload, both in terms of computing and Big Data storage. Anyway, the Desktop version could also leverage GPUs for the training process, enabling the processing of a large amount of data in terms of volume. Finally, the Desktop Client Connector can be easily deployed in any environment, thanks to the use of Docker in order to containerize the Client Connector application. Docker containers ensure a lightweight, standalone, and executable package of the software that includes everything needed to run the Desktop Client Connector: operating system, code, runtime, system tools, libraries, and settings. They are also quite secure since it is possible to limit all capabilities except those explicitly required for any processes (https://docs.docker.com/engine/security/).

Desktop Client
Connector

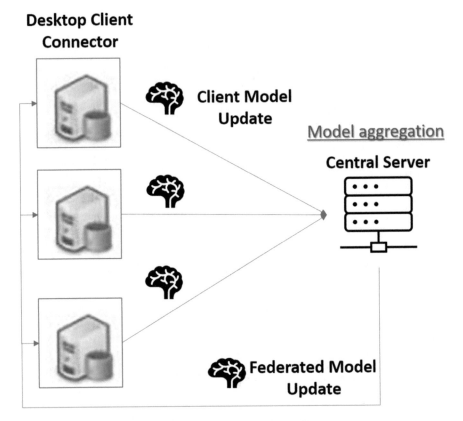

Fig. 3 Federated machine learning through the Desktop Client Connector

Moreover, extra layers of security can be added by enabling appropriate protection systems like AppArmor (https://packages.debian.org/stable/apparmor), SELinux (https://www.redhat.com/it/topics/linux/what-is-selinux), and GRSEC (https://grsecurity.net/), so enforcing correct behavior and preventing both known and unknown application flaws are exploited. Finally, the Docker Engine can be configured to run only images signed using the Docker Content Trust (DCT) signature verification feature.

In this way, the whole Desktop Client Connector application can be easily deployed in a secure sandbox to run on the host operating system of the user.

2.2 Micro-Services

For a viable federated learning platform, trust in the platform is an important requirement. This trust includes privacy protection for sensitive data, which remains on-premise, but also for platform user identities and communications. Ideally, no given user should be able to discover the identity or geographic location of any other user. Additionally, threats from traditional cyber-security attacks should be minimized.

The MUSKETEER server platform, depicted by *Orchestration Services* in Fig. 1, is a collection of cloud-native micro-services. These micro-services manage the life cycle of the federated learning process, using underlying cloud services such as a relational database, cloud object storage, and a message broker.

By employing a brokered architecture, the MUSKETEER server platform enables outbound-only network connections from platform users. Users initiate connections to the platform and do not need to accept connections. This ensures that users are not required to present Internet-facing services, having open ports readily accessible by external, potentially malicious actors. Additionally, all users must register with the platform, by creating a username/password combination account, and all communications use at least TLS 1.2, with server platform certificate validation enabled.

Once registered with the MUSKETEER server platform, each user is assigned a dedicated private message queue, which is read-only. This ensures that only the server platform itself can add messages to the queue but also that only the assigned user has the appropriate privileges to view the contents of their queue. As the server platform is broker based, the Client Connector simply invokes the appropriate procedure to subscribe to the assigned user queue.

As shown in Fig. 4, an important function of the server platform is the routing of messages between participants and aggregators and how the micro-services interact to achieve this. For example, when an aggregator starts a round of training, an initial model may be uploaded to the platform's object storage. During this process, the aggregator obtains write-only privileges to a specific storage location for that model. Upon completion of the upload, the aggregator publishes a message to initiate training, with an included checksum for the model. The platform receives this message and routes it to the queues of multiple users who are part of the federated learning task. Read-only privileges to download the aggregator's model are generated and appended to the message. Multiple participants receive these messages in parallel. They download the model, verify the checksum, and start local training, all via the Client Connector. Upon completion, each participant performs a similar operation to the aggregator, and ultimately, all participant model updates are routed back to the aggregator for model fusion. This routing is deployed within a Kubernetes cluster, leveraging its high-availability features for an always-on, responsive system.

During the fusion process, the aggregator may employ a data contribution value estimation algorithm. Such an algorithm may identify high-value contributions and potentially assign a reward to the originating user, promoting a federated learning data economy. The server platform supports this by providing the capability to the aggregator to store information pertaining to the data value and potential reward. This is discussed in more detail in Sect. 4.

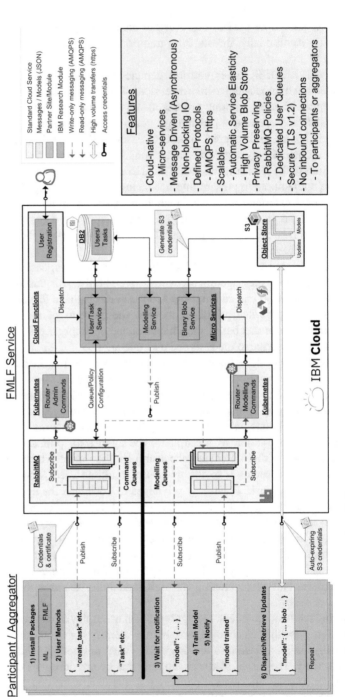

Fig. 4 MUSKETEER micro-services—from [26]

By providing this capability, the server platform is in fact recording each step of the federated learning process. The combination of the recordings at each step, by the end of the federated learning process, enables a view of the complete model lineage for the final model. This lineage includes details such as updates provided per user, when, and of what value.

This architecture is instantiated for use in the MUSKETEER project. The server side (micro-services) is also integrated with *IBM Federated Learning* [22] and is available in the community edition [14]. The community edition supports multiple connection types, one of which is a HTTPS-based connection, using REST, which requires IP addresses to be supplied to participants and aggregators. As previously discussed, there are a number of potential security issues with this approach, which the inclusion of the MUSKETEER option alleviates. Other federated learning platforms also exist, many of which display similar potential security issues due to the direct communication mechanisms employed.

So far, we have described the technological means used to increase the trust of the user on the platform, basically focusing on data/communications security aspects and data confidentiality protection provided by the federated learning approach. In what follows, we provide a legal perspective about the trust required in any data space by further explaining the regulatory data protection (compliance with GDPR principles). Finally, we will focus on the description of several data valuation mechanisms potentially leading to objective credit assignment and reward distribution schemes that further increase the end user trust on the data space operation.

3 Industrial Data Platform, a Legal Perspective

3.1 The Broader Policy Context

Driven by the significant benefits that the use of Big Data analytics technologies (including machine learning) can have for our society, the European Union ("EU") has in the past decade taken several steps toward creating favorable conditions for what it calls a "thriving data-driven economy" [6] and a "common European data space" [7]. Key in these steps is the objective to foster access to and availability of large datasets for re-use for innovation purposes [9]. This is confirmed in the most recent Communication from the European Commission a "European Strategy for Data," where the Commission announces its intention to establish "EU-wide common interoperable Data Spaces in strategic sectors" [10, p. 16]. These spaces, the European Commission goes on, will include "data sharing tools and platforms" [10, p. 17].

3.2   Data Sharing Platforms

Industrial data platforms were already mentioned by the Commission in its earlier guidance on the sharing of private sector data [8, p. 5]. In the aforementioned guidance, the Commission identifies industrial data platforms as one of the modes through which data can be shared among businesses, and it describes these as "platforms dedicated to managing regular data interactions with third parties [and which] offer functionalities when it comes to data exchange [...] storage inside the platform and [...] additional services to be provided on top of the data (based on data analytics)" [8, p. 5].

In academic literature, [28, p. 10] similarly describe data sharing platforms as entities providing "the technical infrastructure for the exchange of data between multiple parties." These scholars discuss several core functions of data sharing platforms and identify the "creation and maintenance of trust [among data users and data suppliers]" as one of their key functions [28, p. 14]. Indeed, they point out that, in order for the platform to achieve its main goal which is to match suppliers of data with users thereof, it is essential that suppliers trust that the data they supply will not be used illicitly and that users trust that the data supplied is fit for use [28, pp. 13–14]. As correctly remarked by these scholars, technology can be a key enabler for trust among users and suppliers of a data platform [28, p. 17].

Aside from a possible lack of trust in the data, users, and suppliers thereof, there may be legal reasons inhibiting the sharing of data among businesses. Crucially, when it comes to the sharing of personal data among businesses, the latter will often qualify as a processing of personal data falling under the scope of application of the General Data Protection Regulation ("GDPR"). Although the GDPR does not prohibit the sharing of personal data among businesses as such, it does impose a number of conditions under which such sharing is allowed to take place.

3.3   Federated Learning as a Trust Enabler: Some Data Protection Considerations

Federated learning has recently been emerging as one of the technologies aimed at overcoming some of the trust and, more specifically, data protection concerns, related to the sharing of personal data. Indeed, federated learning differs from traditional centralized machine learning paradigms, since it does not require that the raw data used to train a machine learning model are transferred to a central server for the training to occur. Instead, under the federated learning paradigm, the machine learning model is trained locally, i.e., on the premises of the data suppliers, under the coordination of a central server. Therefore, under a basic federated learning process, only the local updates to the machine learning model leave the premises of the data suppliers and are sent to the central server for aggregation.

As implicitly recognized by several data protection authorities [2, 4] and the German Data Ethics Commission [5, p. 120], federated learning can facilitate compliance with some principles of the GDPR. Indeed, as pointed out by the Norwegian Data Protection Authority, federated learning helps reducing the amounts of data needed for training a machine learning model [4, p. 26]. Therefore, if the training data qualifies as personal data, federated learning can help complying with the principle of data minimization set forth in article 5.1(c) GDPR. This principle requires personal data to be "adequate, relevant and limited to what is necessary in relation to the purposes for which they are processed." Moreover, since under the federated learning paradigm the training data is not transferred to a central server, the possibilities of such data being re-purposed by that server are also reduced. If the training data qualify as personal data, this means that federated learning could also facilitate compliance with the principle of purpose limitation set forth in article 5.1(b) GDPR. This principle requires personal data to be "collected for specified, explicit and legitimate purposes and not further processed in a manner that is incompatible with those purposes [. . .]." Federated learning can hence be considered as a technique that helps implementing the principle of data protection by design, contained in article 25.1 GDPR. This principle requires controllers of personal data to "[. . .] implement appropriate technical and organizational measures [. . .] which are designed to implement data-protection principles, such as data minimization, in an effective manner [. . .]."

Despite the advantages that federated learning presents from a data protection perspective, it is not, as such, a silver bullet. We name some of the reasons for this. First, as also remarked by [2], the updates that data suppliers share with the central server could, in certain cases, leak information about the underlying (personal) training data to the central server or a third party [23, para. 1.2]. It is hence important to combine federated learning with other privacy-preserving technologies, such as multi-party computation, differential privacy [21, p. 11], and homomorphic encryption [32, pp. 3–4]. Second, "federated learning has by design no visibility into the participants local data and training" [1, para. 1]. This may render federated learning vulnerable to (data and model) poisoning attacks by training participants [17], which could, in turn, in some instances, impair the performance of the final machine learning model. Therefore, the use of federated learning may require an increased attention to not only technical but also organizational accountability measures. The latter may include a careful due diligence investigation into the training participants' compliance with the GDPR (and other relevant legislation) and contractually binding protocols specifying (among other requirements mentioned in the aforementioned EC Guidance on sharing of private sector data [9]), which quality requirements the training data should meet in light of the purpose of the final machine learning model and the population to which it will be applied.

Another key point to consider is about the quality requirements the training data should meet in light of the purpose of the final machine learning model and the population to which it will be applied. To this purpose, we will describe in the next section several data value estimation approaches that can be used to assess the quality of the data provided by each participant, so that the platform is ultimately able to reward every participant proportionally to the contribution to the final model.

The availability of such data value estimations is key to the deployment of a true data economy.

4 Industrial Data Platform, Objective Data Value Estimation for Increased Trust in Data Spaces

As already mentioned, another key requirement for a secure Industrial Data Platform is to measure the impact of every data owner on the accuracy of the predictive models, thus allowing to monetize their contributions as a function of their real data value.

Today data has become the new gold, as it serves to power up advanced Artificial Intelligence (AI) models that form the core of an unlimited number of highly profitable processes, ultimately generating a potentially enormous business value. The importance of collecting large amounts of data as a way to obtain increasingly complex (and therefore accurate) AI models without the problem of overfitting (i.e., complex models that perform well in the presence of input patterns never seen before) is out of the question.

For example, everywhere we are witnessing a struggle to capture as much information as possible from users in the context of mobile applications, to be used or resold for different purposes without any reward for data producers. In this well-known example, the users give their consent (very often inadvertently) for their data to be used by third parties when they install and accept the terms and conditions of a certain application. A fairer scenario would be the one where users[1] are aware of their potential valuable data and agree to share it hoping to receive some compensation in return. It is currently debated that users should be paid for their data in a fairly direct way to foster the data exchange and ultimately improve many AI models. Many economists and politicians believe that data should be treated as an asset, with the possibility of protecting its specific use by third parties and the right of the data owner to sell it for different purposes, like any other "physical" good [31]. In economic terms, data is "non-rival" in the sense that it can be unlimitedly used multiple times for different purposes, unlike other physical goods, which can only be used once [16]. The current situation tends to be the opposite of the desired one, since in most cases large companies accumulate and have the rights over an increasing amount of data, to the detriment of the users who generated them.

The concept of an ideal data market has been studied in [16] where different alternatives (companies own data, people own data, and data sharing is not allowed) have been compared against an optimal economic model administered by a benevolent ruler. As a conclusion of this research, it appears that the situation closest to the ideal reference model is the one in which users handle their own data. On the

---

[1] In what follows, we will refer as "user" to any entity, either person or organization, that has some data of potential interest to a given process.

other hand, the case (more common nowadays) in which companies own the data, the privacy of users is not respected and the data is not shared efficiently with other companies. Finally, when data is not shared at all, economic growth tends to come to an end. Therefore, a reasonable approach would be to allow users to retain the ownership and control over their data and get a revenue whenever they contribute to any machine learning or AI model. The question still to be answered is how to adequately estimate that reward.

As discussed in [15], there are several families of pricing (rewarding) strategies, such as "query-based pricing," which sets the price according to the number of data views [19], "data attribute-based pricing" which fixes prices according to data age or credibility [13], and "auction-based pricing" which sets prices based on bids among sellers and buyers [20]. The aforementioned methods, although potentially useful in certain contexts, have a significant drawback, in the sense that prices (rewards) are set independent of the task to be solved or of the actual utility of the data for the model to be trained. In what follows, we will restrict ourselves to the data value concept that is linked to a real value for a given task, usually the training of a machine learning or AI model.

This data value estimation process is of great interest in a wide range of scenarios with different data granularity. On the one hand, we may have situations where every user provides a unique training pattern (e.g., a person offers data from the clinical record) and a potentially very large number of participants are needed to train a model (millions of people?). On the other side, we have scenarios where a reduced number of entities (organizations, companies, groups) offer a relatively large amount of data (e.g., several companies try to combine their efforts to improve a given process by joining their respective accumulated experience). The first type of scenarios can be associated with the concept of a Personal Data Platform (PDP), where users are individuals who offer their own data for commerce. This is the kind of scenario illustrated in the pioneering work by Google [25] and others in the context of mobile phones [18, 27]. The latter example is associated with the concept of Industrial Data Platform (IDP), where the number of participants is not that high (context also known as enterprise federated learning [32]), but each provides a good amount of training samples. The MUSKETEER platform is oriented toward the latter, and it aims at becoming an IDP offering a variety of possible confidentiality/privacy scenarios, named as Privacy Operation Modes (POMs).

If we assume a scenario where a total amount of reward is to be distributed among the participants (data providers), according to the actual contribution of their respective data to the final model quality/performance, then it is possible to formulate the task as a "profit allocation problem." This type of situation has been studied extensively in the context of cooperative game theory, and the most popular solution is provided by the Shapley value estimation scheme [11, 29]. This approach offers some attractive features: it is task-dependent, the data is valued only if it allows to improve the performance of the model, the reward is fully distributed among the participants, equal data contribution means equal reward, and the addition of several contributions gets a reward equal to the sum of the individual rewards. The calculation of Shapley values is quite simple. If we consider that N is the number of participants, S is a subset of players and $U(S)$ is the utility function that measures the performance of the model produced with the data from users in the set S, then the Shapley value s_i for user i is defined as:

$$s_i = \sum_{S \subseteq I \setminus \{i\}} \frac{1}{N \binom{N-1}{|S|}} [U(S \cup \{i\}) - U(S)] \tag{1}$$

According to the expression in 1, the Shapley value is computed as the average utility gain obtained when player i is added to any other[2] group of participants. Despite the relatively simple definition of the Shapley values, their computation requires an exponential number of different utility computations (each one of them usually requiring to train a brand new model). Therefore, Shapley's approach poses some computational challenges if we opt to use a brute force approach. Some works indicate that it is possible to reduce the exponential computational cost to a linear or logarithmic scale by benefiting from a knowledge transfer between trained models, exploiting some peculiarities of a given machine learning model [15], or using Monte Carlo estimations of the utility values [24].

All the abovementioned optimized methods assume we have an unlimited access to the training data and that we can run the training procedures an unlimited number of times, a situation which is rarely found in real-world situations. Even so, gathering large amounts of data in the same place faces many barriers, such as the growing number of regulations that limit the access/sharing of the information, with the ultimate intention of protecting the privacy and property rights of users (e.g., GDPR [3] or HIPAA [12]).

As already presented in the previous sections, various architectures have emerged in an attempt to circumvent these data exchange restrictions and ultimately facilitate the training of models with increasing amounts of data while preserving the data privacy/confidentiality. For many years, the field of privacy-preserving machine learning (a.k.a. privacy-preserving data mining) has produced solutions relying on different security mechanisms (secure multi-party computation or cryptography, among others). It is obvious that the data value estimation in these scenarios has an additional degree of complexity, sometimes unaffordable. Lately, the federated learning paradigm has emerged as a less complex approach to the problem of training models while preserving data confidentiality. In a federated learning context, we face many restrictions on accessing training data, and the training process is typically only run once. Therefore, the traditional data value estimation methods cannot be used directly in this context.

An interesting approach is the one presented in [30], where the interchanged values (models, gradients) during the federated learning process are used to reconstruct the variety of models needed to estimate Shapley values using 1. In this way, we can calculate estimates of the different models that would be obtained

---

[2] All possible combinations must be considered.

if different combinations of datasets were used, without the need to train them from scratch. Obviously, an exact reconstruction of all models is not possible and we only get estimates, but it is shown in [30] that good approximations are possible.

The procedure is as follows. It is assumed that there is a validation set available in the aggregator, so that for each possible model trained with a subset S of the training data, it is possible to calculate the corresponding utility $U(S)$ needed to estimate the Shapley values. We also assume that the aggregator has access to the following information:

- The initial global (epoch 0) model weights $M^{(0)}$
- The global model weights at epoch n, $M_{all}^{(n)}$
- The model increments[3] contribution from participant m at epoch n, $\Delta_m^{(n)}$

Taking into account all this information, in [30], two approaches are proposed for Data Shapley value estimation. The first one estimates at epoch n the model trained with the datasets from the set of users in set R^4, M_R^n, as the cumulative update from the initial model, i.e.:

$$M_R^{(n)} = \sum_{i=0}^{n} M_g^{(0)} + \sum_{m \in R} \Delta_m^{(n)} \tag{2}$$

and using these model estimates, the corresponding utilities and Data Shapley values in 1 can be calculated, averaging the estimates across all epochs. This approach is prone to divergences from the real model, since the accumulation takes place with respect to the initial (random) model.

The second approach is based on updating the global model $M_{all}^{(n-1)}$ obtained at every step $n - 1$ with the contributions from all participants, so the different submodels are estimated using updates with partial data. For example, the model trained with the datasets from the set of users R at epoch n, M_R^n, is estimated as:

$$M_R^{(n)} = M_R^{(n-1)} + \sum_{m \in R} \Delta_m^{(n)} \tag{3}$$

such that more accurate submodel estimates are obtained, but they are influenced by the contributions from other participants, since $M_R^{(n)}$ is calculated using information from all contributors.

Notwithstanding the restrictions mentioned above, both methods appear to provide reasonable data value estimates in a federated learning environment, as evaluated in [30]. Note that under the approaches described above, the Shapley values are calculated exactly but are based on model estimates. Therefore, the

---

[3] If model weights are exchanged instead of gradient updates, the increments can be obtained as a difference between models.

[4] R can be set to S or $S \subseteq I \setminus \{i\}$, as needed.

quality of those estimates will determine the precision of data value estimates according to Shapley principles.

Various MUSKETEER privacy modes of operation (POM) do not exactly follow federated learning principles and use other security/privacy mechanisms (secure multi-party computation, homomorphic encryption), and it remains to be analyzed how to extend the procedures described above to adapt them to the new scenarios.

The above-described approach is perfectly valid under "honest but curious" security assumptions, where the participants are assumed not to act outside of the defined protocols (which is the case of the MUSKETEER platform), and therefore they can fully trust the aggregator in the sense that they are confident in that it will always declare the correct (estimated) credit allocation values.

However, in some other situations, the aggregator could act maliciously and, after using participant data for a given task, could declare a lower value than actually estimated. In this different security scenario, a different approach would be needed. Also, it would be of great interest to be able to estimate the Data Shapley values *before* actually training any model, so that preliminary data negotiation can be established before actually participating in the training process.

We are exploring the extent to which the data value can be estimated using a collection of statistics calculated on each participant, but which do not contain enough information to train the global model. In the MUSKETEER context, we are interested in answering the following questions (and hence we are investigating in that direction):

- To what extent is it possible to estimate the data values before actually training the model, based on locally pre-calculated statistical values?
- To what extent can the incremental approach proposed in [30] be extended to scenarios other than federated learning, where other privacy mechanisms are used (two-party computation, homomorphic encryption, etc.)?

5 Conclusion

In this chapter, we described an Industrial Data Platform (IDP) for federated learning offering high standards of security and other privacy-preserving techniques (MUSKETEER). Our approach shows how trust respectful of privacy can be enforced from an architecture point of view but also how the techniques used can support the compliance with certain GDPR principles from a legal perspective. Besides, leveraging more data on such data platforms requires incentives that fairly reward shared data; thereby, we also discuss different strategies of data value estimation and reward allocation in a federated learning scenario.

Acknowledgments This project has received funding from the European Union's Horizon 2020 Research and Innovation Programme under grant agreement No 824988.

References

1. Bagdasaryan, E., Veit, A., Hua, Y., Estrin, D., & Shmatikov, V. (2020). How to backdoor federated learning. In *International Conference on Artificial Intelligence and Statistics* (pp. 2938–2948). PMLR.
2. Binns, R., & Gallo, V. (2020). Data minimisation and privacy-preserving techniques in ai systems [www document]. https://ico.org.uk/about-the-ico/news-and-events/ai-blog-data-minimisation-and-privacy-preserving-techniques-in-ai-systems/.
3. Council of European Union. (2016). Council regulation (EU) no 269/2014. http://eur-lex.europa.eu/legal-content/EN/TXT/?qid=1416170084502&uri=CELEX:32014R0269.
4. Datatilsynet. (2018). Artificial intelligence and privacy.
5. Daten Ethik Kommission. (2019). Opinion of the data ethics commission.
6. European Commission. (2014). Towards a thriving data-driven economy.
7. European Commission. (2018). Guidance on sharing private sector data in the european data economy.
8. European Commission. (2018). Guidance on sharing private sector data in the european data economy (no. swd(2018) 125 final).
9. European Commission. (2018). Towards a common european data space.
10. European Commission. (2020). A european strategy for data.
11. Ghorbani, A., & Zou, J. (2019). Data shapley: Equitable valuation of data for machine learning. In *Proc. 36th International Conference on Machine Learning* (vol. 97, p. 2242–2251). PMLR.
12. Gunter, K. (2002). The hipaa privacy rule: practical advice for academic and research institutions. *Healthcare Financial Management: Journal of the Healthcare Financial Management Association, 56*, 50–54.
13. Heckman, J., Boehmer, E., Peters, E., Davaloo, M., & Kurup, N. (2015). A pricing model for data markets. In *Proc. iConference'15*.
14. IBM. (2020). IBM Federated Learning. https://github.com/IBM/federated-learning-lib.
15. Jia, R., Dao, D., Wang, B., Hubis, F., Gurel, N., Li, B., Zhang, C., Spanos, C., & Song, D. (2019). Efficient Task-Specific Data Valuation for Nearest Neighbor Algorithms. http://arxiv.org/abs/1908.08619.
16. Jones, C. & Tonetti, C. (2020). Nonrivalry and the economics of data. *American Economic Review, 110*, 2819–2858.
17. Joseph, A. D., Laskov, P., Roli, F., Tygar, J. D., & Nelson, B. (2013). Machine learning methods for computer security (dagstuhl perspectives workshop 12371). In *Dagstuhl Manifestos* (vol. 3). Schloss Dagstuhl-Leibniz-Zentrum fuer Informatik.
18. Konecny, J., McMahan, H.B., Ramage, D., & Richtarik, P. (2016). Federated optimization: Distributed machine learning for on-device intelligence. http://arxiv.org/abs/1610.02527.
19. Koutris, P., Upadhyaya, P., Balazinska, M., Howe, B., & Suciu, D. (2012). Query-based data pricing. In *Proc. of the ACM SIGACT-SIGMOD-SIGART Symposium on Principles of Database Systems* (vol. 62).
20. Lee, J., & Hoh, B. (2010). Sell your experiences: a market mechanism based incentive for participatory sensing. In *Proc. IEEE International Conference on Pervasive Computing and Communications (PerCom)*.
21. Li, T., Sahu, A.K., Talwalkar, A., & Smith, V. (2020). Federated learning: Challenges, methods, and future directions. *IEEE Signal Processing Magazine, 37*(3), 50–60.
22. Ludwig, H., Baracaldo, N., Thomas, G., Zhou, Y., Anwar, A., Rajamoni, S., Ong, Y., Radhakrishnan, J., Verma, A., Sinn, M., et al. (2020). Ibm federated learning: an enterprise framework white paper v0. 1. Preprint. arXiv:2007.10987.
23. Lyu, L., Yu, H., & Yang, Q. (2020). Threats to federated learning: A survey. Preprint. arXiv:2003.02133.
24. Maleki, S., Tran-Thanh, L., Hines, G., Rahwan, T., & Rogers, A. (2013). Bounding the estimation error of sampling-based shapley value approximation with/without stratifying. http://arxiv.org/abs/1306.4265.

25. McMahan, B., Moore, E., Ramage, D., Hampson, S., & y Arcas, B. A. (2017). Communication-efficient learning of deep networks from decentralized data. In *Procs. of AISTATS* (pp. 1273–1282).
26. Purcell, M., Sinn, M., Simioni, M., Braghin, S., & Tran, M. N. (2020). D3.2 - Architecture design – Final version. https://ec.europa.eu/research/participants/documents/downloadPublic?documentIds=080166e5cf9bc07f&appId=PPGMS.
27. Ramaswamy, S., Mathews, R., Rao, K., & Beaufays, F. (2016). Federated learning for emoji prediction in a mobile keyboard. http://arxiv.org/abs/1906.04329.
28. Richter, H., & Slowinski, P. R. (2019). The data sharing economy: on the emergence of new intermediaries. *IIC-International Review of Intellectual Property and Competition Law, 50*(1), 4–29.
29. Shapley., L. S. (1953). A value for n-person games. In *Annals of Mathematical Studies: contributions to the Theory of Games* (vol. 28, p. 307–317). Princeton University Press.
30. Song, T., Tong, Y., & Wei, S. (2019). Profit allocation for federated learning. In *Proc. 2019 IEEE International Conference on Big Data (Big Data)*, Los Angeles, CA, USA.
31. Walsh, D. (2019). How Much Is Your Private Data Worth - and Who Should Own It? Insights by Stanford Business. https://www.gsb.stanford.edu/insights/how-much-your-private-data-worth-who-should-own-it.
32. Yang, Q., Liu, Y., Chen, T., & Tong, Y. (2019). Federated machine learning: Concept and applications. *ACM Transactions on Intelligent Systems and Technology (TIST), 10*(2), 1–19.
33. Zillner, S., Bisset, D., Milano, M., Curry, E., García Robles, A., Hahn, T., Irgens, M., Lafrenz, R., Liepert, B., O'Sullivan, B., & Smeulders, A., (eds.): Strategic Research, Innovation and Deployment Agenda - AI, Data and Robotics Partnership. Third Release. BDVA, euRobotics, ELLIS, EurAI and CLAIRE (2020)

KRAKEN: A Secure, Trusted, Regulatory-Compliant, and Privacy-Preserving Data Sharing Platform

Silvia Gabrielli, Stephan Krenn, Donato Pellegrino, Juan Carlos Pérez Baún, Pilar Pérez Berganza, Sebastian Ramacher, and Wim Vandevelde

Abstract The KRAKEN project aims to enable the sharing, brokerage, and trading of personal data including sensitive data (e.g., educational and health records and wellbeing data from wearable devices) by returning its control to both data subjects/data providers throughout the entire data lifecycle. The project is providing a data marketplace which will allow the sharing of personal data and its usage for research and business purposes, by using privacy-preserving cryptographic tools. KRAKEN is developing an advanced platform to share certified information between users and organizations by leveraging on distributed ledger technology, promoting the vision of self-sovereign identity solutions (ensuring users' consent and data control in a privacy-friendly way), preserving security, privacy, and the protection of personal data in compliance with EU regulations (e.g., GDPR). The feasibility of the KRAKEN solution will be tested through two high-impact pilots in the education and healthcare fields.

S. Gabrielli
Fondazione Bruno Kessler, Trento, Italy
e-mail: sgabrielli@fbk.eu

S. Krenn · S. Ramacher
AIT Austrian Institute of Technology, Vienna, Austria
e-mail: stephan.krenn@ait.ac.at; sebastian.ramacher@ait.ac.at

D. Pellegrino
TX - Technology Exploration Oy, Helsinki, Finland
e-mail: donato@tx.company

J. C. Pérez Baún (✉) · P. Pérez Berganza
ATOS Spain S.A., Madrid, Spain
e-mail: juan.perezb@atos.net; pilar.perez@atos.net

W. Vandevelde
katholieke Universiteit Leuven, Leuven, Belgium
e-mail: wim.vandevelde@kuleuven.be

© The Author(s) 2022
E. Curry et al. (eds.), *Data Spaces*, https://doi.org/10.1007/978-3-030-98636-0_6

Keywords Privacy-preserving data · Privacy-aware analytic methods · Trusted and secure sharing platforms · Self-sovereign identity · User-centric design · GDPR compliance

1 KRAKEN Overview

The KRAKEN(brokerage and market platform for personal data) project[1] is an innovation action funded by the EU H2020 program (under grant agreement no. 871473) with the main objective to develop a trusted and secure personal data platform with the state-of-the-art privacy-aware analytics methods, guaranteeing metadata privacy and query privacy and returning the control of personal data back to users.

The KRAKEN chapter mainly relates to the technical priorities of data protection, data analytics, and data management of the European Big Data Value Strategic Research and Innovation Agenda [1]. It addresses the horizontal concerns on privacy, data analytics, and data management of the BDV Technical Reference Model. It addresses the vertical concerns on cybersecurity, marketplaces for personal data platforms, and data sharing.

The main challenge to achieve this goal is to empower the citizens on the control of their own personal data, including sensitive data, and motivate the user to share this kind of data.

With this objective KRAKEN is investigating data processing mechanisms working in the encrypted domain with the aim to increase security, privacy, functionality, and scalability for boosting trust.

The first challenges KRAKEN is facing are the loss of control over data and the use of centralized identity management systems. In this sense KRAKEN is returning the control of personal data back into the hands of data subjects and data providers and its subsequent use, which includes the user consent management. Additionally, in contrast to identity management systems which follow centralized approaches involving dependencies, KRAKEN is advocating for a decentralized self-sovereign identity (SSI) management and user-centric access control to data, where the data provider has the control over their data.

Other important challenges this project is addressing are related to individual privacy and security requirements. KRAKEN will develop easy-to-understand privacy metrics and usable interfaces for end users and data subjects, and also privacy-preserving analysis based on advanced cryptography.

A basic aspect to cover when personal and sensitive data are managed and shared is the fulfillment of regulatory framework. KRAKEN addresses this regulatory challenge through General Data Protection Regulation (GDPR) [2] and eIDAS compliance, following standards for compatibility and interoperability and promoting best practices.

Furthermore, in order to motivate the user to share their data, the development of fair-trading protocols and incentive models is envisaged. KRAKEN is handling this

---

[1] https://www.krakenh2020.eu/

business challenge by establishing economic value and innovative business models for "personal Data Spaces" supporting the Digital Single Markets' data economy and engaging SMEs. In this way users can receive some incentive pushing them to share their data.

With the aim to generalize the KRAKEN experience to other economic sectors, KRAKEN will be demonstrated in two high-impact pilots on health and educational domains, in realistic conditions, with legal compliance, considering usability and transparency. In this sense, KRAKEN contributes to the European strategy for data, namely, the boost of the common European Data Spaces by leveraging the SSI paradigm and the cryptographic techniques. These technologies facilitate the fair management of the user data, making them available to be used by several economic domains.[2]

Additionally, the KRAKEN chapter relates to knowledge and learning enablers of the AI, Data and Robotics Strategic Research, Innovation, and Deployment Agenda [3], which can impact the future activities in AI and data.

In summary, KRAKEN is addressing all these challenges providing a sharing data marketplace that is relying on SSI services and cryptographic tools for covering the security, privacy, and user control on data. At the end KRAKEN will provide a highly trusted, secure, scalable, and efficient personal data sharing and analysis platform adopting cutting-edge technologies and leveraging outcomes from the CREDENTIAL[3] and MyHealthMyData[4] projects.

At this moment the high-level KRAKEN architecture (Fig. 1) is provided considering the three main pillars:

- The SSI paradigm providing a decentralized user-centric approach on personal data sharing. The SSI pillar comprises the SSI mobile app for storing verifiable credentials (VCs) and key material, the legal identity manager for issuing an identity of VC leveraging the eIDAS eID network and signing this VC, and the KRAKEN Web Company Tool (KWCT) web tool for VC management.
- A set of different analytics techniques based on advanced cryptographic tools that will permit privacy-preserving data analysis. The cryptographic pillar provides functional encryption (FE) and secure multi-party computation (SMPC) for protecting the sharing of data on the marketplace, a backup service for a secure key material cloud storage, and zero-knowledge proof (ZKP) protocols and proxy re-encryption (PRE) mechanisms for privacy and secure data exchange.
- A data marketplace which will allow the sharing of personal data preserving privacy when Artificial Intelligence/machine learning analysis is made. The marketplace pillar is basically built by a decentralized and distributed processor and a blockchain network for business logic management by using smart contracts.

The health and education domains were selected to demonstrate how SSI and cryptographic technologies can improve the security and privacy of personal data,

---

[2] https://digital-strategy.ec.europa.eu/en/policies/strategy-data

[3] https://credential.eu/

[4] http://www.myhealthmydata.eu/consortium/

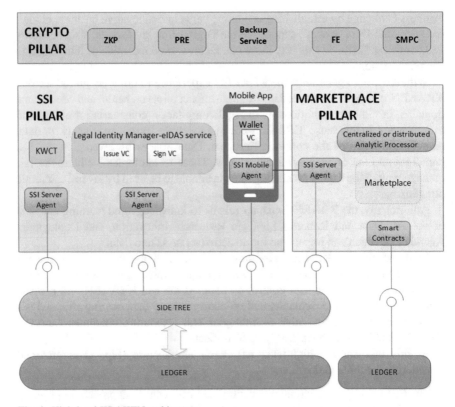

Fig. 1 High-level KRAKEN architecture

including sensitive data, when shared in a marketplace. The health scenario involves sensitive data such as biomedical and wellbeing data, which implies the use of powerful privacy-preserving techniques assuring the data are protected at all times. The education scenario involves personal data such as grades, courses, or diplomas, which can be provided to a third party in a privacy-preserving way. In both cases the use of SSI and cryptographic technologies eases the shared use of these data, assuring the data are protected and the owner has the control over the use of the data.

2 Architectures for Data Platform

2.1 *KRAKEN Data Platform Architecture Overview*

The design of the KRAKEN architecture is based on decentralization, cryptography, and self-sovereign identity (SSI) [4].

The architecture reflects of the user requirements related to the different data products that can be published on the platform. In the KRAKEN marketplace the users are divided into two categories: data providers and data consumers. The data providers are the users whose interest is to publish data products on the platform and earn money by granting access to data consumers. The data consumers are the users whose interest is to buy access to data products.

The data products are divided into three categories: batch data, real-time data, and analytics. Based on the type of data product, the requirements of the users change. One of the requirements that is common between all the three kinds of data products is the eligibility of the data consumer. Data providers are willing to provide their personal data only to data consumers that passed through an eligibility check. In KRAKEN this is accomplished by exploiting blockchain [5] and SSI technology.

The blockchain is the decision-making component of the platform. Through decentralization, the KRAKEN marketplace is able to provide an incorruptible mean whose duty consists in granting access to data products to eligible consumers and keep track of all the transactions in a distributed immutable ledger.

The ledger is also used to store also the policies set by the data providers to instruct the blockchain on how to filter data consumer requests. These policies are checked also against SSI verifiable credentials. To perform this check, the architecture includes an SSI agent. The SSI agent is used to check the validity of the credentials of the users that contain the needed information to be checked against the policies.

One of the requirements of the users is to be in total control of their own personal data. For this reason, the KRAKEN marketplace does not store any data product-related resources (such as the dataset files). However, data consumers need to be able to access the data. To do so, KRAKEN relies on cloud storage systems. Every data provider can choose any cloud storage system available nowadays to store their data. Once they provide the location of the data to the KRAKEN marketplace, such location is shared only with data consumers to let them download the data.

The privacy-preserving analytics data product specifically enables users to share analytics on their personal data without revealing the original data to data consumers and to any third party performing the analytics computation. The element of the architecture that makes this possible is the secure multi-party computation (SMPC) [6] network. SMPC is a technology that allows the establishment of a decentralized network capable of communicating with users exploiting a secret-sharing mechanism. This mechanism consists in encrypting the message in a way that prevents the network from obtaining the original message, but allows the network to perform computation on it and generate an encrypted result that can be decrypted only by the data consumer, still through the same secret-sharing mechanism.

The real-time data product consists of a stream of real-time messages from data providers to data consumers. This needs to happen in a decentralized manner that does not put trust in any middleman. To do so, the KRAKEN marketplace is interfaced with Streamr [7]: a peer-to-peer network for real-time data sharing that aims to become decentralized. In this specific data product, KRAKEN acts as a permission layer to filter the eligible buyers of the data product.

To interact with KRAKEN marketplace users can access the KRAKEN marketplace website. The backend server is used to store the metadata about data products that are fetched by the frontend to allow users to browse through them. The frontend is the tool used by users to perform operations on KRAKEN such as publication and purchase. Exploiting the frontend, users are able to set up policies, present VCs using an SSI wallet, and perform cryptographic processing of their datasets locally.

Payments on KRAKEN are performed using Streamr's DATA coin. DATA coin is a token available on the Ethereum[5] blockchain and on the xDai[6] blockchain. The blockchain used by the KRAKEN marketplace to run the payment smart contract is the xDai blockchain.

Data access in the KRAKEN marketplace is time based by default. The subscription to any of the data products has a parameter that specifies for how much time the data consumer can access the data product. After this time limit, access is automatically revoked by the marketplace.

An overview of the entire architecture involving data flow and analytics is shown in Fig. 2.

2.2 Enabling Decentralized Privacy-Preserving Decision-Making Using Permissioned Blockchain Technology and SSI

In the KRAKEN marketplace the selection of eligible buyers for data products is performed on a blockchain. The specific technology adopted is Hyperledger Fabric [8]. Hyperledger is an open-source community producing blockchain related software. One of them is Fabric: a technology to develop permissioned blockchain solutions. The features provided by Fabric are diverse; the ones that are specifically exploited by the KRAKEN marketplace are the permissioned consensus, the smart contracts, and the distributed immutable ledger.

Fabric is not a public blockchain; this means that nobody outside of the Fabric network is able to access the information inside the distributed ledger. The members of the network are well known and, because of the permissioned nature of the blockchain, are granted permission to participate in the network only by the already existing peers.

The feature of Fabric that enables the decision-making in the KRAKEN marketplace are the smart contracts. Data providers need to be able to declare a set of policies that need to be checked against the SSI verifiable credentials of the buyers. To enable this process, the decision-making in the KRAKEN marketplace

---

[5] https://ethereum.org/en/

[6] https://www.xdaichain.com/

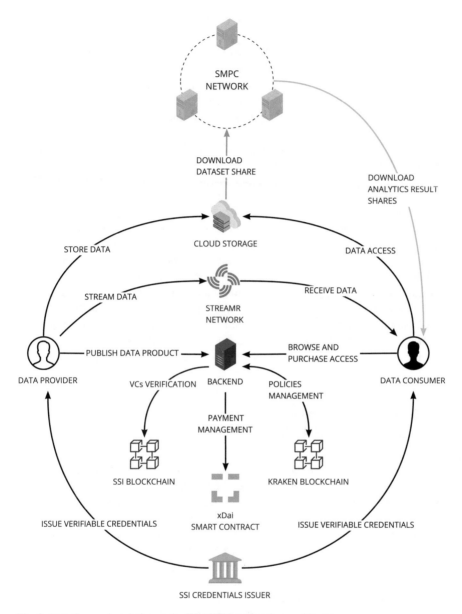

Fig. 2 Data flow and analytics on the KRAKEN marketplace architecture

is programmed using smart contracts. Because of the decentralized nature of the system, this decision-making does not depend on a single party but on a set of organizations that constitute the KRAKEN consortium. In this way the corruptibility of the system is substantially decreased if we compare it to centralized solutions.

The need of the system to have a decentralized decision-making must be joined with the possibility of storing the transactions in a way that nobody, in a later moment, is able to modify or delete it. The ledger is the storage place for information. All the transactions happening on the blockchain are stored on the ledger, including the data product publication and the purchases of eligible buyers. The ledger is not only private but also distributed and immutable. Because of its immutability, it represents the best fit for the purposes of anti-tampering and auditability.

The decentralized decision-making process needs another element to be secure. The information provided to the system has to be verifiable and this needs to happen in a way that preserves the privacy of the users. This need in the KRAKEN marketplace is fulfilled by the SSI technology. Through SSI, users are able to provide verifiable information to the system in the form of a VC.

VCs, in the scope of self-sovereign identity, are certificates released by institutions and organizations to state a specific characteristic of a person, for example, nationality, affiliation to a company or organization, or the fact that one is not underage. This kind of credential is made with the scope of revealing only a specific characteristic of an individual and nothing more, for example, the affiliation to a company does not necessarily also reveal the role that a person has in the company. The credentials are checked using the SSI blockchain. In this way, the privacy of buyers is also protected against its own organization that cannot know when and how the credential is used and cannot block it if not by revocation.

3   Real-Time Data Sharing Using Streamr: A Decentralized Peer-to-Peer Network

One of the data products of the KRAKEN marketplace is the real-time data product. This product consists of streams of real-time messages published by the data provider and received by the data consumers. The streams are enabled by the Streamr network: an open-source peer-to-peer network.

Streamr is a project that aims to realize a decentralized worldwide network for real-time data sharing. In its current state, Streamr is not fully decentralized yet, but it is already a peer-to-peer publish-subscribe network for real-time data transfer. It works with IoT devices, applications, and anything with an Internet connection that can run the Streamr client software.

The network is formed by a set of broker nodes. These nodes are intended to be installed on always-on computers connected to other nodes to route the traffic. The governance of the network is performed by a smart contract on the Ethereum blockchain. All the information regarding coordination, permissioning, and access control of data streams is saved on this smart contract. The actual transfer of data happens off-chain on the Streamr network that benefits from the "network effect" as with the increasing number of nodes, the scalability increases as well.

Through Streamr, users can publish streams of data and not worry about establishing an infrastructure to reach the subscribers. The subscribers can subscribe to the streams in a decentralized way by paying with cryptocurrencies like DATA coin. All of this happens on the Streamr marketplace, but while Streamr successfully performs a selection of buyers based on the payment, it cannot select them based on the eligibility criteria set by the data providers. Here is where KRAKEN gets into action. In addition to providing the other two kinds of data product, in the case of stream data, the KRAKEN marketplace acts as a filter in the already existing pub-subsystem implemented in Streamr where the selection of buyer does not depend solely on the payment but also on the matching of the policies set by the data providers with the VC provided by the data consumer.

4   Privacy, Trust, and Data Protection

In the following we will provide a high-level overview of the cryptographic measures taken by the KRAKEN architecture to guarantee the privacy of the user's data while simultaneously offering high authenticity guarantees to the data consumer. The interplay of all cryptographic primitives discussed in the following is also illustrated in Fig. 3.

Multi-party Computation Secure multi-party computation (SMPC), introduced by Yao [6], has become an interesting building block for many privacy-preserving applications. SMPC allows a group of nodes to jointly perform a computation on secret inputs, without revealing their respective inputs to the remaining nodes in the network or any other third party. More precisely, SMPC guarantees that for a node following the protocol specification, even potentially malicious other parties cannot infer anything about the node's input, except for what can already be inferred from the output of the computation and the malicious parties' inputs. Furthermore,

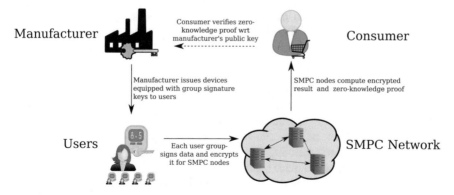

Fig. 3 Overview of the KRAKEN cryptographic architecture

the correctness of the computation can be guaranteed as long as a sufficiently large fraction of the nodes behave honestly.

However, while giving high privacy guarantees to the data provider, classical approaches to secure multi-party computation do not directly fulfill all relevant requirements in the context of KRAKEN. On the one hand, SMPC cannot give authenticity guarantees for the inputs provided by data providers. On the other hand, classical approaches to secure multi-party computation do not directly enable the data provider to verify the correctness of the computation. In the following we will briefly discuss KRAKEN's approaches to solve these two interdependent issues.

End-to-End Authenticity For many application scenarios, the value of data and analytics performed by the KRAKEN platform are highly dependent on the authenticity of the results delivered to a buyer. A natural way to achieve authenticity would be to let the users sign their input data before handing it over to the SMPC nodes. For sensitive data, this signature could be issued directly by a sensor or device owned by the user, which would then guarantee that only data certified by trusted devices (e.g., from a certain manufacturer) would be processed by the SMPC nodes. However, this straightforward approach might violate the users' privacy: verifying the authenticity of input data using the corresponding public key reveals data belonging to the same user and might also allow to identify a user. To avoid this re-identification problem, KRAKEN deploys so-called group signatures [9]: such signatures allow a user to sign messages on behalf of a group while remaining anonymous. That is, the verifier will only be able to check that the message has been signed by some member of group, but not to identify the specific signer. Group membership is controlled by a group manager, with whom any user wishing to join the group needs to execute a registration process. In our context, device manufacturers could now provide each device with a group signature key, which is used to sign, e.g., sensor data. The SMPC nodes as well as the data consumer can now verify the correctness of the signatures using the group manager's public key to verify the authenticity of the input data, without compromising the user's privacy.

On a technical level, it is worth noting that group signatures come with a so-called opening functionality, which allows a predefined third party to identify the signer in case of abuse. To avoid any privacy bottleneck, all key material will be sampled in a way that disables this functionality under standard complexity theoretical assumptions, resulting in a scheme akin to Intel's enhanced privacy ID (EPID) signatures [10].

Correctness With SMPC and group signatures, KRAKEN can give high authenticity guarantees to the data consumer, as long as sufficiently many SMPC nodes are trusted. However, the approach discussed so far neither allows one to drop this assumption, nor does the data consumer have cryptographic evidence about the correctness of the data, meaning that the results could not credibly be presented to any third party. Again, a naive solution could be to let the SMPC nodes sign their respective outputs together with the evaluated function, enabling the data consumer to forward results to third parties, as long as sufficiently many SMPC nodes are assumed to be honest. The approach taken in KRAKEN is different, such that any trust assumptions on the SMPC network can be dropped with regard to the authenticity of the results. Namely, KRAKEN will attach so-called non-interactive zero-knowledge proofs of knowledge [11, 12] certifying the correctness of the provided outputs. Such cryptographic proofs allow one to prove the correctness of a claim without revealing any information than what is already revealed by the claim itself. For KRAKEN, the zero-knowledge proofs will thus cryptographically prove that, starting from private input values which have been signed using a group signature scheme, the function provided by the data consumer has been correctly computed.

Trust Assumption Overall, KRAKEN minimizes the trust assumptions to the best extent possible. Regarding privacy, no user data is revealed to any single entity in the architecture, and also the number of collaborating SMPC nodes necessary to break privacy can be adjusted. Any other ways to break privacy would require compromising communication channels or group signature schemes, for which formal security proofs exist. On the other hand, regarding authenticity, the necessary trust of the data buyer is minimized by the use of group signature schemes and zero-knowledge proofs, and all guarantees can be based solely on the security of the initial signatures on the user's data. For a more detailed discussion about the cryptographic architecture underlying KRAKEN and a detailed privacy analysis following the LINDDUN framework, we refer to Koch et al. [13].

5   Sharing by Design, Ownership, and Usage Control

The most widely deployed approach for data sharing in the cloud, e.g., Google Drive, allows users to upload and share data with others, but beyond the trust put into the cloud provider, no security guarantees can be achieved. While secure communication channels are used between users and the cloud provider, these systems are unable to ensure end-to-end security between users. In an ideal scenario, however, the data owner has complete control over the data and cryptographic schemes to ensure confidentiality of the data with respect to anyone except authorized users. Importantly, this also means that the data is protected against adversarial access by the cloud provider and others. Such strong security guarantees are nontrivial to implement in a cloud-based document and data sharing setting. Approaches based on the use of public-key encryption quickly turn into non-scalable solutions due to the complexity of the involved key management. The use of more advanced techniques such as proxy re-encryption [14] or identity-based encryption [15] often runs into issues when deployed in practice. With these techniques key management remains a crucial part of the system and requires users to constantly interact to exchange key material.

KRAKEN follows a different approach for data sharing that leverages SMPC techniques and the SMPC nodes that are deployed as part of the effort to enable privacy-preserving computation on data. To some extent, data sharing can be seen as a special case of computation on encrypted data. By leveraging the SMPC network, the key management issues can be solved by handling these tasks via the SMPC network [16]. Thereby, the SMPC networks give rise to a scalable system for user-controlled data sharing with end-to-end security. Users are only required to trust one of the SMPC nodes to execute the protocols honestly while keeping their data safe from cloud providers and potential attackers.

The most interesting aspect of running the key management inside the SMPC network is the user's ability to define access policies that after initial verification by the blockchain network are verified and enforced by the SMPC nodes. Similar to a domain-specific language for trust policies [17], users will be able to express their access policies within a policy language designed within KRAKEN [18]. Before sharing data with the receiver, the MPC nodes evaluate if the data receiver satisfies this policy and only then produce the corresponding keys for accessing the data. In comparison to approaches based on encryption schemes with fine-grained access control, users are not required to be online for processing keys within the SMPC network. Additionally, the SMPC network can be extended to provide accountability proofs that give data owners a way to check that the SMPC network validated the access policy [19].

For sensitive data, users are empowered to run their own SMPC node. Especially when dealing with eHealth data, hospitals may host one SMPC node on their own infrastructure. In this case, users do not need to put any trust into any of the other SMPC nodes. Thereby, all trust issues are alleviated. For users unable to host SMPC nodes themselves, privacy-focused organizations may help to distribute trust assumptions and requirements, thereby reducing the risk of data compromise.

5.1   User-Centric Data Sharing

The widespread adoption of the KRAKEN platform depends on new types of user behavior. Users need to understand the value of their personal data [20], the extent to which they are able to control their use, and how they are able to do that. The vision leading the KRAKEN design is to empower users in their ability to control their own data and to promote knowledge about the management of personal data and the distribution of value generated from data. The issue of user adoption has to do with the quality of user experience with the working platform provided; that is why in KRAKEN designers apply user-centric design approaches to understand and assess the needs and preferences of potential data consumers and data providers, in order to realize working prototypes fitting those needs. However, in KRAKEN we are also aware that in order to fully realize the innovation potential of our solution, we need to attract broad masses of users to join and engage with our platform. This requires more than a just a usable working solution, since the key mechanisms explaining how the blockchain and a privacy-preserving data sharing platform work may not be self-explanatory to a standard user. The aim is therefore that of favoring a gradual adoption of our platform, by supporting and gaining user's

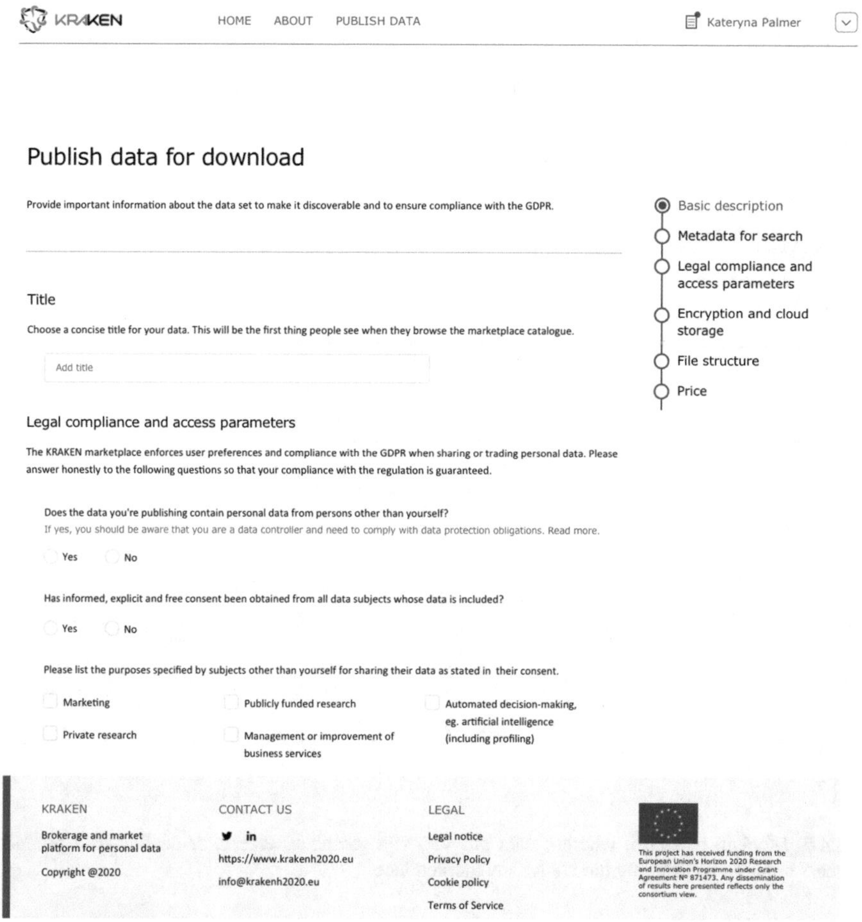

Fig. 4 Mockup of screen enabling a data provider to specify legal compliance and access parameters while creating a data product

trust through the provision of specific built-in privacy-preserving features able to foster increased utilization and sustained data sharing behavior over the long term [21]. The KRAKEN platform will incorporate easy-to-use and easy-to-learn privacy metrics as well as interfaces enabling data sharing through the platform in the most effective and efficient way, ensuring at the same time privacy and safety in its use. Providing to data providers the possibility of fully controlling access to their data by third parties, for example, by specifying legal compliance and access parameters to a data product (Fig. 4) as well as by being able to accept or decline access requests to a data product (Fig. 5), will help to eliminate users' concerns about privacy controls [22].

The KRAKEN platform and marketplace will enforce these consumer-centered features and contribute to educate users on how to best keep control of access

Fig. 5 Mockup of screen where a data provider can visualize, accept, or decline data product access requests received by the KRAKEN marketplace

to their data. It is likely that consumers' willingness to share their data is also affected by factors such as the end purpose of the third party (i.e., making money or research purposes); therefore, enabling mechanisms to support decision-making by the data providers will sound more appealing to a wider audience of potential users of the platform. More reflection is also needed on how to further incentivize data sharing through our platform, by taking into account that some categories of data providers, in the biomedical or healthcare domain for instance, might place greater value on receiving non-monetary forms of compensation (e.g., free treatment, shared research results) instead of value tokens or cryptocurrencies. These examples of design options stress the importance of understanding and monitoring the needs and preferences of KRAKEN users to enable a more successful coevolution and adoption of the data sharing platform, by optimizing and better deploying the advanced technical capabilities of our solution with their users' behaviors.

6 Compliance with Data Protection Framework

This section will give a short overview on the approach of KRAKEN regarding compliance with the relevant data protection framework (i.e., the GDPR). The focus therefore lies on the application of the GDPR [2], even though there are several other frameworks that apply to the KRAKEN platform and accompanying technologies (e.g., the eIDAS Regulation, eCommerce Directive, and future Data Governance Act). In order to ensure that the data subject is adequately protected, and their data are processed fairly and securely, KRAKEN goes beyond a minimum application of the relevant rules by applying a proactive approach toward compliance. This is achieved by considering and integrating important data protection principles and concepts from the outset rather than as an afterthought. Such an approach enhances trust in, and acceptance of, the KRAKEN platform, allowing citizens to benefit from the sharing of their own personal data.

6.1 Data Protection Principles and Their Implementation

The data processing activities in the context of the KRAKEN platform can be divided into two main categories: data processing activities by the KRAKEN platform for the purpose of providing the KRAKEN platform service (i.e., the processing of account data[7]) and data processing activities by the data consumer for their own specific purposes (i.e., processing of content data[8]). This is an important distinction for the application of the GDPR because, as a result, the KRAKEN platform acts as a controller for the processing of account data, while the data consumer acts as a controller for the processing of content data. The implementation of the data protection principles of article 5 GDPR ("principles relating to processing of personal data") will therefore differ depending on the context of the data processing activities. The following overview will mainly focus on the processing of content data by the data consumer since the application of the data protection principles to the processing of account data by the KRAKEN platform is more straightforward in nature.

6.1.1 Lawfulness, Fairness, and Transparency

Lawfulness The principle of lawfulness imposes that all processing activities must comply with the law and must rely on a legitimate legal basis found in article 6

---

[7] Account data refers to data relating to the user profile necessary to provide the KRAKEN platform service (e.g., name, e-mail address, country of residence, etc.).

[8] Content data refers to data that is published on the KRAKEN platform for sharing with data consumers (e.g., educational data or health data).

GDPR ("lawfulness of processing"). In the context of KRAKEN, the processing of content data by the data consumer always relies on the valid consent of the data subject. Consequently, in order to share personal data on the KRAKEN platform, it is necessary to have first obtained valid consent from the data subject.

According to the GDPR, consent is only considered valid if it is (a) *freely given*, (b) *specific*, (c) *informed*, and (d) *unambiguous:*

- *Freely given:* the data subject must have a genuine and free choice; there should be no imbalance of power between the parties involved and the data subject must be able to exercise their free will.
- *Specific:* consent should be given in relation to one or more specific purposes, providing the data subject with a degree of control and transparency. There should be granularity in the consent request and relevant information should be layered in a way that separates it from other information. The data subject should always be able to understand for which specific purpose consent is given.
- *Informed*: the data subject must be properly informed in an intelligible way, using clear and plain language before giving their consent. This should include information about the controller, processing activities, specific purposes, data subject rights, and more.
- *Unambiguous*: consent must constitute a clear affirmative action and must show an unambiguous indication of the data subject's wishes; silence, pre-ticked boxes, and inactivity do not constitute valid consent [23].

In the context of the KRAKEN platform, valid consent is obtained through the user interface and dynamic consent management tool. In the scenario where an institution publishes personal data of data subjects on the KRAKEN platform (e.g., a hospital), they must first confirm that valid consent has been obtained from the data subjects related to the dataset. If an institution wishes to share personal data for different purposes than was included in the original consent, they must obtain new valid consent from the data subjects before proceeding with the publication of the dataset.

In the scenario where a data subject publishes their own personal data on the KRAKEN platform, they are guided through the user interface that allows them to give consent in a *free, specific, informed,* and *unambiguous* manner.

Firstly, the data subject has a real choice and control over whether or not to publish their personal data using the KRAKEN platform. Consent is in no way a non-negotiable condition that is tied to other agreements and the data subject can freely exercise their own will.

Secondly, the data subject is able to select the types of actors that can access and process the data (e.g., public research centers, private companies, governments, etc.) and the specific purposes of processing (e.g., marketing, private research, public research, etc.) in a granular way. Different from a more traditional processing context, it is the data subject that determines the permissions for data processing (incl. specific purposes) when publishing personal data (Fig. 5). Data consumers must also specify and confirm their own intended processing purposes, which are then compared with the specified permissions of the data subject to see whether there is a match. This gives the data subject the necessary control and transparency as to the specific purposes of processing. In order to further safeguard the purpose limitation principle, blockchain technology is used to only allow access to data products by eligible data consumers. In case a data consumer is considered to be ineligible based on a mismatch between the specified permissions, they can still request access to the data product which the data provider can then accept or decline (Fig. 5).

Thirdly, the data subject will be properly informed about the types of processing actors, purposes of processing activities, the possibility to withdraw consent at any time without detriment, and their data subject rights. This information is provided by, in addition to a clear privacy policy, the inclusion of disclaimers and references to additional information throughout the data publication process. In line with the transparency principle, the interface and related information are presented in a concise, transparent, intelligible, and easily accessible form, using clear and plain language. Furthermore, the dynamic consent management tool allows the data subject to manage and modify their consent preferences at any time. Consent can therefore be changed or withdrawn according to the will of the data subject.

Lastly, providing consent on the KRAKEN platform requires multiple affirmative actions by ticking boxes and progressing through the data publication process.

Fairness This principle determines that personal data must not be processed in a way which unreasonably infringes upon the fundamental right to the protection of personal data of the data subject. Processing can therefore be lawful, but still considered unfair with respect to the means foreseen and the reasonable expectations of the data subject. It is essential that the envisioned processing activities, specific purposes, and data subject rights are always clear to the data subject [24].

Transparency As a core data protection principle, transparency applies to all stages of the processing lifecycle. The GDPR makes clear that all information and communications on the processing of personal data should be provided to the data subject in a concise, transparent, intelligible, and easily accessible form while using clear and plain language. The aim is to ensure that data subjects are exhaustively aware of the processing activities and extent of processing relating to their personal data. Thus, the principle of transparency is closely linked to concepts such as valid consent, fairness, information obligations, and the data subjects' rights provided by the GDPR.

The principles of fairness and transparency are also largely implemented by the measures mentioned above, with a special focus on ensuring that the envisioned data processing activities and purposes are in line with the reasonable expectation of the data subject. Additionally, the KRAKEN platform will include easy-to-use privacy metrics that enable the data subject to be aware of their privacy risks at all times.

6.1.2   Purpose Limitation, Data Minimization, and Storage Limitation

Purpose Limitation   This principle states that personal data may only be collected for specified, explicit, and legitimate purposes and not further processed in a manner that is incompatible with those purposes. Purposes should therefore be sufficiently specific and not merely based on broad or vague concepts or notions. They must also be made explicit to the data subject in a clear and intelligible way before any processing activity takes place (cfr. the principle of transparency).

As noted before, it is the data subject that determines the permissions for data processing (incl. specific purposes) when publishing personal data on the KRAKEN platform. It is then up to the data consumers to specify and confirm their own intended processing purposes, which must match with the purposes specified by the data subject. The data consumer, acting as a controller under the GDPR, has to comply with their obligations under the GDPR, including the principle of purpose limitation. Consequently, they may only process the acquired data in accordance with the purposes specified by the data subject.

Data Minimization   The data minimization principle means that personal data must be adequate, relevant, and limited to what is necessary in relation to the purposes for which they are processed. In essence, this principle asks whether the same purpose can be achieved with a more limited collection of personal data. It is therefore intrinsically linked to the purpose limitation principle, as it is an application of the principle of proportionality in relation to the specified purposes.

With regard to the processing of content data, this principle must be complied with by the data consumer that acts as a controller. This can be achieved by only requesting access to strictly necessary data and periodically reviewing whether the personal data they process are still adequate, relevant, and limited to what is necessary for the specified purposes. If the answer is negative, unnecessary personal data should be deleted and incorrect or incomplete data should be rectified. With regard to the processing of account data, the KRAKEN platform only processes what is strictly necessary to provide the KRAKEN platform service in a secure and privacy-friendly way. This encompasses the processing of personal data such as the name, e-mail address, country of residence, etc.

Storage Limitation   According to this principle, which is closely linked to the principles of purpose limitation and data minimization, personal data must be kept in a form which permits identification of data subjects for no longer than is necessary for the purposes for which the personal data are processed. Consequently, once personal data are no longer necessary for the specified purposes, they must be removed from storage or irreversibly de-identified.

Similar to the application of the data minimization principle, it is up to the data consumer acting as a controller to conduct periodic reviews and establish storage, retention, and deletion policies prior to data collection. The KRAKEN user interface allows for the specification of storage periods by the user, which the data consumer must comply with.

6.1.3 Accuracy, Integrity, and Confidentiality

Accuracy The principle of accuracy says that personal data should be accurate and, where necessary, kept up to date. With regard to content data, the data consumer that acts as a controller should keep data accurate at all stages of the processing lifecycle, taking every reasonable step to erase or rectify inaccurate personal data without delay. This can be achieved through review mechanisms and the exercise of the data subject's right to rectification and erasure. With regard to account data, the KRAKEN platform should aim to keep the relevant account details accurate and up to date.

Integrity and Confidentiality This principle states that personal data must be processed in a manner that ensures appropriate security of the personal data. The aim is to protect personal data against unauthorized or unlawful processing, accidental loss, destruction, or damage.

The data consumer that acts as a controller in relation to content data should take steps to implement appropriate technical and organizational measures, such as clearly defined access policies, systemic quality controls, and technical features against data breaches. The level of security should be periodically reviewed to ensure constant protection of personal data. The KRAKEN platform, on the other hand, should also aim to secure the integrity and confidentiality of account data.

Additionally, in order to secure the storage and transfer of personal data, the KRAKEN project introduces appropriate security measures. Because no data products are stored on the KRAKEN platform, but rather by external cloud service providers, strong end-to-end encryption is in place. The use of privacy-preserving analytics also safeguards the integrity and confidentiality of personal data by enabling users to share analytics on their personal data without revealing the initial data. Finally, the use of blockchain technology as a decision-making component allows KRAKEN to only allow access to data products by eligible data consumers. The same blockchain technology stores policies set by the data provider which are checked against SSI VCs of the data consumer by making use of smart contracts.

6.1.4 Accountability

The principles of accountability relate to all previous principles by stating that the controller is responsible for, and must be able to demonstrate compliance with, the other data protection principles.

This means that the controller is responsible for actively implementing appropriate technical and organizational measures in order to promote and safeguard the protection of personal data and to be able to demonstrate that the processing activities are conducted in accordance with the GDPR. In this context, the controller is obliged to keep records of processing activities under its responsibility in order to promote and demonstrate compliance. This also applies to the legal basis of consent, which the controller should also be able to demonstrate according to article 7 GDPR ("conditions for consent"). For these reasons, it is important that the data consumer that acts as a controller implements record-keeping systems for possible audits and inspections. The KRAKEN platform also contributes to the accountability of data consumers by storing evidence of consent through the dynamic consent management application and the tracking of transactions through the blockchain. KRAKEN also informs data consumers about their obligations under the GDPR and provides a system that allows data consumers to clearly stay within the boundaries of valid consent, such as the purposes specified by the data provider.

6.2 The Exercise of Data Subject Rights

Under Chapter III of the GDPR, data subject is entitled to exercise and request their rights vis-à-vis the responsible controller. In the context of KRAKEN, the exercise of data subject rights has two dimensions: vis-à-vis the KRAKEN platform in relation to account data and vis-à-vis the data consumer that acts as a controller in relation to content data. Data subjects are informed about their rights under the GDPR at several points, for example, at profile creation and publication of a data product, in addition to the privacy policy.

With regard to the exercise of data subjects' rights vis-à-vis KRAKEN, data subjects may request their rights by using the KRAKEN contact details and communication channels provided to them. The right to erasure of personal data can be exercised through a profile deletion process, which erases their personal data held by KRAKEN.

For the exercise of data subject rights vis-à-vis the data consumer that acts as a controller, KRAKEN provides data subjects with the appropriate contact details and communication tools. In this context, KRAKEN acts as a communication channel in order to exercise data subject rights, but the requests must be granted by the data consumer. In any case, the possibility to exercise specific data subject rights is subject to the conditions and exceptions of the GDPR, which must be assessed by the data consumer.

6.3 The KRAKEN Approach Toward Data Monetization

Under the EU framework, there does not yet exist legislation that explicitly regulates the monetization of personal data. However, existing legislation applicable to the processing of personal data (i.e., the GDPR) may provide some initial guidelines. From a GDPR point of view, the discussion on the monetization of personal data is quite straightforward. The GDPR does not make specific mention of the monetization of personal data, but since these activities are in fact processing activities in the form of personal data transfers between parties, in exchange for a monetary reward, the GDPR applies as if it would to any other processing activity. The lack of an explicit prohibition means that the monetization of personal data is, in principle, allowed under the GDPR, provided that all principles and provisions are complied with. The question of whether the monetization of personal data is allowed under the GDPR thus becomes a question of compliance. Additionally, when personal data has been fully de-identified through anonymization, the processing of this data will fall outside the scope of the GDPR, which means that the accompanying legal obligations do not have to be complied with.

One of the main objectives of KRAKEN is to enable data subjects to benefit from the processing of their own personal data (e.g., a monetary reward) while still leaving data subjects in control over those data. The KRAKEN platform offers the possibility for data consumers to find relevant personal data for specific processing activities in exchange for compensation. It is important to note that transactions on the KRAKEN platform do not rely on a transfer of ownership rights over personal data (i.e., a transfer of data ownership). The data subject still remains the "owner" of their personal data and they are merely compensated for providing permission to the data consumer to process their personal data for predefined purposes and within the limits of the informed consent given by the data subject. In this sense, the KRAKEN platform merely facilitates the coming together of data providers and data consumers, with the added value of compensating the data provider.

7 Business Challenges

The KRAKEN project aims to release the marketplace of reference for sharing, brokerage, and trading personal data, based on the self-sovereign principle to ensure a user-centered approach for the management of sensitive data. From a business perspective such marketplace needs to generate value for the data providers by offering them mechanisms to evolve toward self-sovereign identity on one hand and by offering added-value services to let them generate revenues on the other hand.

In a digital world the use of digital credentials is required for a huge variety of services, from those provided by public administration including education, health, mobility, and tax declaration to those provided by private organizations such as financial, entertainment, and other services which need to verify the source and integrity of those credentials.

Digital identity is experiencing growing relevance over the last years, changing the way that citizens interact with public institutions and by extension with the private sector as well. There are market drivers that have been stimulating the development and adoption of digital identity in recent years such as the increasing number of online services (related to mobility, smart cities, digital governance etc.) which entails protective supervision of digital certification systems to properly guarantee data security and muster citizenship trust.

This scenario has brought the development of the self-sovereign identity (SSI) that states the right of individuals to control their own data without the involvement of a third party. Therefore, a new paradigm with three main stakeholders emerges:

the individual who owns and manages their digital identity, the issuer who is able to certify a specific attribute of the individual, and the verifier who requests some of these attributes.

Blockchain is the technology which has allowed to take digital identity one step further. Thanks to the immutability, dis-intermediation, and transparency of blockchain, the self-sovereign identity (SSI) paradigm has become a reality allowing users the control and portability of their data securely.

Now, individuals have the control of a huge amount of data of greatest interest for public and private institutions that can be directly or indirectly monetized through personal data marketplaces. In this context a variety of stakeholders from companies and research institutions to citizens and public administration can exchange data in a secure way and obtain a reward (monetary or not monetary). This business model releases a value proposition for all stakeholders involved by enabling the decentralized exchange of data using blockchain technologies; on one hand the use of digital identity reduces the clerical work and facilitates the interoperability among different organizations, increasing the efficiency of administrative processes; on the other hand decentralization guarantees control and integrity of data by the data owners which possess their digital wallet and decide how, when, and with whom to share the data.

The KRAKEN project takes the leadership of data marketplace evolution focusing on healthcare and education sectors, although the resulting platform could be extended to a variety of markets and business cases.

Both current healthcare and education marketplace scenarios share many characteristics. There are decentralized options to manage and share data to users but without monetization mechanisms (beyond the fact of accessing the service for free or incentive mechanisms related to gamification), with the companies being able to get revenues from data commercialization. Both types of marketplaces suffer from poor interoperability among services and they need to explore new business models enhancing aspects such as pricing and rewarding strategies.

KRAKEN aims to disrupt data marketplace market by releasing a strong value proposition based on providing added-value monetization opportunities both for organizations and individuals, guaranteeing data control by data owners and a secure and GDPR compliance data access. The KRAKEN value proposition also will empower data providers and organizations as data unions by removing the intervention of third parties. With regard to healthcare market, KRAKEN will drive the market one step further in the field of personalized medicine and telemedicine development around the concept of real-world data (RWD) [25] by facilitating data transaction at affordable cost to improve and extend traditional studies in the case of researchers and to foster innovation and AI-based applications in the case of IT companies.

From a business perspective the launch and adoption of data marketplace relies upon two aspects which feed each other: on one hand they need to provide attractive value propositions to engage data providers which will benefit from the platform, and on the other hand they need to develop mechanisms to generate economic value to incentivize stakeholders. KRAKEN addresses both issues by analyzing different

B2C and B2B business models to be applied in different phases of the process able to generate monetary and non-monetary revenues. The engagement activities take place from the very beginning of the project by open KRAKEN deployment to entities including their use case for testing. Additionally, the individual users will be engaged through the "data for services" agreement facilitating the matching between data provision and the access to services and rewards (e.g., discount on insurance premium or access to innovative data-driven services) as well as contribute to aggregated data products getting reimbursement for it. KRAKEN will democratize the data market economy by establishing mechanisms to effectively redistribute monetary revenues among all parties including individuals which are indeed the main data generators.

Acknowledgments The KRAKEN project has received funding from the European Union's Horizon 2020 research and innovation program under grant agreement no. 871473.

References

1. Zillner, S., Curry, E., Metzger, A., Auer, S., & Seidl, R. (2017). *European big data value strategic research & innovation agenda*. Big Data Value Association.
2. Regulation (EU) 2016/679 of the European Parliament and of the Council of 27 April 2016 on the protection of natural persons with regard to the processing of personal data and on the free movement of such data, and repealing Directive 95/46/EC (General Data Protection Regulation), OJ 2016 L 19/1.
3. Zillner, S., Bisset, D., Milano, M., Curry, E., García Robles, A., Hahn, T., Irgens, M., Lafrenz, R., Liepert, B., O'Sullivan, B., & Smeulders, A., (eds) (2020). *Strategic research, innovation and deployment agenda - AI, data and robotics Partnership. Third Release. September 2020, Brussels. BDVA, euRobotics, ELLIS, EurAI and CLAIRE.*
4. Der, U., Jähnichen, S., & Sürmeli, J. (2017). Self-sovereign identity - opportunities and challenges for the digital revolution. *ArXiv*, abs/1712.01767.
5. Nakamoto, S. (2008). *Bitcoin: A peer-to-peer electronic cash system*. Accessed March 31, 2021, from https://bitcoin.org/bitcoin.pdf
6. Chi-Chih Yao, A. (1982). Protocols for secure computations (Extended Abstract). *FOCS* (pp. 160–164).
7. Streamr. (2017). *Unstoppable data for unstoppable apps: DATAcoin by Streamr*. Accessed March 31, 2021, from https://s3.amazonaws.com/streamr-public/streamr-datacoin-whitepaper-2017-07-25-v1_1.pdf
8. Androlaki, E. (2018). *Hyperledger fabric: A distributed operating system for permissioned blockchains*. Accessed March 31, 2021, from https://arxiv.org/pdf/1801.10228.pdf
9. Chaum, D., & van Heyst, E. (1991). Group signatures. *EUROCRYPT* (pp. 257–265).
10. Brickell, E., & Li, J. (2010). Enhanced privacy ID from bilinear pairing for hardware authentication and attestation. *SocialCom/PASSAT* (pp. 768–775).
11. Goldwasser, S., Micali, S., & Rackoff, C. (1985). The knowledge complexity of interactive proof-systems (Extended Abstract). *STOC* (pp. 291–304).
12. Bitansky, N., Canetti, R., Chiesa, A., & Tromer, E. (2012). From extractable collision resistance to succinct non-interactive arguments of knowledge, and back again. *ITCS* (pp. 326–349).
13. Koch, K., Krenn, S., Pellegrino, D., Ramacher, S. (2021). Privacy-Preserving Analytics for Data Markets Using MPC. In: Friedewald, M., Schiffner, S., Krenn, S. (eds) *Privacy and Identity Management. Privacy and Identity 2020. IFIP Advances in Information and*

Communication Technology, vol 619. Springer, Cham. https://doi.org/10.1007/978-3-030-72465-8_13

14. Blaze, M., Bleumer, G., & Strauss, M. (1998). Divertible protocols and atomic proxy cryptography. *EUROCRYPT* (pp. 127–144).
15. Shamir, A. (1984) Identity-based cryptosystems and signature schemes. *CRYPTO* (pp. 47–53).
16. Archer, D. W., Bogdanov, D., Lindell, Y., Kamm, L., Nielsen, K., Pagter, J. I., Smart, N. P., & Wright, R. N. (2018). From keys to databases—real-world applications of secure multi-party computation. *The Computer Journal, 61*(12).
17. Mödersheim, S., Schlichtkrull, A., Wagner, G., More, S., & Alber, L. (2019) TPL: A trust policy language. *IFIP TM* (pp. 209–223).
18. Alber, L., Stefan, S., Mödersheim, S., & Schlichtkrull, A. (2022). Adapting the TPL trust policy language for a self-sovereign identity world. *Open Identity Summit.*
19. Alber, L., More, S., Mödersheim, S., & Schlichtkrull, A. (2021). Adapting the TPL Trust Policy Language for a Self-Sovereign Identity World. In: Roßagel, H., Schunck, C. H. & Mödersheim, S. (Hrsg.), *Open Identity Summit 2021*. Bonn: Gesellschaft für Informatik e.V.. (S. 107–118).
20. Kugler, L. (2018). The war over the value of personal data. *Communications of the ACM, 61,2,* 17–19.
21. Yeratziotis, A., Van Greunen, D., & Pottas, D. (2011). Recommendations for usable security in online health social networks. In *Pervasive Computing and Applications (ICPCA): 2011 6th International Conference IEEE*. Oct 220–226.
22. Daglish, D., & Archer, N. (2009). Electronic personal health record systems: A brief review of privacy, security, and architectural issues. privacy, security, trust and the management of e-Business. World Congress on December 2009.
23. European Data Protection Board. (2020). *Guidelines 05/2020 on consent under Regulation 2016/679*, 7–20.
24. Clifford, D., & Ausloos, J. (2017). Data protection and the role of fairness. CiTiP working Paper 29/2017, KU Leuven Centre for IT & IP Law, 11–20.
25. Lipworth, W. (2019). Real-world data to generate evidence about healthcare interventions. ABR11, 289–298 (2019). doi:https://doi.org/10.1007/s41649-019-00095-1. Accessed March 31, 2021from https://link.springer.com/article/10.1007/s41649-019-00095-1

Connecting Data Spaces and Data Marketplaces and the Progress Toward the European Single Digital Market with Open-Source Software

Achille Zappa, Chi-Hung Le, Martín Serrano, and Edward Curry

Abstract In this book chapter, recent advances in the development and implementation of open-source software technologies and information management systems to support the progression of the data economy by means of data operations and data offering descriptions are introduced. The management of controlled registries, mapping of information using metadata aggregation, interfaces among components, links of data and actors, discovery and retrieval of data, compiling smart contracts, and other core operations are introduced. This chapter contributes to the state of the art by providing the definition, creation, and collection of data-driven marketplaces that, by design, look at sharing and exchanging data using a common description framework called i3-MARKET. i3-MARKET's main design objectives are to support the sharing data assets, execute operations, and provide API services and generally all the security-related functionalities relying on data details, credentials, contracts, and pricing. i3-MARKET also uses a sharing semantic model to facilitate the work with people in improving and maintaining the models for the present and future. The i3-MARKET described in the chapter introduces the concept of a backplane architecture as a support tool that aims to innovate the data market industry providing solutions and support in developing building blocks to overcome the barriers to interoperable and integrative data using trusted, federated, and decentralized software components.

A. Zappa · M. Serrano (✉)
Insight SFI Research Centre for Data Analytics, University of Galway, Galway, Ireland
e-mail: achille.zappa@nuigalway.ie; martin.serrano@nuigalway.ie

C.-H. Le
Insight SFI Research Centre for Data Analytics, Data Science Institute, University of Galway, Galway, Ireland
e-mail: chihung.le@nuigalway.ie

E. Curry
Insight SFI Research Centre for Data Analytics, University of Galway, Galway, Ireland
e-mail: edward.curry@nuigalway.ie

E. Curry et al. (eds.), *Data Spaces*, https://doi.org/10.1007/978-3-030-98636-0_7

131

Keywords Data marketplace · Data offering · Semantic vocabulary · Digital economy

1 Introduction

There is a high demand for advancing and promoting the European data market economy by innovating Data Spaces and data marketplace platforms following single digital market strategies and demonstrating with industrial implementations that the growth of data economy is possible [1]. There is a growing need for solutions that provide technologies for trustworthy (secure and reliable), data-driven collaboration and federation of existing and new future data marketplace platforms, and those with special attention on large industrial data [2]. It is well known that despite various research and innovation attempts working on Big Data management, of personal and or industrial data integration and security, there is no broadly accepted trusted and secure data marketplace [3]. The H2020 i3-MARKET project aims to promote data market economy by providing support tools and avoiding to create another new marketplace but implementing a solution in the form of a backplane set of tools introduced as a framework of solutions that allow other data marketplaces and Data Spaces to expand their market functions, facilitating the registration and discovery of data assets and supporting the trading and sharing of data assets among providers, consumers, and owners for a better data sharing and trading processes. The i3-MARKET platform described in this chapter is designed to enable secure and privacy-preserving data sharing across Data Spaces and marketplaces by deploying a backplane across operational data marketplaces. The i3-MARKET Backplane, on the one hand, can be seen as a set of tools that can be deployed and integrated as backend technologies in current running marketplaces facilitating and allowing to add the missing functionalities that current marketplaces lack, and, on the other hand, i3-MARKET acts as baseline technologies for stand-alone reference implementation(s) that facilitates the starting point for a modern data sharing economy. In other words, the i3-MARKET Backplane provides the tools for setting up the foundations of a data marketplace ecosystem.

This book chapter reviews the progression of the data economy through data operations and data offering descriptions. It introduces the most recent advances in supporting, developing, and implementing open-source software technologies and information management systems [4]. As part of the design description, the management of controlled registries, mapping of information using metadata aggregation, interfaces among components, links of data and actors, discovery and retrieval of data, compiling of smart contracts, and other core operations are described [5]. This chapter contributes to the state of the art, providing the definition, creation, and collection of data-driven marketplaces that by design look at sharing and exchanging data using a common description framework called i3-MARKET. The i3-MARKET's main design objectives are to support the sharing data assets, execute operations, and provide API services and generally all the security-related functionalities relying on data details, credentials, contracts, and pricing.

This chapter analyzes the basis of data marketplace design and reviews the best practices on implementation and deployment for developing lacking technologies and solutions for a trusted (secure, self-governing, consensus-based, and auditable), interoperable (semantic-driven), and decentralized (scalability) data marketplace infrastructure [6, 7]. This chapter introduces and explains the principles for enabling federation and semantic data interoperability of the existing and future emerging Data Spaces and marketplaces. Finally, this chapter describes the principles of data monetization being added to formerly closed systems to offer and share data in the form of intelligent data economy services (smart contracts) and lowering the market entry barriers for stakeholders—especially SMEs—to trade their data assets to ignite a flourishing data economy that fosters innovation and business in Europe.

This chapter focuses on explaining the high demand for the design and implementation of technologies for enabling privacy and confidentiality levels that allow data marketplaces to support both legal and user-desired control and at the same time enable transparency for sharing data among data marketplaces. This chapter also focuses on reviewing the current demands on regulatory aspects around sensitive data assets and the more stringent demands on privacy and security using legal frameworks implementing the required security and access control measures that enable secure trading of data, including support for automated contracting and real-time exchange of data.

The organization of this chapter is as follows: Sect. 2 presents challenges in the data marketplace design and data economy in terms of best practices. Section 3 reviews the current state-of-the-art situation and introduces best practices for data exchange approach in three streams, i.e., security, privacy, and trust. Section 4 presents the i3-MARKET's innovations for the data economy that acts as a baseline supporting the information interoperability approach to the current state of the art and describes the features for a federated data marketplace and data space ecosystem. Section 5 introduces i3-MARKET Backplane architecture at a glance, including the reference data flow for Data Spaces and marketplaces. Section 6 describes industrial innovations as necessary features and addresses their benefits in industrial marketplace platforms. Finally, Sect. 7 presents the conclusions.

2 Challenges in Data Marketplace Design and Data Economy

The current state-of-the-art analysis reveals barriers for data marketplaces that require attention on security, data privacy, and trust [8, 9]. Persistent challenges blocking progress in data marketplace design and deployment are the following.

2.1 Data Marketplace Openness and Fairness

Data marketplaces are traditionally large, closed ecosystems in the hands of a few established lead players or a consortium that decides on the rules, policies, etc. This approach limits the number of data offers that can be included and, at the same time, creates barriers to a more open data economy. For example, the business interest of data marketplace established members' interests often conflict with those from new applicants. Thus, the request from those who want to join an ecosystem might be denied. This is a significant barrier for growing a competitive data economy, where all stakeholders can freely participate under fair conditions.

In addition, the rights for using the data are often defined in contractual agreements among the involved parties. Therefore, companies with a stronger negotiating position usually obtain preferred rights to use the data. Unfortunately, this often results in smaller companies being excluded from gaining access to the data. This legal disadvantage is particularly detrimental to small- and medium-sized enterprises (SMEs) in a weaker negotiating position in such negotiations [3]. Yet, the main barrier of the European data economy is the fact that current Data Spaces and marketplaces are "silos," without support for data exchange across their boundaries. These wall gardens significantly limit market competition and overall data economy growth [10].

2.2 High Demands on Security and Privacy

In a large data-driven economy, sharing sensitive personal or industrial data assets demands high-security standards. This demand is applicable for Data Spaces and marketplace platforms and especially necessary over those interfaces and special hooks developed for data owners to control with whom their data is exchanged. In addition, the new European privacy regulation (GDPR) requires an unprecedented level of transparency and control for end-users, which, by design, must be implemented before personal data can be exchanged. Moreover, Data Spaces and marketplaces also lack adequate legal frameworks that address questions like how the data source is identified and verified, which type of data reuse is appropriate, who is responsible if data are leaked, etc. Those legal barriers prevent all stakeholders (SMEs, large enterprises, and any other organization or individual) from fully exploiting the opportunities of the data economy [2].

2.3 Data Marketplace Interoperability

The lack of interoperability due to missing standards, common APIs, and data models makes it difficult for data consumers to discover relevant data assets, access the data, and uniformly integrate the data, especially when combined from multiple providers. This is even challenging among data providers and consumers participating in the same data space or marketplace. The lack of extended common data models typically requires developers of a consumer application or service to consult the data provider first to learn how to interpret and use the data. This is a very timely and cumbersome task and thus restricts the sharing of data assets and potential innovation significantly. In addition, the missing semantic interoperability is an obstacle to providing solid, consistent, and objective models of data quality [11]. Thus, transparent pricing and quality assurance strategies are essential for successful data markets. However, the main barrier of today's data economy is that there is no trusted backbone for Data Spaces and marketplaces to share data assets across the "silos." The lack of a self-governing and transparent backbone thus hampers the growth of the data economy in Europe and beyond.

3 Advancing the State of the Art on Security, Privacy, and Trust

To address current data marketplace challenges and leverage the full potential of the data economy, there is a need for developments in security, privacy, and trust as they are analyzed as follows.

3.1 Security

Self-sovereign identities are a must, independently from a single/central provider, to support global-scale identity management and enable the ability to associate personal data with end-users in a trusted and traceable manner. This is also the basis for identity-based access control and user consent management.

Accessibility to deployed data services is necessary. Therefore, there is a need to develop secure access for data consumers to access sensitive personal or commercial data directly and be compliant with data provider services. The use of application programming interfaces (APIs) is widespread, and this feature has the advantage of enabling data services to be connected and exposed. At the same time, the data can remain on the provider infrastructure and does not need to be stored or passed through a central infrastructure. However, APIs require a well-documented approach and a fully specified and published method, allowing existing data marketplaces or providers to adopt this. Thus, the Open-API approach is the primary best practice, allowing multiple stakeholders to participate in the driven data economy.

Data wallets and data economy require the adoption of emergent digital technologies, and the provisioning of a novel smart wallet framework enables data owners (i.e., end-users in the case of personal data or companies in the case of industrial data) to directly interact with smart contracts related to their data to give or revoke their consent for the anticipated data exchange. Therefore, in conjunction with the novel type of smart contracts, smart wallets will be a key component toward addressing user privacy following the GDPR and thus promoting the data-driven economy.

3.2   Data Privacy

There is a need for decentralized storage and access to semantic descriptions of the offered as data assets to enable data discovery across today's data marketplace silos. This requires enabling federation among the individual Data Spaces and marketplaces, without the need for central control or coordination that has to be trusted by all parties in the federated data marketplace ecosystem.

Transparency in Data Spaces and marketplaces is a feature that requires policy commitments and technology. Cryptocurrency/token provides a transparent, cost-efficient, and fast payment solution for trading data assets among the participating Data Spaces and marketplaces. As a result, the cryptocurrency/token will incentivize data providers to offer their data assets and thus accelerate the European data economy. Furthermore, the solution will be designed so that the participating Data Spaces and marketplaces can also use the tokens as an internal payment medium.

There is a requirement when sharing data that needs to be secured, mainly if a federated ecosystem is designed. A Secure Semantic Data Model Repository is a feature that enables data consumers to efficiently discover and access data assets (due to precise semantic queries) and integrate the data into their applications/services (based on a common understanding of the meaning of the data). This allows completely independent data providers and consumers to exchange and use data in a meaningful way—without prior information exchange. The availability of common data models is a key enabler for establishing a scalable data economy.

3.3   Trust

Secure and trusted APIs are required to allow Data Spaces and marketplace providers to obtain identities, register data assets, fetch their semantic descriptions, create, and sign smart contracts, make payments, etc. This ensures complete openness, i.e., any data space or marketplace provider can connect its local ecosystem with the global data market ecosystem.

Immutable and auditable smart contracts are necessary to trade data assets across data space and marketplace boundaries. All stakeholders, namely, data providers (for confirmation of the offer and its conditions, e.g., license, price, and SLAs), data consumers (for agreement of the contract conditions), and data owners (for consent to the data exchange), must sign these contracts. In addition, individual marketplaces can also adopt this advanced solution for handling local contracts.

Legal frameworks are obstacles that need to be removed by designed and implemented following the requirements (e.g., contractual basis for smart contracts and crypto-token) and innovative business models for incentivizing the sharing and trading of data assets and the operation of the decentralized backplane by the marketplace providers.

4 The i3-MARKET Backplane Innovations for the Data Economy

The i3-MARKET Backplane is an initiative/project that addresses the growing demand for a single European data market economy by innovating marketplace platforms demonstrating industrial implementations of the data economy. I3-MARKET implements features and requirements in the form of backend tools. The i3-MARKET Backplane implements reference components that can be used under different Data Spaces and marketplaces to satisfy common needs. i3-MARKET provides technologies for trustworthy (secure and reliable), data-driven collaboration, and federation of existing and new future marketplace platforms; special attention on industrial data and particularly on sensitive commercial data assets from both SMEs to large industrial corporations is taken.

4.1 Privacy and Data Protection

The i3-MARKET framework implements a trusted and secured backplane offering privacy preservation and fine-grained access control using an identity access management system (IAM) for data owners and consumers. Also, based on new types of smart contracts and a secure data access/exchange API to enable sharing of sensitive personal data and commercial/industrial data, i3-MARKET will ensure transparency and control. The i3-MARKET project will allow data producers to register their data offers (with all the necessary metadata to describe the offer). Data consumers will use data APIs to discover data descriptions as available information in the marketplace and thus start commercializing relations in a secure and controlled manner. i3-MARKET's strong focus on trust and security has the potential to remove the fear of data owners, to start sharing and trading their sensitive data, which are kept close as of today.

4.2 Trust and Security Platform

The i3-MARKET platform main target is to develop the missing building blocks for building a trusted, interoperable, and decentralized European data market, as well as an integrated platform (the i3-MARKET Backplane) that allows the federation (via integration) of currently emerging but yet isolated Data Spaces and marketplaces.

Besides addressing the privacy concerns of data owners to share their data in a user-controlled and transparent manner and the security concerns of companies to trade-sensitive/non-sensitive industrial data, the i3-MARKET project focuses predominantly on developing technologies, best practices, and reference design approaches that create trust. Based on distributed ledger technologies and their blockchain-based decentralization, consensus-based ruling, and auditability [12], the i3-MARKET Backplane aims to be a fully scalable and trusted reference platform to power the overall European data economy, where every stakeholder can participate under fair conditions.

4.3 Secure Sharing of Personal Data and Industrial Data

The i3-MARKET platform addresses the increasing need to share sensitive data (i.e., industrial data or personal). This is key to growing the European data economy beyond "open data" and leveraging the potential of data in commercial settings. For this, i3-MARKET has developed the missing building blocks for trusted data-driven collaboration (interoperability) and trading platforms (economy) for sensitive commercial data assets.

Concerning privacy risk and threat methods to protect industrial data assets, the i3-MARKET platform uses stochastic models/algorithms that have been tested in previous platforms like AGORA and VITAL-IoT, and the aim is to use them for the identification of privacy attacks promoting secure and scalable trading of proprietary/commercial data assets with support for automated detection. The method for securing data sharing will be smart contracts, including the required legal framework(s), and enabling data exchange. The focus of i3-MARKET is on industrial data; however, the provided data access and protection frameworks will guarantee protection for both personal and industrial data, enforce user-desired privacy levels, and allow end-users to control by consent who can access their data.

4.4 Large-Scale Federated Data Platform

The i3-MARKET Backplane and its APIs act as a reference design and implementation for existing and emerging Data Spaces and marketplaces to federate and trade data assets across the existing ecosystem boundaries. Moreover, i3-MARKET introduces the concept for cross-domain data sharing and, employing federated tools, incentivizes opening formerly closed systems and offers their data assets via the i3-MARKET Backplane and lowers the market entry barriers for stakeholders (especially SMEs) to ignite a common, federated data market in Europe. In particular, i3-MARKET will address the interoperability challenge for trading data assets across independent stakeholders through a common, standard data access API and a shared data model repository, allowing data providers to semantically describe their data assets (metadata) and data consumers to access and integrate them in a uniform and standard manner (based on the metadata).

4.5 Policy and Regulation for Data Marketplace Backplane

The i3-MARKET project will address not only the latest policy and regulatory requirements in terms of data protection and privacy, e.g., the need for flexible and easy-to-use controls for data access (GDPR), but also the lack of interoperable data access APIs and data monetization support to enable the data exchange and incentivize data economy.

The i3-MARKET data market backplane, with its Open APIs and easy-to-use SDKs for developers to integrate their Data Spaces and marketplaces, makes sharing and trading of data assets across the participating Data Spaces and marketplaces a straightforward task for developers. Thus, i3-MARKET reduces the lack of ICT and data skills needed to grow the European data economy and increases the capacity of Europe to respond to the digitalization challenges.

Through the project results and their dissemination and exploitation, i3-MARKET will also increase the number of human ICT capacities with the required skills and know-how for industry digitalization in general and the relevant European regulations in particular.

5 i3-MARKET Backplane at a Glance

The i3-MARKET project innovates industry solutions by developing building blocks to overcome the barriers discussed above. As depicted in Fig. 1, we integrate them into a trusted, interoperable, and decentralized data backplane. In the same way, other marketplaces can be integrated to enable secure privacy-preserving data sharing across Data Spaces and marketplaces.

5.1 i3-MARKET High-Level Architecture

To validate the solution, the i3-MARKET Backplane is a deployment that will work across operational data marketplaces. Firstly, Atos and SIEMENS operate two marketplaces for data, Bridge.IoT [13] and AGORA [14].

These marketplaces allow data providers and consumers to share or trade data in an open and fair (every organization can participate under equal conditions) and interoperable manner. First, the BIG IoT marketplace and APIs are being transferred to the open-source community by establishing a new project called Bridge.IoT [12] within the Eclipse Foundation, where all future extensions could be maintained. Secondly, Atos operates a data marketplace for the automotive industry sector in Spain. This full-scale data marketplace prototype, AGORA, runs on the trusted Linked Data platform called with the same name and developed and owned mainly by Atos.

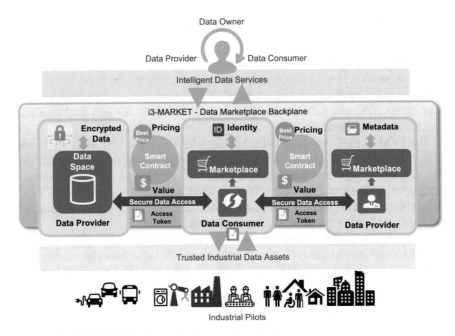

Fig. 1 The i3-MARKET Backplane for federated marketplace ecosystem

However, even though many use cases call for and would directly benefit from sharing or trading data across the different marketplace instances, there is still no scalable, trusted, and interoperable solution that enables sharing or trading data assets across individual marketplace instances. Moreover, today's data marketplace platforms, including Bridge.IoT or AGORA, still lack the capability for exchanging sensitive industrial data assets as the required levels of security and privacy (in accordance with the GDPR) are not yet supported—especially not across ecosystem boundaries. We consider those problems to be solved using blockchain-type technologies and part of the building blocks for the i3-MARKET Backplane. Furthermore, they will help the participating marketplace operators, who collectively run the consensus-based, self-governing, and decentralized backplane. As a result, the data owners, providers, and consumers trust the overall solution.

5.2 i3-MARKET Data Flow as Reference Implementation

The project's overall goal is to develop the i3-MARKET Backplane, defined as a software framework that provides the lacking technologies trusted, interoperable, and decentralized marketplaces for industrial data [15]. Figure 2 gives an overview of the i3-MARKET data flow to achieve that goal. The i3-MARKET Backplane

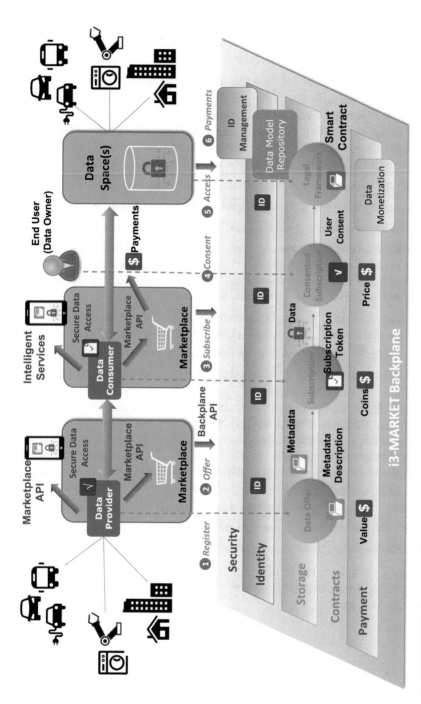

Fig. 2 i3-MARKET reference data flow

platform builds upon the state-of-the-art design principles [16–18] and the following assets:

1. Two data marketplace platforms: Bridge.IoT and AGORA. Both platforms will be operational during the project lifetime, allowing data providers and consumers to register **1** and then sign up via subscription **3** and secure access identification **ID** and offer **2** or consume data in a protected manner. The Bridge.IoT platform has been developed by Atos, SIEMENS AG, NUIG, and UPC in the EU project BIG IoT [14] and is maintained in an Eclipse Foundation project. i3-MARKET aims to be extensible and integrative; thus, the solution can also be extended to other marketplaces, including other domains.

2. A secured data access API enables data providers secured registration **v** and consumers verification **☑** to access **5** and/or exchange data in a peer-to-peer fashion once the contracts **▣** and security mechanisms for identity management **ID** have been confirmed and executed. This improves scalability and avoids the need for data providers to share their data assets with intermediaries (e.g., a marketplace provider). This data access API also uses semantic data descriptions to access available types of data assets in the marketplace.

3. A marketplace APIs **/** for all the communication between data providers and data consumers to the marketplace. Data providers use this API to register their data offers **2** (with the necessary metadata **▣** to describe the offer). Data consumers use the APIs **/** to register their demand and find the matching offerings they can then subscribe to.

4. A set of IoT-related vocabularies and data models to semantically describe data offers and demands. This is the key for enabling the trading of data assets across domains and stakeholder boundaries, without the need for developers of an application (data consumer) to learn about the meaning of the data from the data provider or through manual analysis or experimentation with the data and also for the owner of the data to consent **4** access.

5. A set of intelligent services for data monetization is defined first by the pricing model **$** the provider of the data assigned and second by the activity and interactions in the marketplace following dynamic pricing models **$**.

Figure 2 gives an overview of the designed Data Spaces and marketplace data flow, including the architectural overview of the i3-MARKET Backplane approach as described above.

6   Industrial Innovation for a Data-Driven European Ecosystem

The ability to exchange or monetize data and gain rewards for offering data for various purposes while supporting data owners' privacy and security demands has great potential to incentivize and grow the European data economy and top the international competition in a socially acceptable manner [19].

The i3-MARKET Backplane aims to become one of the key enablers for embracing the latest European Commission Digital Single Market strategy, which mentioned according to a recent announcement that "it is time to make the EU's single market fit for the digital age—tearing down regulatory walls and moving from 28 national markets to a single one. This could contribute 415 billion euros per year to our economy and create hundreds of thousands of new jobs."

6.1   Data Sharing/Brokerage/Trading Build on Existing Computing Platforms

Despite various research and innovation projects working on Big Data management and large amounts of data integration and security, there is no broadly accepted trusted and secure data marketplace. At the core of the i3-MARKET is the definition and implementation of a data market backplane able to facilitate trusted and secure sharing and trading operations of proprietary/commercial data assets in a secure way and at the same time be capable of unifying data access across heterogeneous platforms (i.e., different types of Data Spaces and marketplaces).

Significant limitations and challenges when managing industrial data enable and facilitate trusted and secure sharing and trading mechanisms of the data assets. In i3-MARKET, we address those challenges by aggregating a backplane with security tools that will automatically establish robust and scalable controls of data protection over the activity occurring in the marketplace(s) and with legal compliance when data assets are to be discovered, shared, and then exchanged as part of a commercial transaction (brokerage or trading). Furthermore, we follow an innovative lifecycle that ensures that the security controls are compliant with the legal right and fair remuneration to the data owners.

We also acknowledge that several industrial applications require cross-domain use cases involving both large amounts of industrial data (Big Data), and based on those facts, our approach called i3-MARKET is also looking at scalability and efficiency levels concerning time response to preserve utility metrics for data analysis enabling intelligent services from the data and the marketplace to be easy to understand for the owner of the data and other stakeholders and at the same time contribute back the value in the form of data monetization.

6.2 Data Privacy in Industrial Data Marketplace Platforms

The i3-MARKET identity access management (IAM) approach is based on self-sovereign identities and new types of smart contracts to exchange and trade data in a privacy-preserving manner (in accordance with the GDPR) and with the desired level of control by the data owners. The blockchain-based decentralized backplane, with its support for smart contracts and crypto-tokens, is the basis for incentivizing data owners to share their assets. Based on the i3-MARKET crypto-token, we will incentivize especially early adopters of the i3-MARKET technology to overcome the common challenge of data markets, namely, to reach the initial liquidity level needed to achieve the network effect of marketplaces. Furthermore, addressing the aspects of data confidentiality and privacy is fundamental to i3-MARKET, as the project targets use case scenarios where personal and industrial data are shared or traded among the involved stakeholders. We will use self-sovereign identities and new smart contracts, which the data providers and consumers must sign and the data owners (e.g., end-users or corporations) to ensure that all parties consent to the data exchange. i3-MARKET will also support data encryption on data access interface to ensure that only the involved parties can see the data.

6.3 Industrial Data Marketplace Platforms

Three i3-MARKET use cases are implemented in the form of industrial pilots. The three are selected based on the involvement of multiple stakeholders (e.g., manufacturers, suppliers, as well as leasing and financial companies) and also from a cross-domain nature (e.g., manufacturing (Industry 4.0) and human-centric, as well as automotive sector) to demonstrate i3-MARKET's capability of integrating heterogeneous data platforms, solving the interoperability and integrative challenge, and providing the backbone for a single European data market. i3-MARKET builds on proven deployed and functional platforms toward extending them by providing publicly accessible and extensible services for secure and protected data assets.

7 Conclusions

This chapter addresses best practices for data space and data marketplace design and their implementation identified from the state-of-the-art analysis. These challenges are tested and validated in the context of an H2020 European Data Ecosystem called i3-MARKET. The best practices leverage the full potential of Big Data, IoT, and AI applications in data marketplaces and identify a need for further improvements in other streams supporting scaling-up applications.

The main objective of the i3-MARKET Backplane is to overcome hurdles in the current and new design Data Spaces and marketplace approaches by developing the lacking building blocks (in the form of a software framework called i3-MARKET Backplane) for data providers and consumers, and thus incentivize and enable the creation of a more trusted European data market economy.

The i3-MARKET Backplane software platform addresses the interoperability and integration challenges for trading data assets across independent stakeholders using secured transactions based on data annotation (semantics) as well as a trusted data trading platform and will provide a network of decentralized and economy-driven and scalable data repositories that can be extensible for enabling the deployment of intelligent industrial data services fostering innovation and business opportunities.

The i3-MARKET Backplane aims at enabling the federation of data markets and targets to become a key enabler for embracing the latest European Commission Digital Single Market strategy, incentivizing the industrial data market economy. The impact of the exploitation of the i3-MARKET will be tackled, overall and individually, by exploitation and business models that will benefit all stakeholders in the data economy and take into account the information societies in Europe.

This book chapter analyzed the basis of data space design and data market-places discussing the best practices for data privacy, data protection, and data sharing/exchange alike, introduced concepts for data economy, and illustrated the i3-MARKET Backplane tools to enable semantic interoperability of the metadata using an open-source reference architecture and following an agile methodological innovative approach.

Acknowledgments This work has been partially supported by the H2020 i3-MARKET Project—Intelligent, Interoperable, Integrative, and deployable open source MARKETplace backplane with trusted and secure software tools for incentivising the industry data economy (www.i3-market.eu)—which is co-funded by the European Commission under H2020 framework program, contract number H2020-ICT-2020-871754-i3-MARKET. It is also partially supported by Science Foundation Ireland under grant number SFI/12/RC/2289_2.

References

1. *European Initiative on Digital Single Market*. Accessible online October 2021, from https://eufordigital.eu/discover-eu/eu-digital-single-market/
2. *A European Strategy for Data*. Accessible online October 2021, from https://digital-strategy.ec.europa.eu/en/policies/strategy-data
3. https://www.digitalsme.eu/data-economy-open-data-market-unleashing-untapped-potential-smes/
4. Brynjolfsson, E., & Mcafee, R. The business of artificial intelligence. *Harvard Business Review (HBR)*, July 17.
5. Fernandez, R. C., Migliavacca, M., Kalyvianaki, E., & Pietzuch, P. (2014). Making state explicit for imperative big data processing. In *USENIX ATC*.
6. Serrano, J. M., Serrat, J., & Strassner, J. Ontology-based reasoning for supporting context-aware services on autonomic networks. 2007 IEEE/ICC International Conference on Communications, 24–28 June 2007, Glasgow, Scotland.

7. Muan Sang, G., Xu, L., & De Vrieze, P. (2016). *A reference architecture for big data systems* (pp. 370–375). doi:https://doi.org/10.1109/SKIMA.2016.7916249.
8. Serrano, M., Strassner, J., & ÓFoghlú, M. A formal approach for the inference plane supporting integrated management tasks in the future internet. *1st IFIP/IEEE ManFI International Workshop, In conjunction with*11th IFIP/IEEEIM2009, 1–5 June 2009, at Long Island, NY.
9. Zillner, S., Curry, E., Metzger, A., Auer, S., & Seidl, R. (Eds.). (2017). *European big data value strategic research & innovation agenda.* Big Data Value Association. Retrieved from http://www.edwardcurry.org/publications/BDVA_SRIA_v4_Ed1.1.pdf
10. Curry, E., Metzger, A., Zillner, S., Pazzaglia, J.-C., & García Robles, A. (2021). *The elements of big data value.* Springer International Publishing. https://doi.org/10.1007/978-3-030-68176-0
11. Curry, E. (2020). *Real-time linked dataspaces.* Springer International Publishing. https://doi.org/10.1007/978-3-030-29665-0
12. Carson, B., Romanelli, G., Walsh, P., & Zhumaev, A. *Blockchain beyond the hype: What is the strategic business value?*", June 2018.
13. https://projects.eclipse.org/proposals/eclipse-bridge.iot
14. http://big-iot.eu and http://www.automat-project.eu/
15. Pääkkönen, P., & Pakkala, D. (2015). *Reference architecture and classification of technologies, products and services for big data systems.* Big Data Res. 2, 4 (December 2015) (pp. 166–186). doi:https://doi.org/10.1016/j.bdr.2015.01.001.
16. Le-Phuoc, D., Nguyen-Mau, H. Q., Parreira, J. X., & Hauswirth, M. (2012). A middleware framework for scalable management of linked streams. *Journal of Web Semantics, 16*(November 2012), 42–51.
17. Neumeyer, L., Robbins, B., Nair, A., & Kesari, A. (2010). *S4: Distributed stream computing platform.* In *2010 IEEE International Conference on Data Mining Workshops* (pp. 170–177).
18. Serrano, J. M. (2012). *Applied ontology engineering in cloud services, networks and management systems*, March 2012. Springer. Hardcover, 222 pages, ISBN-10: 1461422353, ISBN-13: 978-1461422358.
19. Zillner, S., Bisset, D., Milano, M., Curry, E., Hahn, T., Lafrenz, R., et al. (2020). *Strategic research, innovation and deployment agenda - AI, data and robotics partnership. Third Release* (Third). Brussels: BDVA, euRobotics, ELLIS, EurAI and CLAIRE.

AI-Based Hybrid Data Platforms

**Vassil Vassilev, Sylvia Ilieva, Iva Krasteva, Irena Pavlova,
Dessisslava Petrova-Antonova, and Wiktor Sowinski-Mydlarz**

Abstract The current digital transformation of many businesses and the exponential growth of digital data are two of the key factors of digital revolution. For the successful meeting of high expectations, the data platforms need to employ the recent theoretical, technological, and methodological advances in contemporary computing and data science and engineering. This chapter presents an approach to address these challenges by combining logical methods for knowledge processing and machine learning methods for data analysis into a hybrid AI-based framework. It is applicable to a wide range of problems that involve both synchronous operations and asynchronous events in different domains. The framework is a foundation for building the GATE Data Platform, which aims at the application of Big Data technologies in civil and government services, industry, and healthcare. The platform implementation will utilize several recent distributed technologies such as Internet of Things, cloud, and edge computing and will integrate them into a multilevel service-oriented architecture that supports services along the entire data value chain, while the service orchestration guarantees a high degree of interoperability, reusability, and automation. The platform is designed to be compliant with the open-source software, but its open architecture supports also mixing with commercial components and tools.

Keywords Vertical layering · Semantic technologies · Horizontal integration · Service-oriented architectures · Cloud · Application containerization · Service orchestration

V. Vassilev (✉) · W. Sowinski-Mydlarz
Cyber Security Research Centre, London Metropolitan University, London, UK

GATE Institute, Sofia University, Sofia, Bulgaria
e-mail: v.vassilev@londonmet.ac.uk; w.sowinsky-mydlarz@londonmet.ac.uk

S. Ilieva · I. Krasteva · I. Pavlova · D. Petrova-Antonova
GATE Institute, Sofia University, Sofia, Bulgaria
e-mail: sylvia@gate-ai.eu; iva.krasteva@gate-ai.eu; irena.pavlova@gate-ai.eu;
dessislava.petrova@gate-ai.eu

© The Author(s) 2022
E. Curry et al. (eds.), *Data Spaces*, https://doi.org/10.1007/978-3-030-98636-0_8

147

1   Introduction

Europe is home to more than 50 Big Data Centers of Excellence (CoE), participating in the European Network of National Big Data Centers of Excellence [1]. Big Data for Smart Society Institute at Sofia University (GATE) is building the first Big Data CoE in Eastern Europe. Its advanced infrastructure and unique research ecosystem aim to create data services and analytical and experimentation facilities to deal with the challenge of contemporary digital revolution. The GATE Data Platform will enable high-quality research with wide scope and big impact along the entire data value chain. The platform will also support data-driven innovations and will serve the needs of multiple projects within different application domains—Future City, Smart Industry, Intelligent Government, and Digital Health. As a by-product of these activities, the GATE Data Platform will create an advanced and sustainable ecosystem for both application developers and nontechnical businesses to exploit the full potential of the available services and acquired knowledge and data. For this purpose, the GATE Data Platform will also enable creating Data Spaces with high-quality pre-processed and curated data sets, aggregating and semantically enriching data from heterogeneous sources. The acquired knowledge for management and usage of data will be made available through reusable intelligent cross-domain data models and data processing services. The GATE Data Platform will enable start-ups, SMEs, and large enterprises, as well as other organizations in a wide range of societal sectors, to build advanced data-driven services and vertical applications. This way, the GATE Data Platform will become a focal point for sharing data, services, technology, and knowledge that eases the creation of an ecosystem of diverse stakeholders, adds value to the businesses, and facilitates creation of new business and commercial models for digital transformation of the industry and the society.

These ambitious goals can be achieved effectively only with wider employment of the achievements of contemporary computing and data science and technologies. To utilize their potential, the data platforms must adopt a hybrid approach in order to address the data processing from theoretical, technological, engineering, and organizational standpoint. Artificial Intelligence allows to utilize many powerful concepts, to build complex models, to automate difficult tasks, and to manage the complexity of technical projects through knowledge representation and problem solving, decision making, and action planning, execution monitoring, and explanation. This article presents a framework for developing a hybrid data platform, which embodies many AI techniques adding intelligence along the entire data value chain.

The chapter relates to the *data management*, *data processing architectures*, *data analytics*, and *data visualization* technical priorities of the European Big Data Value Strategic Research and Innovation Agenda [2]. It addresses the respective horizontal concerns of the BDV Technical Reference Model as well as the vertical concerns of the development perspective— *engineering and DevOps*, *cybersecurity*, and *data sharing*. The chapter also relates to the *data, knowledge, and learning*, *reasoning and decision making*, *action, interaction, and explainable AI*, and *systems, hard-*

ware, methods, and tools enablers of the recent AI, Data and Robotics Strategic Research, Innovation and Deployment Agenda [3].

The rest of the chapter is organized as follows. Firstly, it reviews some of the relevant reference architectures, component models, and data platforms, existing within the European Big Data space. Next, it presents the requirements for the GATE Data Platform considering the complex role of GATE CoE as an academic and innovation hub as well as business accelerator in several domains. After these preliminaries, the next section presents the multi-layer approach to hybridization, adopted for building the GATE Data Platform. In a subsequent section, the implementation of this concept is discussed and the final chapter presents one of the flagship projects of GATE, which will leverage from the GATE Data Platform. Finally, conclusions and directions for future work are presented.

2 Brief Overview of Architectures, Frameworks, and Platforms

This section provides a brief overview of some of the most prominent reference models and component frameworks for data processing across Europe. Several examples of platforms operating at other European Big Data centers are also presented.

2.1 Reference Architectures for Big Data Processing

The European Big Data Value Association (BDVA) has proposed a reference architecture for Big Data systems [2]. It has a two-dimensional structure with components structured into horizontal and vertical concerns (Fig. 1). The horizontal concerns cover the entire data processing value chain together with the supporting technologies and infrastructure for data management, analysis, and visualization. The main sources of Big Data, such as sensors and actuators, are presented along the horizontal dimension. The vertical concerns include cross-cutting issues, relevant to all horizontal concerns—data types and formats, standards, security, and trust. Communications, development, and use are other important vertical concerns which add engineering to the vertical concerns.

This reference model provides a very high-level view of data processing without imposing any restrictions on the implementation, regardless of the area of applicability. There are more specific reference architectures developed with particular application areas in focus, such as the hierarchical model Industrie 3.0 and the three-dimensional model Industrie 4.0, which account for more detailed relationship between the business processes, but they are focused entirely on the needs of industry.

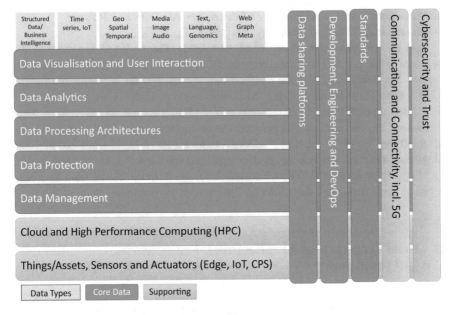

Fig. 1 Big Data Value Association reference architecture

One of the more generic enhancements of the BDVA reference architecture has been developed under the EU Horizon 2020 project OPEN DEI [4]. It aligns the reference architecture of BDVA with the requirements of open platforms and large-scale pilots for digital transformation. The OPEN DEI reference architecture framework (RAF) is built upon six fundamental principles which are generally applicable to digital platforms for data-driven services:

- Interoperability through data sharing
- Openness of data and software
- Reusability of IT solutions, information, and data
- Security and privacy
- Avoiding vendor lock-in
- Supporting a data economy

OPEN DEI reference architecture is three-dimensional (Fig. 2), with the third dimension providing directions for implementation according to the underlying philosophy of the framework. The horizontal layers include Field Level Data Spaces, Edge Level Data Spaces, and Cloud Level Data Spaces in which data is shared. The Smart World Services included in the Field Level Data Spaces enable interaction with IoT, automation systems, and humans. The Edge Level Data Spaces provide services for data acquisition, brokering, and processing. Cloud Level Data Spaces include different operations on the cloud such as data storage, data integration, and data intelligence. These Data Spaces offer the services to the main orthogonal dimension of the RAF—the X-Industry Data Spaces. The X-Industry Data Spaces

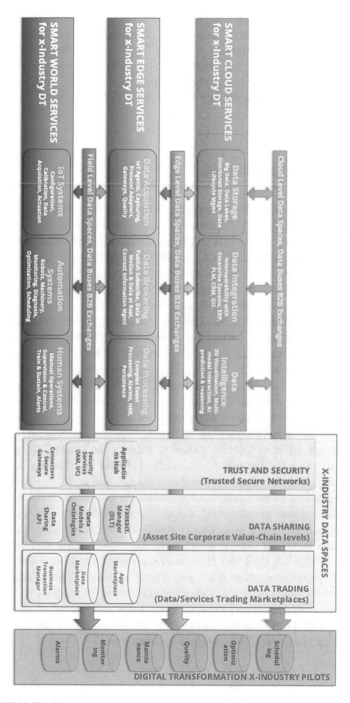

Fig. 2 OPEN DEI reference architecture

provide trustful and secure communication, data sharing, and data trading through appropriate technical infrastructure and development frameworks. All these Data Spaces support the implementation of Digital Transformation X-Industry Pilots for specific business scenarios. The main enhancement of the BDVA reference architecture by OPEN DEI RAF is in embedding the innovation and commercialization directly into the architecture through the concepts of *Data Spaces*, *smart services*, and *industry pilots*.

2.2 Component Frameworks

One of the most significant efforts to provide support for building data platforms has been undertaken by FIWARE Foundation. The FIWARE framework comprises open-source software components which can be assembled together and with other third-party components to accelerate the development of smart solutions [5]. The FIWARE component model is *broker*-based and provides an API for utilizing the functionality of the components. For this purpose FIWARE offers the so-called `Generic Enablers`, which provide support for common and specific operations for interfacing with data sources, processing, analysis, and visualization of context information, as well as usage control, publishing, and monetizing of data. The key enabler is the `Context Broker` which integrates all platform components and allows applications to update or consume the context information in a highly decentralized and large-scale manner. The `Context Broker` is the only mandatory component of any platform or solution which builds upon the FIWARE platform. A number of applications in the areas of smart agriculture, smart cities, smart energy, and Smart Industry are built upon FIWARE components.

More recent effort to provide technical support for assembling applications based on existing components is undertaken by the International Data Space Association (IDSA). Its component model elaborates further the broker architecture by standardization of two additional elements of the broker pattern—the *data provider* and the *data consumer* [6]. On the data provider side IDSA architecture is based on the concept of `Data Space` together with an associated `Connector`, while on the consumer side it operates through `DataApps` and `AppStore`. Similarly to FIWARE, the IDSA framework has an open architecture and supports large-scale system integration. IDSA has mainly in focus B2B industrial channels with extensive communications or distributed networks of institutions involved in collaborative work such as national and international government or public systems. As a member of IDSA Gate Institute considers building its own platform so that it can expose and integrate IDSA-compatible components.

2.3 Data Platforms

Platforms operating at other European Big Data Centers of Excellence include general-purpose as well as application-specific ones. The most relevant to the task of developing the GATE Data Platform, built specifically to provide support for Big Data and AI projects regardless of their application domain, are briefly presented here.

Teralab is an open Big Data and AI platform hosted by Institut Mines-Telecom (IMT)—a leading French institute of technology [7]. The platform aims to support and accelerate projects in Big Data and AI by providing technical, legal, and infrastructure tools and services as well as an ecosystem of specialists in those fields. The main asset toward providing various services is the diverse expertise hold by the Teralab teams that elaborate the best solution for each specific project.

ICE, the infrastructure and cloud research and test environment, is hosted by the RISE research institutes of Sweden and provides technical infrastructure, research data, and expert support [8]. As part of its services, ICE offers a tailor-made data platform that supports Big Data analytics and ML. The platform provides both Big Data services and customized development environment.

The Swiss Data Center has implemented RENKU platform as an open-source standalone solution with the aim of making the collaboration in data science teams more effective, trustful, and easy [9]. The RENKU platform can be deployed on a Kubernetes cluster within an organization. It supports versioning of data and code and allows customization of the environment. It enables traceability and reusability of all the artifacts developed in a data science project.

The discussed background provides a steppingstone for designing the GATE Data Platform and in particular for specifying the requirements for the platform, which are outlined in the next section. Presented reference architectures and component frameworks for data processing are designed to be general enough to support various usage scenarios and application domains and to provide common understanding of the architectural components and connections between them. By adhering to these reference architectures, the GATE platform will ensure high level of reusability of artifacts and processes, as well as of standardization and interoperability. On the other hand, the presented data platforms are a good example of different styles of utilization to be followed—from standalone instances, through service-oriented mode, to customized solutions. In addition, they demonstrate how various technologies provide support for the vast landscape of Big Data and AI projects.

3   Requirements for the GATE Data Platform

The complex role of GATE CoE as a scientific and innovation hub and business accelerator in several domains leads to multidimensional requirements:

- To empower the research on AI and Big Data technologies conducted within GATE CoE
- To enable development and exploitation of data-driven innovations
- To support the education and training activities on MSc, PhD, and professional level
- To facilitate creation of a Big Data ecosystem within the country, in the region, and in Europe

To reflect these objectives, the platform requirements were considered to be *holistic, symbiotic, open, evolving*, and   *data-driven* [10], which fully aligns with the fundamental principles of BDVA and OPEN DEI reference architectures. Here we are briefly specifying them.

3.1   Requirements from Research Perspective

To support simultaneous work on research projects across the entire data value chain in different application areas, the following is required:

RR1 Vertical hybridization: Combining symbolic, statistical, and numerical AI methods with semantic technologies and Knowledge Graphs to derive value from domain knowledge

RR2 Horizontal integration: Combining multiple technologies to provide flexibility in the implementation of data services

RR3 Modularization and reusability: Integration of generic domain-independent components and data with domain-specific and problem-specific components and data for enabling the use and reuse of third-party components and APIs, such as the Fireware and Geospatial components

RR4 Synchronization and orchestration: Control over the execution of data services to support simultaneous use of the resources when working on different projects while executing the individual data services in an isolated and safe environment

RR5 Robustness and security: Coping with a wide range of problems, caused by human errors, technical faults, or external interventions

RR6 Multilevel explainability: Transparency of both the data and the operations in mapping the solutions to the problems by uncovering the circumstances and dependencies behind decisions, plans, processes, and events and thus explaining the specific results during data processing

3.2 Data-Driven Requirements

The specific requirements toward the data are:

DR1 Integration of diverse data sources: Mixing data coming from multiple data sources over different transport protocols

DR2 Support for data variability: To ensure possibility for processing data in structured, unstructured, and semi- structured formats

DR3 Multi-mode data processing: Support for different modes of data processing—batch, messaging, and streaming in both discrete and continuous flows

DR6 Scalability: Scaling out for processing large amounts of data without compromising the performance

DR5 End-to-end velocity: Capability to handle data through processing in parallel and mitigating bottlenecks and latency within the existing infrastructure

The produced datasets and metadata will be integrated into Data Spaces. A key step in realizing GATE data space is data acquisition, including public and private data. Data sharing will be realized by adhering to FAIR (findability, accessibility, interoperability, and reuse) principles:

FR1 Findability: Support of rich machine-readable metadata for automatic discovery of datasets and data services

FR2 Accessibility: Strict mechanisms for control, based on consumer profiling for both data and metadata

FR3 Interoperability: Well-defined data models, common vocabularies, and standardized ontologies for data processing

FR4 Reusability: Clear usage of licenses, detailed provenance information, and domain-relevant community standards

3.3 Service Provisioning Requirements

Currently, there is a wide variety of open-source and commercial products which can be used to implement the platform [11]. They need to be chosen in accordance with the service provisioning objectives and integrated to achieve the following:

SR1 Openness: Building on open standards, providing APIs and public data

SR2 Integration of open-source and commercial technologies: Exploiting open-source solutions as a cheaper alternative, providing for better customization and extendability, but also leveraging mature concepts, established methodologies, and stable commercial technologies to minimize migration and to foster quick innovation and commercialization

SR3 Technological agnosticism: Through relying on proven industrial and open-source solutions which support modularization, isolation, and interoperability without dependence on the underlying technology of implementation

SR4 Explainability: Through dedicated services to be able to provide for rational explanation at different level of operation, abstraction, and granularity

3.4 Data Governance Requirements

The Big Data is seen as an asset that needs to be effectively managed. This requires governance for the decentralized and heterogeneous data sharing and data processing. It should also facilitate building trust in AI as a key element for Data Spaces as defined by the recent European AI Data and Robotics Partnership [3]:

GR1 Data sovereignty and privacy: By implementing data connectors with guaranteed level of control following various compliance requirements such as GDPR, RAF, IDS, etc.

GR2 Non-repudiation and auditability: Enabling responsible development through maintenance of versioning, tracing, and auditing at all levels of operation

GR3 Trustfulness: Building trust between organizations in partnership and collaboration through enhanced data sovereignty and privacy, transparency, explainability, auditability, security, and control of the access and operation

As a conclusion, we can say that the GATE Data Platform must be open, extendible, and very flexible to improve the comprehensiveness of the different processes and enhance the transparency of its operation at theoretical, methodological, technological, and engineering levels. The architecture which can support all these requirements will need to strike a fine balance which cannot be achieved by simply endorsing the reference architectures or repeating the experience of other Big Data CoE.

4   Hybridization of Data Platforms

This section presents the theoretical, technological, and methodological choices behind the hybrid approach adopted for the GATE Data Platform.

4.1   Multilevel and Service-Oriented Architectures

Traditionally, AI as an area of advanced research and development has been divided into several sub-domains: knowledge representation, automated inference, problem solving, decision making, machine learning, etc. Although most of them are relevant to data processing, only a few are directly present in data platforms. There is an urgent need to bridge the different AI sub-domains on theoretical, methodological, technological, and engineering levels in order to add intelligence to data processing

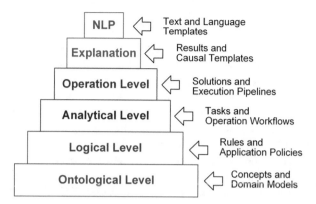

Fig. 3 Vertical layering of the GATE Data Platform

along the entire value chain and on all levels. Our approach, multilevel conceptualization, allows for a seamless integration of several AI technologies as shown
in Fig. 3. The backbone of the architecture is the mapping between the levels.
The *ontological, logical, analytical, operation, explanation,* and *linguistic* levels
are based on a common foundation—the theory of situations and actions, which
allows to model both the statics and the dynamics in a single framework [12].
Technological support for the mapping between the layers comes from the "layered
cake" of the Semantic Web serialized languages. The software implementation is
based on the service-oriented architectures (SOA), which utilize the containerization
and orchestration capabilities of contemporary cloud technology.

4.2 Levels of Intelligence

The multilevel architecture of the platform can enhance Big Data projects through
adding intelligence on several levels:

Ontological level: Models the metadata, the domain knowledge, the methods,
  and algorithms for data processing as well as the parameters of the processes
  they generate during execution. Fundamental concepts on this level in our
  approach are the *situations*, which describe the static state of affairs; *events*,
  which formalize the asynchronous results of completing the data processing; and
  actions, which model the synchronous data operations. On this level the model
  is represented as an OWL ontology.
Logical level: Specifies the heuristics which control the execution of data management operations, referring to the concepts and individuals on ontological
  level. They are modeled using SWRL rules.
Analytical level: Integrates the two previous levels into operational workflows,
  modeled as RDF graphs. These workflows are much simpler than BPEL work-

flows as there is no need to use the web services API for remote orchestration of
the services.

Operational level: Controls the workflows for executing data processing operations such as collection, pre-processing, transportation, aggregation, storing,
analyzing, and interpretation of the data, together with the events associated with
the pathways during executing the data processing workflows. Each operation
will be implemented as a software component, configured and deployed to the
cloud containers using the metadata from the ontology, and controlled by the
workflow monitoring tools available on the platform.

Explanation level: Generates rational explanation based on the representations
of the causal dependencies and their logic on ontological, logical, and analytical
levels, plus the results of data processing on operational level. It is based on a
separate ontology, which can be extended to different domains and problems with
domain-specific and problem-specific concepts, roles, and heuristics especially
for explanation [13].

Linguistic level: The attributes of the various ontological concepts form a case-
based grammar. It can be used for template-based generation of the text narrative
of the explanation [14].

The top concepts of the ontological hierarchies are the OWL classes `Problem`,
`Data`, and `Solution`, which represent the domain-specific and problem-specific
information. The taxonomies of `Infrastructure` and `Resource` classes
describe the available software and hardware components and tools for data
processing on the platform. On logical level this model can be expanded further
with domain-specific, problem-specific, and application-specific heuristics. From
the OWL and SWRL representations, we can extract information to build a pure
RDF graph, which forms the basis for the models on the next levels (Fig. 4).

Figure 5 illustrates the analytical model of a possible scenario for data analysis in
the form of such an AND-OR graph. It can be expanded further on analytical level
based on additional logical analysis and heuristic decisions. Later on, this graph
can be used to control the execution of the workflow operations. The above multi-
layer model of the data platform has been designed to allow seamless integration
of knowledge and data representation, problem solving, data analytics, and action
planning in a single conceptual framework. On the other hand, it splits the domain-
specific from problem-specific and application-specific logics, which supports high
modularization, interoperability, and reusability on all levels of operation. Our
recent research also shows the possibility to integrate decision-making components
with it based on stochastic process control, thus adding further capability to the
framework [16]. The use of explicit representation of knowledge in OWL and
SWRL also allows to address the problem of explainability, which is important for
presenting the work of the platform in a rational way to both professionals and non-
specialists [3].

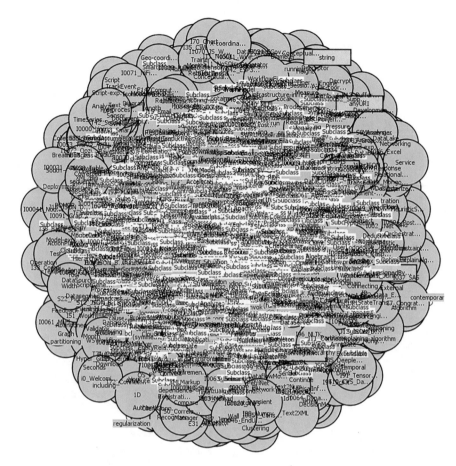

Fig. 4 Ontology of data processing on the platform

4.3 System Architecture

Many of the requirements for data platform can be met by contemporary SOA (Fig. 6). The cloud-based infrastructure for such an architecture is shown in Fig. 7. Its distinctive feature is the horizontal integration of application components through containerization and their orchestration using the service control languages of the container management system. This perfectly meets the requirements for the GATE Data Platform, supporting multiple different projects and activities in several application areas on different level and using a variety of technologies.

The system architecture shown in Fig. 7 is based on public domain software. However, this is not a limitation of the platform. Although it has been designed from the ground up using public domain software in mind, it does not exclude the use of commercial software. Nowadays the Big Data software vendors (IBM, Oracle,

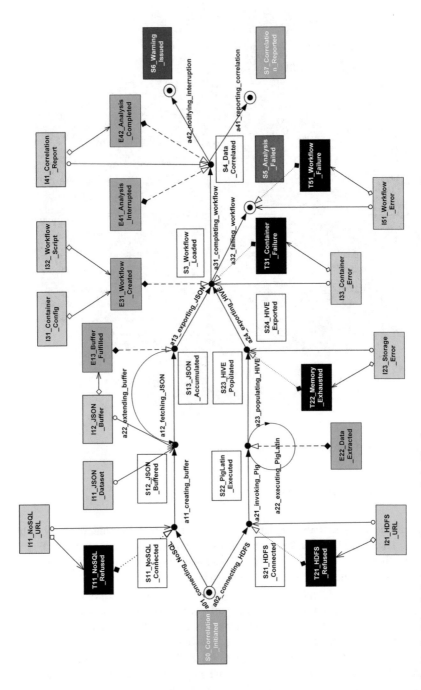

Fig. 5 Graph of the platform operations on analytical level

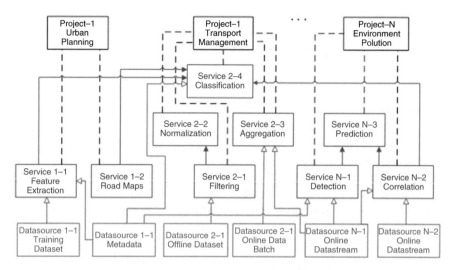

Fig. 6 Horizontal integration of the platform services

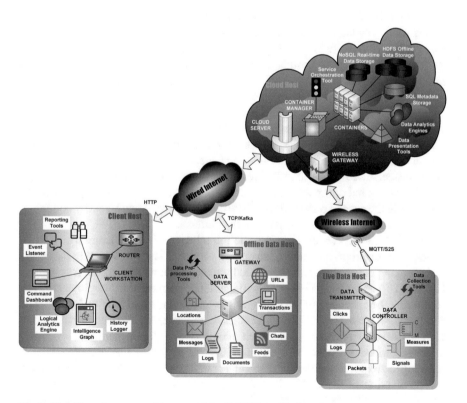

Fig. 7 Cloud-based system architecture of the GATE Data Platform

Microsoft, Hortonworks, Cloudera, MapR, etc.) also develop their own software based on open platforms and this allows compatibility between open-source and commercial technology stacks.

5 Implementation

The platform as presented can support a wide range of problems which involve both synchronous operations and asynchronous events. Such problems are typical in most of the application themes of GATE and especially in future cities and Smart Industry where the potential projects will need to deal with production line fault recovery, critical infrastructure protection, urban planning, public safety management, etc.

5.1 Enabling Technologies

Contemporary cloud technology relies on several cornerstones—application *containerization*, container *isolation*, process *synchronization*, and service *orchestration*. Cloud deployment is especially attractive for data platforms due to the support for SOA. Initially cloud hosting was pushed by big software vendors like Amazon, Google, Microsoft, and Oracle, which introduces dependence on the service providers. Hosting data platform on public cloud might not be feasible for project-oriented organizations such as GATE due to the large running costs. Fortunately, nowadays this computing paradigm can be implemented on the premises using open-source software [15, 17], which allows to untangle the dependence from the service providers. Of course, this introduces additional requirements for the maintenance. At the same time it gives more opportunities for system integration, software reuse, and optimization of the costs. Since the cloud service provision does not differ in the case of public from private cloud hosting, it can be easily combined, which would combine the benefits of both solutions.

The GATE Data Platform implementation relies on cloud software which exists in both vendor-proprietary and open-source versions. The two scripting languages for managing cloud resources supported by most container management systems—YAML [18] and CWL [19]—are sufficient for specification, deployment, and controlling the execution of the data services on the cloud and their orchestration in data processing workflows. The control can be implemented using cloud dashboards such as Apache Airflow [20], which monitors the deployment and execution of containerized software components. Such an approach has the advantage of reducing the requirements for the client and significantly simplifies the maintenance.

The layering of the GATE Data Platform allows additional automation of the component deployment, service configuration, and workflow orchestration on the cloud. The scripts needed can be generated directly from the OWL ontology, the SWRL heuristics, and the Knowledge Graphs created on the previous levels. When the general task requires a local workflow of data processing operations, which has to be executed directly within the data management system, it can also be generated in the command language supported by it, like Oozie in Hadoop [21], JavaScript in NoSQL, or stored procedures in SQL databases.

5.2 Data Services

The data services must support the data processing operations along the entire Big Data value chain and will be the main focus of the research and innovation projects of GATE. The process of developing containerized components for data analysis on the cloud based on two popular methods for ML—SVM and NN—is described in [22]. The GATE Data Platform does not impose any restrictions on the programming languages, libraries, and tools, but will require parameterization to support the interoperability and reusability.

The data services will run within separate containers under the control of the container management system of the cloud. Each containerized component will be developed according to a common methodology, which will be based on the use of templates for configuration, deployment, and controlling the execution. This will increase the productivity of the development and will support additionally the automation of the deployment.

5.3 Engineering Roadmap

The major pathways supported by GATE Data Platform are the following:

Data warehousing and analysis of data at rest. Big Data requires powerful infrastructure capable of running HDFS-compatible data management system such as Hadoop [23], installed on cloud-enabled container management systems and executing services within containers. The enterprise tools for transporting the data are Kafka [24] and NiFi [25]. From platform perspective the analytical engines, including machine learning tools, are parametrized black boxes and their specifications will become part of the top-level ontology of the platform. The analysis will be performed by containerized applications organized in data processing workflows under the control of powerful tools such as Spark [26] or ad hoc programs which include ML software libraries such as TensorFlow in Python [27] or Deeplearning4j in Java [28].

Data streaming, integration, and analysis of data in motion. Using simple IoT controllers such as ESP32 [29] or Arduino [30], over wireless protocols such as MQTT [31], the sensor data can be transported and pre-processing in real-time on the fly, can be stored to NoSQL databases [32] for later analysis using suitable data analytics and machine learning tools.

Conceptual modeling of the explanation.   Using ontological editors such as Protege the analysts and domain experts can develop problem-solving, machine learning, decision-making, and operation models for explanation. They can be then stored in graph databases such as GraphDB [33] for later use during explanation of the entire data processing from conceptual, theoretical, technological, and computational viewpoint.

Data services.   As a by-product the cloud-based GATE Data Platform can also support various data services offered to third parties—downloading of datasets, broadcasting of live data streams, online and offline data pre- and post-processing, data analytics on demand, etc.

To employ the GATE Data Platform, the project teams must complete several steps:

1. Develop domain- and problem-specific ontologies to extend the ontology of data processing.
2. Specify the problem-solving heuristics for solving particular problems on logical level and index them against the ontologies.
3. Generate the working scenarios and extend them with decision-making heuristics to control the execution on analytical level.
4. Develop domain- and problem-specific ontologies for explanation of the methods, algorithms, tasks, solutions, and results.
5. Develop software components implementing specific methods for data management, data analysis, and data insight for various tasks on operational level.
6. Generate the containerization and orchestration scripts needed to deploy the software components using the ontological representation and metadata.

The operations required for setting up a project may look demanding, but the SOA of the platform allows to use design patterns. To leverage on this GATE considers adopting a suitable model-centric methodology of working and dedicated team training. On the other hand, the multi-layered architecture allows focusing on a specific vertical level and/or horizontal component which will lower the staff requirements.

After completing some project tasks the more universal components can be incorporated in the library of platform services for further use. Such an incremental development will allow the platform to grow and expand over time without the need for changing its core.

6   City Digital Twin Pilot

This section presents one of the flagship projects of GATE which will benefit from the use of data platform supporting hybrid AI. At the same time, it is a testbed of the philosophy behind it. City Digital Twin is a large interdisciplinary pilot project that aims at developing a Digital Twin platform for designing, testing, applying,

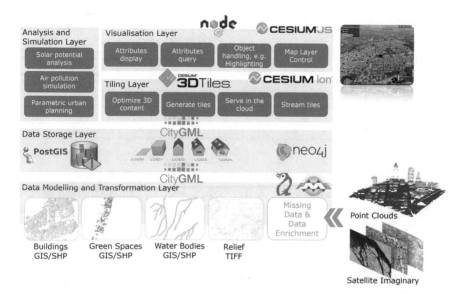

Fig. 8 Multi-layer framework of city Digital Twin pilot

and servicing the entire lifecycle of the urban environment. The project is focused on a spatially resolved exploration of a broad range of city-specific scenarios (urban climate, mobility of goods and people, infrastructure development, buildings' energy performance, air and noise pollution, etc.). The core of the platform is a semantically enriched 3D model of the city. Simulation, analytical, and visualization tools will be developed on top of it enabling the basic idea of the Digital Twin "design, test, and build first digitally." The technological framework used for initial implementation of the platform is shown in Fig. 8.

The development has started with implementation of a CityGML-compliant 3D model at ontological level, covering District Lozenets of Sofia. CityGML is an official international standard of the Open Geospatial Consortium (OGC). It is implemented as a GML application schema [34]. Because CityGML is based on GML, it can be used with the whole family of GML-compatible web services for data access, processing, and cataloguing, such as Web Feature Services, Web Processing Services, and Catalogue Services. It allows to model the significant objects in a city taking into account their 3D geometry, 3D topology, semantics, and visual appearance. Explicit relationships and component hierarchies between objects are supported and thus the 3D model is applicable to urban planning, environment analysis, 3D cadastres, and complex simulations [35]. Table 1 presents a mapping between the multi-layer framework of the GATE Data Platform and the technological layered framework of city Digital Twin platform. The city digital framework spans over several data pathways of the GATE Data Platform: (1) data at rest, (2) conceptual explanation, (3) data services, and in the future—(4) data in motion.

Table 1 Mapping of pilot layers to GATE Data Platform levels

	Modeling and transformation	Data storage	Analysis and simulation	Tiling	Visualization
Explanation	✓				✓
Operation		✓	✓	✓	✓
Analytics	✓		✓		
Logics	✓				
Ontology	✓				✓

The 3D model is based on three main data sources: cadastral data, covering most of thematic modules of CityGML standard, such as buildings, green spaces, relief, road network, etc.; high-resolution satellite image; and point cloud data. The satellite imaginary and point cloud data are used for semantic enrichment of the 3D model as well as for urban analysis, such as cadastre validation and urban change detection. Currently, the 3D model covers `Building` and `Relief` thematic modules in CityGML 2.0, including information about the buildings addresses as well as their intersection with the terrain. It is mainly developed using FME software, which allows to create and reuse data integration workflows. Additional transformations, related to the integration of the buildings and terrain, are performed, using MathLab. At operational level, the 3D model is stored in a 3D City Database [36], which can be implemented on either Oracle Spatial/Locator or PostgreSQL/PostGIS. PostGIS database is chosen for the current implementation of the platform.

Regarding the second data pathway, a domain-specific city data model and a corresponding ontology will be elaborated at ontological level for the purpose of urban planning. Thus, the full potential for mixing symbolic and graphic representation of information in Knowledge Graphs using graph databases, such as Neo4j [37] or GraphDB [33], will be exploited at the operational level. The domain model is needed to establish the basic concepts and semantics of the city domain and help to communicate these to GATE stakeholders. NGSI-LD [38] is chosen for its implementation, since it allows for specification of rules, which control the execution of data management operations at the logical level. NGSI-LD supports both the foundation classes which correspond to the core meta-model and the cross-domain ontology. The core meta-model provides a formal basis for representing a "property graph" model using RDF/RDFS/OWL. The cross-domain ontology is a set of generic, transversal classes which are aimed at avoiding conflicting or redundant definitions of the same classes in each of the domain-specific and problem-specific ontologies.

The third data pathway is realized through sharing the 3D model for user interaction through the web. The 3D model is currently uploaded to a Cesium ion platform, which optimizes and tiles it for the web. Cesium ion serves the 3D model in the cloud and streams it to any device. A web application is developed for visualization of the building model (Fig. 9) which will become part of the explanation level. It is currently hosted on a Node.js web server. Cesium.js is used

Fig. 9 3D model visualization of Lozenets District in Sofia

for its implementation due to its extensive support of functionality, such as attribute display and query, object handling, highlighting, map layer control, etc.

Several use cases in a process of implementation demonstrate the potential of the GATE Data Platform in urban setting. The first one is related to urban planning. The main idea behind it is to develop a tool for parametric urban design, supporting urban planning by taking into account neighborhood indicators related to population, green areas, transport connectivity, etc. The logic of the parametric urban design and its implementation using genetic algorithms fit within the logical, analytical, and operation level, respectively. The second use case deals with analysis and simulation of air quality, focusing on pollution dispersion independent of the wind direction and velocity, as well as the geometry of the buildings. In collaboration with researchers from Chalmers University, the wind flow in an urban environment is explored by applying computational fluid dynamics. The simulations themselves are implemented on the operational level, while their visualization corresponds to the explanation level of the GATE Data Platform.

The fourth, real-time data pathway of the GATE Data Platform will be fully operational after GATE City Living Lab is completed. The Living Lab will generate data for air, weather, and noise monitoring and will continuously deliver data to the cloud for real-time data analysis. A pilot implementation for processing of data about air pollution, generated by open air sensors across the city, is already on the way. When the GATE Data Platform is fully operational, we plan to transfer the entire project to it, which would allow us to reuse various components in other projects related to the analysis of the urban environment.

7   Conclusion and Future Work

The data platform under development for the GATE CoE is unique in the way it combines theoretical, technological, and applied aspects in a simple but powerful multi-layered hybrid framework, based on AI and empowered by the cloud technologies. The separation of domain-specific from problem-specific knowledge at ontological, logical, and analytical levels allows detaching the tasks for specification and modeling from the technical tasks for processing the data, which is the cornerstone of the interoperability of both data and operations. At the same time, it facilitates explanation on different level and with different granularity. Furthermore, thanks to the containerization and the orchestration of data services, the platform adds high degree of automation, reusability, and extendibility. Still, this hybrid platform can be used in a uniform way, regardless of the mode of working—locally, remotely, over network, or on the cloud.

The possibility for vendor-independent implementation of the platform, based on open software, very well supports both academic teaching and professional training practices, which is an additional advantage for GATE. In order to leverage the full advantages of AI technologies for data processing presented in this chapter, the software development for research, innovation, and commercialization requires conceptual, methodological, and technological discipline which will gradually become a practice at the GATE CoE.

GATE has already completed the development of the ontological level of its data platform and currently proceeds with formulation of heuristics, which will guide its operations. The immediate follow-up plans include developing of an ontological model of explanation, which will complete its conceptual framework as an explainable AI framework. GATE is relying on both its academic partners and its industrial supporters to build the technical infrastructure needed to implement this solution, and it is expected that by the end of the year, the GATE Data Platform will be fully operational. Its belief is that this framework can be of interest to other organizations with similar goals within the European Big Data space.

Acknowledgments This research work has been supported by GATE project, funded by the H2020 WIDESPREAD-2018-2020 TEAMING Phase 2 programme under grant agreement no. 857155, by Operational Programme Science and Education for Smart Growth under grant agreement no. BG05M2OP001-1.003-0002-C01, and by London Metropolitan University Transformation project no. 0117/2020, funded by UK HEIF. GATE is also grateful for the continuing support of the companies Rila Solutions and Ontotext. The understanding, the concepts, and the claims formulated in this material are of the authors only and should not be attributed to the official policy of any of these organizations.

References

1. KNOW-CENTER GmbH: European Network of National Big Data Centers of Excellence. Retrieved March 9, 2021 from https://www.big-data-network.eu/map/
2. Zillner, S., Curry, E., Metzger, A. et al. (Eds.). (2017). *European big data value strategic research & innovation agenda*. Big Data Value Association.
3. Zillner, S., Bisset, D., Milano, M., Curry, E. et al. (Eds.). (2020). Strategic research, innovation and deployment agenda—AI, data and robotics partnership. Third Release. September 2020, Brussels. BDVA, euRobotics, ELLIS, EurAI and CLAIRE. Retrieved March 9, 2021 from https://ai-data-robotics-partnership.eu/wp-content/uploads/2020/09/AI-Data-Robotics-Partnership-SRIDA-V3.0.pdf
4. OpenDei Project: Reference Architecture for Cross-domain Digital Transformation. Retrieved March 9, 2021 from https://www.opendei.eu/wp-content/uploads/2020/10/
5. Fiware Foundation, e.V.: FIWARE-NGSI v2 Specification. Retrieved March 9, 2021 from http://fiware.github.io/specifications/ngsiv2/stable/
6. International Data Spaces Association: Reference Architecture Model Version 3.0 (2019). Retrieved March 9, 2021 from https://internationaldataspaces.org/publications/
7. Institut Mines-Telecom: Data Science for Europe. Artificial intelligence and big data platform. Retrieved March 9, 2021 from https://www.teralab-datascience.fr/?lang=en
8. RISE Research Institutes of Sweden: ICE Data center. Retrieved March 9, 2021 from https://www.ri.se/en/ice-datacenter
9. Swiss Data Science Center: Multidisciplinary Data Science Collaborations Made Trustful and Easy. Retrieved March 9, 2021 from https://datascience.ch/renku/
10. Petrova-Antonova, D., Krasteva, I., Ilieva, S., & Pavlova, I. (2019). Conceptual architecture of GATE big data platform. In *Proc. 20th Int. Conf. on Computer Systems and Technologies (CompSysTech)* (pp. 261–268). ACM.
11. Petrova-Antonova, D., Ilieva, S., & Pavlovam I. (2017). Big data research and application—a systematic literature review. *Serdica Journal of Computing, 11*(2), 73–114.
12. Bataityte, K., Vassilev, V., & Gill, O. (2020). Ontological foundations of modelling security policies for logical analytics. In *IFIP Advances in Information and Communication Technology (IFIPAICT)* (Vol. 583, pp. 368–380). Springer.
13. Chari, S., Seneviratne, O., Gruen, D., et al. (2020). Explanation ontology: A model of explanations for user-centered AI. In J. Pan, V. Tamma, C. d'Amato et al. (Eds.), *Proc. 19th Int. Semantic Web Conference (ISWC)*. LNCS (Vol. 12507, pp 228–243). Springer.
14. van Deemter, K., Theune, M., & Krahmer, E. (2005). Real versus template-based natural language generation: A false opposition? *Computational Linguistics, 31*(1), 15–24.
15. Docker, Inc.: Get Started with Docker. Retrieved February 19, 2021 from https://www.docker.com
16. Vassilev, V., Donchev, D., & Tonchev, D. (2021). Risk assessment in transactions under threat as a partially observable Markov decision process. In *50th Int. Conf. Optimization in Artificial Intelligence and Data Sciences (ODS2021)*, 14–17 Sep 2021, Rome, Springer.
17. Cloud Native Computing Foundation: Building sustainable ecosystems for cloud native software. Retrieved February 19, 2021 from https://www.cncf.io/
18. yaml.org: YAML Ain't Markup Language. Retrieved February 9, 2021 from https://yaml.org/
19. Amstutz, P., Crusoe, M., & Tanic, N. (Eds.). Common Workflow Language, v1.0.2. Retrieved February 19, 2021 from https://w3id.org/cwl/v1.0/
20. Apache Software Foundation: AirFlow. Retrieved February 19, 2021 from http://airflow.apache.org/
21. Apache Software Foundation: Oozie. Retrieved February 28, 2021 from https://oozie.apache.org/
22. Sowinsky-Mydlarz, W., Li, J., Ouazzane, K., & Vassilev, V. (2021). Threat intelligence using machine learning packet dissection. In *20th Int. Conf. on Security and Management (SAM21)*, Las Vegas, USA. Springer.

23. Apache Software Foundation: Hadoop. Retrieved February 19, 2021 from https://hadoop.apache.org/
24. Apache Software Foundation: Kafka. Retrieved February 19, 2021 from https://kafka.apache.org/
25. Apache Software Foundation: NiFi. Retrieved February 19, 2021 from https://nifi.apache.org/
26. Apache Software Foundation: Spark—Unified Analytics Engine for Big Data. Retrieved February 9, 2021 from https://spark.apache.org/
27. tensorflow.org: An end-to-end open source machine learning platform. Retrieved February 9, 2021 from https://www.tensorflow.org/
28. Eclipse Foundation: Deep Learning for Java. Retrieved February 9, 2021 from https://deeplearning4j.org/
29. Espressif Systems: SoCs. Retrieved February 9, 2021 from https://www.espressif.com/en/products/socs
30. Arduino: Arduino Pro. Retrieved February 9, 2021 from https://store.arduino.cc/new-home/iot-kits
31. MQTT.org: MQTT: The Standard for IoT Messaging. Retrieved February 19, 2021 from https://mqtt.org/
32. MongoDB, Inc.: The database for modern applications. Retrieved February 9, 2021 from https://www.mongodb.com/try/download/community
33. Ontotext: GraphDB—The Best RDF Database for Knowledge Graphs. Retrieved February 9, 2021 from https://www.ontotext.com/products/graphdb/
34. Open Geospatial Consortium Europe, CityGML. Retrieved March 8, 2021 from https://www.ogc.org/standards/citygml
35. Gröger, G., Kolbe, T., Nagel, C.. & Häfele, K. (2012). OGC City Geography Markup Language (CityGML) Encoding Standard, Wayland MA: Open Geospatial Consortium.
36. Kolbe, T., Nagel, C., Willenborg, B. et al. 3D City DB. Retrieved March 3, 2021 from https://www.3dcitydb.org/3dcitydb/d3ddatabase/
37. Neo4j, Inc.: Introducing Neo4J. Retrieved March 1, 2021 from https://neo4j.com/
38. Context Information Management (CIM) Industry Specification Group (ISG), NGSI-LD API. Retrieved March 3, 2021 from https://www.etsi.org/deliver/etsi_gs/CIM/
39. Vassilev, V., Sowinski-Mydlarz, W., Gasiorowski, P. et al. (2020) Intelligence graphs for threat intelligence and security policy validation of cyber systems. In *Advances in Intelligent Systems and Computing (AISC)* (Vol. 1164, pp. 125–140). Springer.

Part II
Deployment

A Digital Twin Platform for Industrie 4.0

Magnus Redeker, Jan Nicolas Weskamp, Bastian Rössl, and Florian Pethig

Abstract In an Industrie 4.0 (I4.0), rigid structures and architectures applied in manufacturing and industrial information technologies today will be replaced by highly dynamic and self-organizing networks. Today's proprietary technical systems lead to strictly defined engineering processes and value chains. Interacting Digital Twins (DTs) are considered an enabling technology that could help increase flexibility based on semantically enriched information. Nevertheless, for interacting DTs to become a reality, their implementation should be based on open standards for information modeling and application programming interfaces like the Asset Administration Shell (AAS). Additionally, DT platforms could accelerate development and deployment of DTs and ensure their resilient operation.

This chapter develops a suitable architecture for such a DT platform for I4.0 based on user stories, requirements, and a time series messaging experiment. An architecture based on microservices patterns is identified as the best fit. As an additional result, time series data should not be integrated synchronously and directly into AASs, but rather asynchronously, either via streams or time series databases. The developed DT platform for I4.0 is composed of specialized, independent, loosely coupled microservices interacting use case specifically either syn- or asynchronously. It can be structured into four layers: continuous deployment, shop-floor, data infrastructure, and business services layer. An evaluation is carried out based on the DT controlled manufacturing scenario: AAS-based DTs of products and manufacturing resources organize manufacturing by forming highly dynamic and self-organizing networks.

Future work should focus on a final, complete AAS integration into the data infrastructure layer, just like it is already implemented on the shop-floor and business services layers. Since with the standardized AAS only one interface type would then be left in the DT platform for I4.0, DT interaction, adaptability, and autonomy could be improved even further. In order to become part of an I4.0 data

M. Redeker (✉) · J. N. Weskamp · B. Rössl · F. Pethig
Fraunhofer IOSB, IOSB-INA Lemgo, Fraunhofer Institute of Optronics, System Technologies and Image Exploitation, Lemgo, Germany
e-mail: magnus.redeker@iosb-ina.fraunhofer.de https://www.iosb-ina.fraunhofer.de

E. Curry et al. (eds.), *Data Spaces*, https://doi.org/10.1007/978-3-030-98636-0_9

space, the DT platform for I4.0 should support global discovery, data sovereignty, compliance, identity, and trust. For this purpose, Gaia-X Federation Services should be implemented, e.g., as cross-company connectors.

Keywords Industrie 4.0 · Digital Twin · Asset Administration Shell · Platform · Microservices patterns · Interprocess communication · Time series · Industrie 4.0 data space · Industrial data economy · Gaia-X · Digital Twin controlled manufacturing · Collaborative condition monitoring

1 Introduction

In this day and age, increasing manufacturing productivity and sustainability faces challenges regarding interoperability and scalability of technical systems [1, 2]. The rigid operational technology (OT) and information technology (IT) architectures widely used today lead to the establishment of equally inflexible value chains and prevent the implementation of highly dynamic and globally connected value networks that could enable new forms of collaboration and business models [3]. It is the vision of Industrie 4.0 (I4.0) to enable this implementation through the establishment of standards, which should be implemented based on highly scalable and sovereign data infrastructure as an environment of trust and the basis for a global I4.0 data space [4, 5]. Major standards evolving in the German initiative "Plattform Industrie 4.0" currently focus on the implementation of one main technological concept that should foster semantic interoperability and autonomy of technical systems: a Digital Twin (DT) for I4.0 [6, 7]. Benefits expected from the DT for I4.0 include efficient re-configuration of production lines and machines for mass customization in application scenarios like "plug-and-produce" or "order-controlled production," as well as cross-company collaboration across value chains and life cycle stages in order to optimize products and production processes [8–15].

The Reference Architecture Model Industrie 4.0 (RAMI 4.0) includes the concept of an I4.0 component, consisting of asset and Asset Administration Shell (AAS) [16]. The AAS concept "helps implement Digital Twins for I4.0 and create interoperability across the solutions of different suppliers" [17]. The AAS specifies a meta information model for I4.0 that is based on properties standardized according to IEC 61360 [9, 18, 19]. Property specifications in dictionaries like the IEC Common Data Dictionary include formalized semantic descriptions and are an important step toward unambiguous and automatic interpretation of knowledge by I4.0 components [20]. The AAS meta information model is being standardized as IEC 63278 ED1, and first mappings of the meta model to specific technologies, e.g., to the OPC Unified Architecture (OPC UA), exist [21, 22]. In addition to the information model, the first infrastructure components for AAS are being developed as well, e.g., a registry for AAS and runtime environments based on different programming frameworks [23, 24]. AAS services executed in such an environment exchange information via a specified application programming interface (API) or they interact autonomously via a specified language for I4.0 [25, 26]. Additional I4.0 software components and services will be developed within the Industrial Digital Twin Association (IDTA) [27].

This chapter relates to the technical priorities "data management," "data processing architectures," "Big Data standardization," and "engineering and DevOps for Big Data" of the European Big Data Value Strategic Research and Innovation Agenda [28]: It addresses the horizontal concern "IoT, CPS, Edge, and Fog Computing" of the BDV Technical Reference Model as well as the vertical concerns "standards" and "industrial data platforms." Also, the chapter relates to the "systems, methodologies, hardware, and tools" and "action and interaction" enablers of the AI, Data, and Robotics Strategic Research, Innovation, and Deployment Agenda [29].

Accelerating the development, deployment, and resilient operation of DTs by developing the foundation for a DT platform for I4.0 is the main objective of this chapter. To this end, Sect. 2 presents the user stories of platform architect, chief financial officer, platform operator, manufacturing manager, and data scientist. They all share the vision of I4.0, but face individual challenges leading to specific requirements. Section 3 takes these requirements into account and introduces generic patterns for architectural style and interprocess communication. A time series messaging experiment is conducted in order to evaluate architectural styles and synchronous as well as asynchronous communication paradigms and their applicability for different use cases, for example, DT interaction and time series processing.

Section 4 then proposes a DT platform for I4.0, which is composed of specialized, independent, loosely coupled services and structured into four layers: continuous deployment, shop-floor, data infrastructure, and business services. Finally, Sect. 5 demonstrates how DTs of products to be manufactured and manufacturing resources executed on the proposed DT platform for I4.0 can autonomously organize manufacturing by forming highly dynamic and self-organizing networks. The DT of a product instance, for example, comprises, as a passive part, all its AASs and data contained in the platform related to the product instance and, as an active part, those business services executing services for the specific product instance. It actively takes decisions and interacts with the DTs of the manufacturing resources aiming at an efficient manufacturing of the product instance it represents.

2 User Stories

This section presents user stories describing the functionality the Digital Twin platform for Industrie 4.0 shall provide from the perspective of platform architect, chief financial officer, platform operator, manufacturing manager, and data scientist. What they have in common is that they want to implement Industrie 4.0.

	Industrie 4.0 community
I40C-S 1	**Industrie 4.0**
	We want to use the platform to implement Industrie 4.0, in which assets across all hierarchy levels of the former automation pyramid and beyond interact autonomously in an Industrie 4.0 environment and data space in order to enable higher efficiency, flexibility, and new data-driven business models

As a platform architect, my user stories consider encapsulated applications, data transport, storage, sovereignty and discovery, the discovery of services, and the provision of guidelines for development, integration, and consumption of services.

	Platform architect
PA-S 1	**Security**
	I want the platform to be secure regarding authentication, authorization, and communication as well as integrity, confidentiality, and availability of data
PA-S 2	**Encapsulated applications**
	I want the platform to use and reuse encapsulated applications that run isolated and that are scalable and effortlessly maintainable
PA-S 3	**Data transport**
	I want data transported efficiently from, into, and within the platform
PA-S 4	**Data storage**
	I want the platform to be able to store any kind of data generated either within or outside of the platform
PA-S 5	**Data sovereignty**
	I want the platform to guarantee sovereignty over data to the respective data provider
PA-S 6	**Data and service discovery**
	I want the platform to provide a discovery interface providing semantic descriptions and endpoints of any data and service available in the platform, so that data and services can be discovered easily
PA-S 7	**Guidelines**
	I want the platform to provide an extensive instruction to develop, integrate (also third-party software), or consume services

My concern as chief financial officer is of course the platform's ROI.

	Chief financial officer
CFO-S 1	**Return on investment**
	I want the platform to run cost efficiently, increasing the company's competitiveness

As a platform operator, the platform shall be easily maintainable, and the software shall be of high quality and scalable.

	Platform operator
PO-S 1	**Platform management**
	I want the platform to be effortlessly maintainable, flexibly manageable, and well-organized
PO-S 2	**Service quality**
	I want the platform to integrate and offer solely high-quality software guaranteeing effectiveness, efficiency, and reliability
PO-S 3	**Service scalability**
	I want to be able to offer data portal services to at least 100 data scientists in parallel without performance losses, in order to prevent customer complaints or denial of service

As a manufacturing manager, I need a dashboard, and, in the spirit of Industrie 4.0, I want the Digital Twins to control the manufacturing. Furthermore, I appreciate easy design and integration of Digital Twins and semantic descriptions.

	Manufacturing manager
MM-S 1	**Semantics**
	I want the platform to suggest semantic identifications based on free-text descriptions or, if not available, to create and publish them automatically in a repository
MM-S 2	**Digital Twin design**
	I want the design, storage, and deployment of Digital Twins to be effortless
MM-S 3	**Digital Twin decision making**
	I want Digital Twins of products and manufacturing resources to interact and autonomously make arrangements for their assets
MM-S 4	**Digital Twin asset control**
	I want a Digital Twin to have sovereignty over the assignment of tasks to the asset it represents
MM-S 5	**Digital Twin documentation**
	I want Digital Twins of products and manufacturing resources to document manufacturing progresses and the acquisition and execution of assignments
MM-S 6	**Dashboard**
	I want the manufacturing performance to be evaluated in real time and visualized in a dashboard: at present or at a specific time in the past or future

As a data scientist, I value capable data portals simplifying data engineering

	Data scientist
DS-S 1	**Data portal**
	I want to efficiently search for, browse, download, annotate, and upload process data and metadata via a standardized (web-)interface and information model, saving time for data analysis

3 Architectural Style and Interprocess Communication

In Industrie 4.0, assets across all hierarchy levels of the former automation pyramid and beyond interact in order to enable higher efficiency, flexibility, and new data-driven business models. To this end any asset, which according to [16] is any object of value for an organization, is equipped with an Asset Administration Shell (AAS, [16, 19, 25]). The AAS is the core technology for the implementation of I4.0-compliant Digital Twins, i.e., comprising semantically enriched asset descriptions. Both asset and AAS must be globally uniquely identifiable. Together, they constitute an I4.0 component that can become part of an I4.0 system.

I4.0 components can call for and provide services and data. Since supply and demand of the union of all assets will never be constant, I4.0 systems dynamically evolve over time: Digital and physical assets are added, adapted, or removed. Therefore, it seems worth to consider the following two key questions:

1. Which software architecture is best suited for a dynamic DT platform for I4.0?
2. Which communication technologies are appropriate to address the requirements from Sect. 2 like data transport efficiency?

3.1 Architectural Style

According to [30] an architectural style is "formulated in terms of components, the way that components are connected to each other, the data exchanged between components, and finally how these elements are jointly configured into a system."

Since the DT platform for I4.0 will contain highly distributed and frequently changing components, a monolithic approach, in which components are only logically separated and structured as a single deployable unit [30, 31], is practically impossible. Furthermore, the expected high complexity of the platform, the typically long development and deployment cycles of a monolith, and its limited scalability argue against a monolithic approach.

For distributed systems layered, object-based, data-centered, event-based, and microservice architectures are the most important styles [30, 31]. In a layered architecture, only services of underlying layers can be requested directly via Service Access Points (SAP) [30–32]. Since this is in a diametrical opposition to the core idea of I4.0 about the equality of assets and self-organizing networks, the layered architectural style is inept.

In the object-based architecture "components are connected through a (remote) procedure call mechanism," and data-centered architectures "evolve around the idea that processes communicate through a common (passive or active) repository," and in event-based architectures "processes essentially communicate through the propagation of events, which optionally also carry data" [30]. In [31] these three styles are combined in a microservice architecture, where "the components are services, and the connectors are the communication protocols that enable those services to collaborate." "Each service has its own logical view architecture, which is typically a hexagonal architecture" placing "the business logic at the center" and using inbound adapters to "handle requests from the outside by invoking the business logic" and outbound controllers "that are invoked by the business logic and invoke external applications" [31].

For the DT platform for I4.0, significant examples, as depicted in Fig. 1, for inbound adapters would be controllers implementing sets of HTTP/REST (Hypertext Transfer Protocol/Representational State Transfer) or OPC Unified Architecture (OPC UA) endpoints or message broker clients consuming messages of broker topics. On the other hand, major examples for outbound adapters would

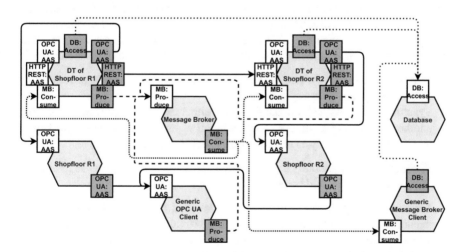

Fig. 1 Most significant adapters of shop-floor resources and their Digital Twins in the Digital Twin platform for Industrie 4.0: Inbound adapters are indicated by white rectangles and dark gray rectangles mark outbound adapters. An outbound adapter of component A invokes the business logic of component B by connecting and sending information to B's inbound adapter. Shop-floor resources and their Digital Twins either interact directly and synchronously via OPC UA or HTTP/REST for Asset Administration Shell (AAS) or indirectly and asynchronously using a message broker as intermediary. Generic OPC UA and message broker clients stream time series data from shop-floor resources via the message broker into databases and Digital Twins

be data access objects implementing operations to access databases, message broker clients producing messages in broker topics, and service triggers invoking external applications via their HTTP/REST or OPC UA endpoints.

A microservice "is a standalone, independently deployable software component that implements some useful functionality" [31]. Services "have their own datastore and communicate only via APIs" (Application Programming Interfaces), which consist of commands and queries: "A command performs actions and updates data," and on the other hand "a query retrieves data" [31]. A microservice architecture "functionally decomposes an application into a set of services" enabling "continuous delivery and deployment of applications" and "easily maintainable, independently deployable, and scalable services," and allows "easy experimenting and adoption of new technologies" [31].

The most significant benefits of the microservice architecture concur with the primary architectural and operational requirements of the DT platform for I4.0: PA-S 2 ("encapsulated applications") and PO-S 1–3 ("platform management," "service quality," and "service scalability"). Consequently, the platform will be developed based upon a microservice architecture.

3.2 Interprocess Communication

In a microservice architecture, in which services address specific challenges, services must collaborate to achieve an overall objective. Suitable interservice or rather interprocess communication (IPC) technologies depend on the use case at hand. According to [31] the various styles can be categorized into two dimensions:

- One to one/directly vs. one to many/indirectly: Each request is processed by either exactly one vs. many services.
- Synchronous vs. Asynchronous: A service either expects a timely response from another service and might even block while it waits vs. a service does not block, and the response, if any, is not necessarily sent immediately.

Within the I4.0 community certain technologies and systems promise to become major standards and should be taken into account.

- Services of shop-floor machines can be invoked by OPC UA, via which the machines also provide data access. OPC UA provides a service-oriented architecture, consisting of several features like security, communication, and information model. In OPC UA the communication is either organized according to the client/server or publish/subscribe paradigm. Clients can establish a connection to a server and interact with its information model, e.g., to read/write data or call methods. Subscribers can subscribe to certain information model elements, e.g., published cyclically by a corresponding publisher.
- The bidding procedure [26, 33] standardizes interactions between I4.0 components bringing together providers and users of physical shop-floor services: These

standardized messages are exchanged in a standardized sequence via MQTT [34].

- Server applications host AASs that can be accessed via HTTP/REST- or OPC UA-APIs [24, 25, 35]. REST operates on textual represented resources, e.g., JavaScript Object Notation (JSON) or Extensible Markup Language (XML). Within the I4.0 community the AAS meta model is typically represented in such a format [19, 25].
- The AAS meta model standardizes the invocation of operations or services of the administered asset, i.e., the "executable realization of a function," which can, e.g., be shop-floor or IT services [19]. While, according to [25], an AAS server's resources can only be manipulated synchronously, operations can also be invoked asynchronously using an operation handle to retrieve the results.

The previous considerations lead to additional requirements for the DT platform for I4.0: An MQTT message broker shall be provided and services of I4.0 components shall be invoked via AAS-APIs of AAS servers (see Fig. 1), either synchronously or asynchronously. AAS or asset data shall be manipulated and queried synchronously.

3.3 Time Series Integration

Data is produced continuously by the assets within the platform, e.g., during operation of shop-floor machines or IT services. In general, an AAS is supposed to be the virtual representation of an asset and consequently would have to contain all asset data. To a certain degree it might be feasible to integrate time series data into an asset's AAS directly. However, there is reasonable doubt that Big Data is manageable in such a structure. As mentioned above, adding data to an AAS in a server has to be done synchronously. Considering that an asset could produce data in nanosecond intervals, it seems unfeasible to integrate it synchronously. Besides this, an AAS server's capacity and performance objectives limit the amount of data that can be integrated: How to deal with unbounded time series data?

For this time series use case asynchronous messaging suits best. A message is a container for data "in either text or binary format" and "metadata that describes the data being sent," such as "a unique message id" or "an optional return address, which specifies the message channel that a reply should be written to" [31].

Messages are exchanged over channels: point-to-point or publish-subscribe. "A point-to-point channel delivers a message to exactly one of the consumers that is reading from the channel. Services use point-to-point channels for the one-to-one interaction styles" [31], like the invocation of operations. On the other hand, a "publish-subscribe channel delivers each message to all of the attached consumers. Services use publish-subscribe channels for the one-to-many interaction styles" [31].

Publish-subscribe messaging enables decoupling of publisher and subscriber in three dimensions: time, space, and synchronicity [36]. Consequently, publisher and subscriber do not need to know each other, do not have to be active at the same time, and do not have to block.

In particular, publish-subscribe channels are suitable for the time series use case. For example, a shop-floor machine might produce drinking cups. One quality requirement might be that the injection molding temperature during production is continuously within a given temperature range. Sensors in the machine measure temperature in nanosecond intervals and publish the time series data to a publish-subscribe channel. Why one-to-many interaction? Because there is most likely more than one consumer service:

- A service integrating the data into a time series database for persistence
- A service monitoring the condition of the machine
- A service deciding whether or not a specific drinking cup meets the quality requirements or must be sorted out

Consequently, in the DT platform for I4.0, Big Data and time series will be exchanged via asynchronous messaging. Endpoints for retrieving the data from a channel or a time series database are added to the AASs of the assets generating the data. The results of the following time series messaging experiment substantiate the previous argumentation.

3.4   Time Series Messaging Experiment

Let us assume that energy consumption data is recorded in second intervals in order to determine peak loads of a machine consuming up to 1 kW power. During an 8-h shift a sensor produces 28,800 records, each consisting of a measurement (the data type is double in the range from 0 to 1000 with 1 digit) and a timestamp (the format is unix timestamp in seconds). The peak loads are defined as the greatest one percent of measurements.

In seven setups the time it took to read the data from the source and to determine the peak loads was measured. In line with the previous considerations regarding architectural styles and interprocess communication schemes, these setups are entitled as:

[*Monolith*] a monolithic approach, in which a software program decodes a compressed file (AASX package format, [19]), deserializes the content as in-memory object, and queries the object, where the file contains an AAS with only one submodel containing:

[*-SMC*] per record a submodel element collection consisting of two properties: the record's measurement and timestamp
[*-Array*] just two properties: arrays of the measurements and timestamps

[*Micro*] a microservice architecture

[*-Sync*] with synchronous one-to-one communication, in which a software program requests from an HTTP-API server a submodel, deserializes the JSON as in-memory object, and queries the object, where the submodel is equal:

[*-SMC*] to the submodel of case *Monolith-SMC*
[*-Array*] to the submodel of case *Monolith-Array*

[*-Async*] with asynchronous one-to-many communication, in which a software program:

[*-Stream*] consumes a data stream containing the records and queries the records in-memory
[*-TSDB-ext*] extracts the records from a time series database and queries the records in-memory
[*-TSDB-int*] applies a built-in function of a time series database to query the records

Table 1 specifies the sizes of the consumed data in the seven setups. This data is consumed, decoded, and deserialized by a software program before the peak load determination begins. For the Monolith setups the data is provided in files, embedded in serialized AAS meta models, and compressed as AASX package format with JSON: In the SMC case the file's size is 672 kB and 139 kB in the Array case. The HTTP-API server in the SMC case returns a JSON of size 19,448 kB and of size 757 kB in the Array case. Please note that the significant difference in sizes of Array and SMC cases results from the duplication of metadata in the SMC cases: Each property containing a measurement or timestamp also contains metadata.

In the Micro-Async-Stream setup data of size 1266 kB is consumed from Apache Kafka [37], and in the case of Micro-Async-TSDB-ext, the consumed data from InfluxDB [38] sizes up to 157 kB. For the execution of Micro-Async-TSDB-int no

Table 1 Sizes in kB of the consumed data in the time series messaging experiment. In the first six setups, a software program must consume this data from the provider and decode and deserialize it prior to determining the peak load. Contrarily, in the seventh setup Micro-Async-TSDB-int the providing database determines the peak load internally

	Size of consumed data in kB	
Monolith-SMC	672	[AASX with JSON; uncompressed: 21,278]
Monolith-Array	139	[AASX with JSON; uncompressed: 759]
Micro-Sync-SMC	19,448	[HTTP with JSON-body]
Micro-Sync-Array	757	[HTTP with JSON-body]
Micro-Async-Stream	1266	[TCP binary]
Micro-Async-TSDB-ext	157	[HTTP with Gzip-body; uncompressed: 703]
Micro-Async-TSDB-int	-	[only TSDB-internal querying]

data must be consumed by an external software program, since InfluxDB determines the peak load internally.

The experiment was executed 1000 times per setup by a computer with an 11th Gen Intel(R) Core(TM) i7-1165G7 processor, 32 GB LPDDR4x 4267 MHz memory, and a 64-bit system architecture. To perform the experiment as close as possible to laboratory conditions, this one computer processed the setups one after the other and ran all their components. It was ensured that no disturbing other network or data traffic biased the results and garbage collection was executed before each iteration. Times were measured in microseconds, separately for data reading (consuming, decoding, deserializing) and top percentile determination. The client software programs executed highly optimized built-in C# and Python functions to determine the top percentile without performance losses, i.e., a built-in function from C# in the case of the Monolith and Micro-Sync setups and Python in the case of the external Micro-Async setups.

Figure 2 and Table 2 display the runtime results. What must be considered is that in the case of setup Micro-Async-TSDB-int, the time could only be measured for the complete process. By far, this setup achieved the best runtime results.

In the other six setups the peak load determination runtimes are approximately on the same low level. The runtimes of their complete processes are, however, dominated by the reading parts, in which their runtimes differ considerably. Surprisingly, Monolith-Array and Micro-Sync-Array are significantly faster than, in ascending order, Micro-Async-TSDB-ext, Micro-Async-Stream, Monolith-SMC,

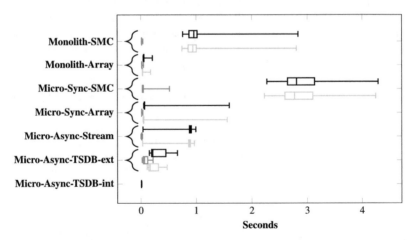

Fig. 2 Boxplots displaying the time it took to read the energy consumption data and determine the peak loads in the seven setups of the time series messaging experiment. Each setup was executed 1000 times. The light gray boxplot figures the reading data part runtime results of a setup, in dark gray the runtime results of the peak load determination, and black displays the complete process runtime results. Lower resp. upper whisker represent the minimum resp. maximum of the measurements. In the case of setup Micro-Async-TSDB-int, the time could only be measured for the complete process

Table 2 Median, mean, and normed standard deviation of the time reading and aggregation parts taken separately and together in the seven setups of the time series messaging experiment. For setup Micro-Async-TSDB-int the time could only be measured for the complete process. Each setup was executed 1000 times. Median and mean are both specified in seconds and the normed variance is determined by $(\frac{1}{999} \sum_{i=1}^{1000} (\frac{x_i - x_{mean}}{x_{mean}})^2)^{1/2}$

	Reading data			Peak determination			Complete process		
	Median	Mean	n-SD	Median	Mean	n-SD	Median	Mean	n-SD
Monolith-SMC	0.9440	0.9710	0.2024	0.0160	0.0169	0.2164	0.9620	0.9880	0.2016
Monolith-Array	0.0340	0.0344	0.1795	0.0210	0.0219	0.1815	0.0550	0.0563	0.1345
Micro-Sync-SMC	2.7695	2.8767	0.1425	0.0310	0.0431	1.0532	2.8060	2.9198	0.1422
Micro-Sync-Array	0.0510	0.0547	0.8835	0.0140	0.0136	0.0711	0.0650	0.0684	0.7175
Micro-Async-Stream	0.8778	0.8773	0.0497	0.0168	0.0165	0.2747	0.8945	0.8939	0.0488
Micro-Async-TSDB-ext	0.1671	0.2196	0.4166	0.0633	0.0830	0.5978	0.2140	0.3026	0.4480
Micro-Async-TSDB-int	–	–	–	–	–	–	0.0085	0.0088	0.1111

and Micro-Sync-SMC. As the sizes of the consumed data sources could suggest, Monolith-SMC and Micro-Sync-SMC performed worst.

In each run of the Monolith setups a compressed AASX file is decoded and its content is deserialized as in-memory object before the peak load determination begins. The file in the Array case is approximately five times smaller than in the SMC case; reading, however, is on average even approximately 28 times faster. Similar observations can be made for the Micro-Sync setups: While the size in the Array case is approximately 26 times smaller than in the SMC case, reading is on average approximately 53 times faster. For reasons of efficiency, time series data should therefore, if actually stored directly in AASs, be highly compressed avoiding duplication of metadata.

Please note that except for the Monolith setups, data provider and consumer in I4.0 are usually executed on separate machines connected via a network. Consequently, time losses at network layer would most likely extend the reading runtimes of the Micro setups.

The payload in the Micro-Sync setups could be compressed, as it is standardized for HTTP, so that the consuming runtime would decrease. Additionally, an HTTP-API server could provide server-side execution of queries, as is typically done by databases like in the Micro-Async-TSDB-int setup. So far, though, such services are not implemented for the AAS meta model.

A monolithic architecture in general and a microservice architecture with synchronous messaging are, due to the requirements of Sect. 2 and as deduced in the previous subsections, not eligible for the DT platform for I4.0 and highly dynamic interprocess communication. However, the runtime results show that such constructs are suitable for end-user applications aggregating, for example, in low-frequency data from AASs in the platform for reporting or documentation purposes.

The competitive runtime results in the three Micro-Async setups, with either data stream or time series database, justify the decision to integrate and consume Big Data or time series via asynchronous messaging in the Digital Twin platform for Industrie 4.0.

4   A Digital Twin Platform for Industrie 4.0

The Digital Twin platform for Industrie 4.0 (DT platform for i4.0) is based on the architectural deductions from Sect. 3 and meets the user requirements formulated in Sect. 2. It facilitates efficient development, deployment, and operation of I4.0 business services (user stories PA-S 2 and PA-S 7: "encapsulated applications" and "guidelines"). To account for scalability, robustness, performance, maintainability, and security (user stories PA-S 1 and PO-S 1–3: "security," "platform management," "service quality," and "service scalability"), I4.0 business services are implemented based on a set of data infrastructure microservices, e.g., for discovery, connectivity, synchronicity, and storage.

A continuous development and integration pipeline facilitates the release of robust microservices that are deployed to and orchestrated by a Kubernetes cluster as a backbone of the DT platform for I4.0 [39]. The main features of the cluster include load balancing and fault management, e.g., by scaling and restarting service instances automatically. Moreover, the operational phase of the platform is supported by monitoring and logdata collection services. Here, Fluentd collects logs and ingests them in Elasticsearch, a scalable search and analytics suite [40–42].

Figure 3 depicts the platform's architecture comprising the following four layers:

- Business Services Layer: As value-added services, I4.0 services are located on the business services layer of the platform, utilizing the data infrastructure microservices on the layer below. I4.0 services are accessible and interact via an API (application programming interface). Additionally, the data portal provides a graphical user interface (GUI).
- Data Infrastructure Layer: The data infrastructure layer contains the microservices supporting the business services layer. They are accessible via an API and some of them are interacting. These services are implemented as generic as possible.
- Shop-Floor Layer: The shop-floor layer contains each component, machine, and manufacturing facility on the physical shop-floor. Each of them is equipped with an OPC UA server.
- Continuous Deployment Layer: The continuous deployment (CD) layer holds the base software for a continuous deployment of microservices into the data infrastructure and business services layers.

Furthermore, the platform comprises two components crossing these layers:

- A Kubernetes cluster orchestrates and monitors the microservices in the data infrastructure and business services layers.
- Cross-company connectors enable cross-company access to all four layers of the platform and can connect to the platforms of collaborating business partners.

The following subsections describe the CD, data infrastructure, and business services layer in detail and discuss how the platform will become part of an I4.0 data space. A deeper insight into the interoperation of business and data infrastructure microservices and the shop-floor assets will follow in Sect. 5.

4.1 The Continuous Deployment Layer

The CD layer contains a pipeline based on GitLab, a container registry, and Kubernetes [43, 44]. Through this pipeline new versions of microservices are developed, tested, built, tagged, registered, and deployed to either the data infrastructure or business services layer of the platform. For each microservice a separate GitLab

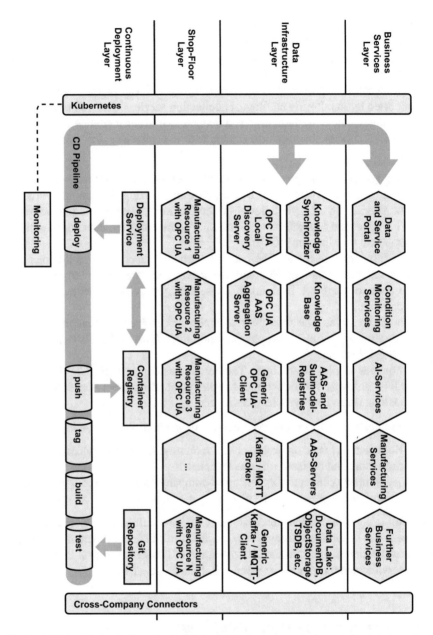

Fig. 3 A Digital Twin platform for Industrie 4.0. It is composed of specialized, independent, loosely coupled services structured into four layers: continuous deployment, shop-floor, data infrastructure, and business services layer. Connectors enable cross-company collaboration

project exists. Triggered by developments in these GitLab projects, the CD pipeline performs the following tasks [43]:

- Build: Following an update to a project's master branch, a container image is built. In particular, the master branch contains the corresponding container configuration file providing all necessary settings. Each new image is tagged with a version number and registered in the GitLab container registry.
- Deployment: As soon as an updated container image is registered in the registry, the corresponding containerized microservice in the data infrastructure or business services layer is also updated by the deployment service.

4.2 The Data Infrastructure Layer

The data infrastructure layer contains the microservices that support the I4.0 business services. The knowledge base is a document-based database storing all relevant metadata of known AAS (Asset Administration Shell) and OPC UA servers as JSON (JavaScript Object Notation) in a predefined mapping (user stories PA-S 6, MM-S 1, and DS-S 1: "data and service discovery," "semantics," and "data portal"). Here, Elasticsearch is used [41]. Furthermore, the knowledge base contains semantic descriptions of the data contained in the platform. A free-text search can be used to find information relevant for the use case at hand. Please find more metadata and knowledge-specific orchestration details in [45, 46].

Data from AAS, in particular submodel templates, as the usual semantic description of submodel instances, as well as those descriptions of semantic identifications of submodel elements that had to be defined manually since no public repository contained proper descriptions, are included in the knowledge base. Parts of this database are accessible from the Internet, so that the authorized partners can, for example, find semantic descriptions for the submodels and elements contained in AASs shared with them.

To ensure that the available data is always in sync, the OPC UA Local Discovery Service (LDS) automatically discovers each available OPC UA server in the edge-cloud ecosystem. Some assets hosting a local OPC UA server, e.g., on a programmable logic controller (PLC), possess additional AASs in AAS servers in the cloud that provide corresponding metadata. In such a case, the knowledge synchronizer service is responsible for linking and collecting AASs and OPC UA servers or rather metadata and data sources and storing them in the knowledge base.

In the case that an asset is equipped with an OPC UA server but not yet with an AAS, neither locally in its OPC UA server nor in the cloud, an OPC UA AAS aggregation server automatically migrates the asset server's information model to the OPC UA companion specification for AAS and serves it [22, 47]. Consequently, the OPC UA AAS aggregation server ensures that all assets with an OPC UA server possess at least one AAS.

AAS servers like [23, 24] with an HTTP/REST- or an OPC UA-CRUD-API (create, read, update, delete) host AASs in the intranet. AAS and submodel registries contain descriptors of AASs and submodels [25]. Each descriptor includes at least identification and endpoint of either AAS or submodel.

An MQTT broker is used for the execution of the bidding procedure for I4.0 [26, 33]. This procedure standardizes the interactions between I4.0 components and brings together providers and requesters of physical manufacturing services, e.g., drilling, transportation, or storage.

On request, the data and service portal on the business services layer establishes or cuts off Apache Kafka data streams (user story PA-S 3: "data transport") [37] from, e.g., OPC UA or AAS servers to the data lake, to the business microservices, or to the cross-company connectors. To this end generic OPC UA and Kafka clients are instantiated and deployed. Fieldbus protocols can be integrated using, e.g., Internet of Things (IoT) gateways.

The data lake contains databases for any kind of (raw) data (user story PA-S 4: "data storage"): structured like rows and columns in a relational database, semi-structured like JSON, or unstructured like files. Due to the knowledge base, the search for process and metadata in streams and lake is quick and efficient. Via the connectors to cross-company data-sharing systems, data can be exchanged and collaboratively enriched with, e.g., partners in a value-adding chain, where data sovereignty (user stories PA-S 1 and 5: "security" and "data sovereignty") is ensured. Please find use case-specific details in Sect. 5.

4.3   The Business Services Layer

The business services layer finally contains the I4.0 business microservices. For example, the data and service portal enable a user to search—freely or using tags—for relevant data and its sources in the platform (user stories PA-S 6 and DS-S 1: "data and service discovery" and "data portal"), including AAS and OPC UA servers, data streams, and lake. The portal provides suitable results and optionally, as mentioned before, sets up data pipelines from selected sources. Please find a detailed description in [45].

Within the data and service portal AASs for assets and dashboards, aggregating data can be designed (user stories MM-S 2 and 6: "Digital Twin design" and "dashboard"). Furthermore, the portal shows the user which Artificial Intelligence (AI) services could be applied to which shop-floor assets, e.g., anomaly detection for conveyor belt drives. For this purpose, the portal automatically matches the semantic identifications of the offered data of a shop-floor asset and the required data of an AI service [15]. Subsequent to the user's design of an AAS or dashboard or the selection of an AI service for a shop-floor asset, the portal deploys the new I4.0 component and sets up necessary data streams from shop-floor asset to AI service and back.

4.4   The Cross-Company Connectors

The DT platform for I4.0 enables cross-company data and service access by adding connector components which cross all four platform layers vertically.

In addition to the obvious need for extended security, e.g., regarding authentication and access control, a cross-company collaboration requires additional solutions, such as a global and well-known AAS registry that synchronizes with the company's internal registries. Strict firewall rules and sub-networks that cannot be reached from the outside make it impossible to establish a connection to the dedicated AAS-API endpoints. Inevitably, such a connector must not only carry the registry to the outside but also establish a kind of AAS gateway at the same time.

Gaia-X [5] anticipates many of these challenges to build a trustful environment, e.g., an I4.0 data space to leverage industrial data economy. The use case "collaborative condition monitoring" (CCM), for example, describes a shift from a bilateral data exchange to a multilateral relationship [48]. The leading idea is that by sharing and using data in a collaborative way, insights can be gained and thus processes and products can be optimized. In Fig. 4 three potential stakeholders of such a use case are depicted: a component supplier, a machine integrator, and a manufacturer, who operates the integrators' machines.

Please find a detailed description in [14] on how the DT platform for I4.0 can be used for CCM. The paper addresses the challenges of interoperability, self-describing and managed data flow, authorization and access control, authentication, and usage control. In perspective, the DT platform for I4.0 shall become part of an I4.0 data space based on the Federation Services of the European cloud initiative Gaia-X.

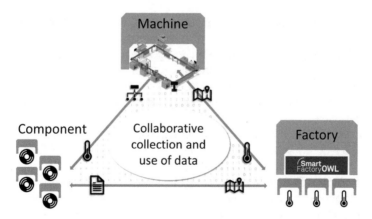

Fig. 4 The three-point fractal sketches the minimum of a multilateral data exchange

5   Digital Twin Controlled Manufacturing

The DT Platform for I4.0 presented in Sect. 4 facilitates the implementation of the main application scenarios for I4.0: mass customization, the resulting need for increased autonomy of manufacturing execution services. Using standardized I4.0 services enables cross-company collaboration of partners in a value chain consisting of, e.g., product developers, manufacturers, vendors, as well as their customers. Vendors could request manufacturers' services to manufacture product instances of certain types, e.g., to cope with high-order volume or to handle product orders that require specific manufacturing services or machinery.

Additionally, the DT platform for I4.0 facilitates product and process optimization by utilizing a peer-to-peer sharing system to collaboratively enrich AASs (Asset Administration Shells) of products with data from specific life cycle stages, e.g., product type development, customization of product instances, documentation of manufacturing processes, or product usage data. This system combines a blockchain for booking who changed an AAS and when, a peer-to-peer distributed file system for encrypting and storing versions of AASs, and a version control system for tracking changes in detail. In combination, these technologies ensure the sovereignty of the collaborators' data, i.e., integrity, confidentiality, and availability. Please find details in [13].

A manufacturer's factory houses working stations and warehouses, each with specific capabilities. These physical assets process their individual work orders successively according to the order in a queue. They, as well as the company's business microservices, all comprise an AAS [19] as standardized interface for I4.0 via which they can interact, offer their capabilities, and trigger each other's services. AASs are either implemented based on OPC UA or AAS server technologies (see Sect. 4) [22–24].

In contrast to the assets of the physical world, the assets of the information world can execute multiple instances of their services in parallel. Also, these I4.0 business microservices each serve their corresponding AAS on their own, enabling authorized consumers HTTP/REST-read access and the triggering of their specific service. For example, the order taking service serves its AAS via the Internet. It can be accessed by collaborating vendors in order to trigger the manufacturing of product instances.

The microservices discover and trigger each other via registries [25], in which they register descriptors of their specific services and endpoints for triggering. Figure 5 visualizes the registering, finding, and triggering process.

In the spirit of I4.0, product types and manufactured instances are themselves considered as assets as well. Here, their corresponding AASs are made available via AAS servers with AAS-APIs (application programming interfaces) for creating, reading, updating, and deleting complete AASs as well as submodels and submodel elements. These product types' and instances' AASs provide submodels containing data necessary for the manufacturing of the product instances and submodels for process documentation. In addition to the collaboration use case described before,

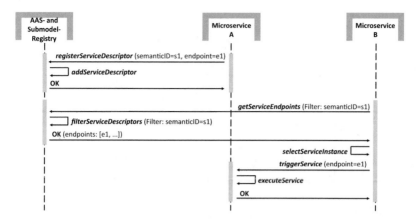

Fig. 5 The mutual discovery and triggering of I4.0 business microservices

an instance's AAS facilitates product optimization based on manufacturing data and enables vendors and customers to track manufacturing progresses.

What the business services of the information world have in common with the working stations and warehouses of the physical world is that they only consider the data contained in their own AAS, as well as the data contained in the AASs of the product instance and its type they are currently working on. This distributed knowledge enables fast decision making and flexible manufacturing management.

Based on the following four business microservices, manufacturing can be executed: order taking service (OTS), manufacturing driving service (MDS), bidding request service (BRS), and recording and notification service (RNS). Authorized vendors can trigger the execution of a manufacturing process run via the OTS.

Figure 6 visualizes the manufacturing process flow.

- A customer customizes and orders a product instance in a vendor's webshop. For each product instance to be manufactured, the vendor creates an AAS containing the product instance's and type's identification, customization details, as well as an endpoint via which the vendor wants to be notified of manufacturing progresses. The vendor shares its AAS with the manufacturer via the peer-to-peer sharing system.
- When triggered, OTS (order taking service)—the starting point of each manufacturing process run—puts the vendor's AAS of a product instance into an AAS server, extends it with submodels providing information required for the execution and submodels for the documentation of the manufacturing process, and adds descriptors of AAS and submodels to the registries. OTS completes with triggering RNS (recording and notification service) and MDS (manufacturing driving service) providing each with the identification of the AAS and, in addition, RNS with "Manufacturing accepted."

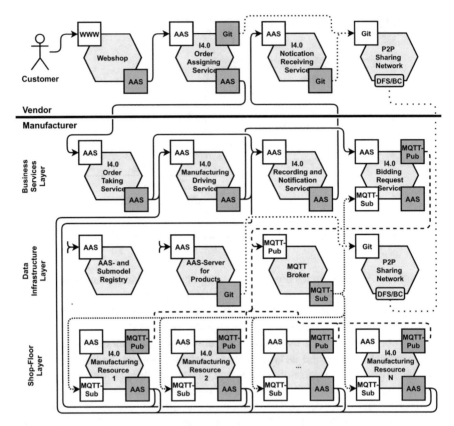

Fig. 6 Digital Twin controlled manufacturing in the manufacturer's Digital Twin platform for Industrie 4.0. Each manufacturing process run begins with a trigger of the manufacturer's order taking service via its AAS-API by an order assigning service of a vendor, which itself can be triggered by, e.g., a customer order in a webshop. The components' inbound adapters are indicated by white rectangles, dark gray rectangles mark outbound adapters, and the rounded light gray rectangles in the peer-to-peer sharing network components identify the distributed file system via which AASs are exchanged securely across the value-adding chain. For the sake of a clearer view, Apache Kafka and the data lake are not shown in the data infrastructure layer. Please add mentally also that every business service and every manufacturing resource interact with the AAS and submodel registry and the AAS server for manufactured products: either to register or find business services and AASs and submodels of product types and instances, to write or read manufacturing instructions, or to document manufacturing progresses

- RNS, when triggered, records the respective AAS's current server state in the peer-to-peer sharing system, notifies the corresponding vendor transmitting the status message, and terminates.
- MDS drives the manufacturing process of a product instance. It is triggered either by OTS for the first manufacturing step or by a working station or warehouse that completed a step. It determines, from the process documentation in a product instance's AAS, the step to be executed next, triggers BRS (bidding request

service) providing as parameter the identification of the submodel describing this step, and terminates.

- BRS publishes a call for proposal, using the interaction element from the specified submodel [26, 33] in the designated MQTT brokers. It collects the received proposals and selects the one suiting the manufacturer's guidelines—top priority instances, as soon as possible; lowest priority instances, as inexpensive as possible—best, confirms it to the proposer, documents it in the respective submodel, triggers RNS with the message "Manufacturing step <step-number> scheduled," and terminates.

- The winning proposer (working station or warehouse) of a bidding procedure is responsible for the completion of the respective manufacturing step. Its responsibility also includes the picking up of the product instance and the documentation of the step in the corresponding submodel. When completed, it triggers RNS with "Manufacturing step <step-number> completed" and, again, MDS.

- When MDS determines that the manufacturing of a product instance is completed, it adds to the corresponding AAS a submodel describing the instance's nameplate [49], triggers RNS with "Manufacturing completed," and terminates—the endpoint of each manufacturing process run.

The bottom line is that the Digital Twin of a product instance in the DT platform for I4.0 can be identified, as visualized in Fig. 7, as the union:

- Of a passive part composed of all data contained in the platform related to the product instance, that is, all AAS versions as well as process and metadata contained, e.g., in AAS servers, data streams, data lake, and peer-to-peer sharing network

Fig. 7 A Digital Twin of a manufactured product instance

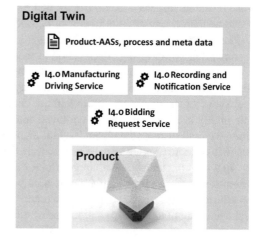

- Of an active part comprising those instances of the I4.0 manufacturing driving, I4.0 recording and notification, and I4.0 bidding request services that execute services (in past, present, or future) for the product instance

Jointly, a product instance and its Digital Twin compose an active composite I4.0 component: The Digital Twin actively takes decisions and interacts with I4.0 manufacturing resources aiming at an efficient manufacturing of the product instance.

6   Conclusion and Outlook

This chapter developed a Digital Twin (DT) platform for Industrie 4.0 (I4.0) enabling efficient development, deployment, and resilient operation of DTs and I4.0 business microservices, each of which comprises a software asset and an Asset Administration Shell (AAS).

User stories of platform architect, chief financial officer, platform operator, manufacturing manager, and data scientist laid the foundation for a comprehensive analysis and evaluation of architectural styles. Based on the result of analysis and evaluation, the DT platform for I4.0 was developed applying microservices patterns. A time series messaging experiment conducted in the course of this evaluation led to the conclusion that specialized, independent, and loosely coupled DTs, I4.0 microservices, and I4.0 manufacturing resources on the platform should in general communicate asynchronously via their AAS-APIs, channels, streams, or time series databases. Particularly, time series integration, consumption, and processing should be implemented in an asynchronous fashion via streams or time series databases. Besides higher performance, blocking and crashed services can be prevented this way. Seldom executed end-user applications, like reporting or documentation services, could consume time series directly from AASs in files or API servers in the platform. In this case, however, time series should be highly compressed avoiding, e.g., duplication of metadata.

The platform consists of four layers: continuous deployment (CD), shop-floor, data infrastructure, and business services layer. The services in the CD layer, the backbone of the platform, ensure that the data infrastructure and business services are highly available and can be deployed continuously. Data infrastructure services integrate and manage data from the I4.0 manufacturing resources on the shop-floor layer in order to provide semantically enriched information to I4.0 business services.

In a manufacturer's actual application of the DT platform for I4.0, I4.0 business microservices accept manufacturing orders from trusted vendors, control manufacturing processes, dynamically interact with I4.0 manufacturing resources to schedule manufacturing execution, record the manufacturing of product instances, and notify vendors about the progress. Each I4.0 business microservice hosts its own AAS. They discover and trigger one another by submitting proper input parameters to proper AAS endpoints found in AAS and submodel registries. The DT of, e.g., a manufactured product instance can be identified as the union of a passive part comprising all data related to the product instance contained in the platform (AASs in files and servers, process data and metadata in streams and data lake) and of an active part composed of those instances of I4.0 microservices executing services (in past, present, and future) for the considered product instance.

For each physical manufacturing step that must be executed, a call for proposals is issued and the received proposals from I4.0 manufacturing resources are evaluated by the product instance's DT. Subsequently, the DT assigns the step to the capable working station proposing the best execution—priority-dependent this might either be the fastest or the most economic proposal. After the execution and documentation of an assigned step, the working station in charge hands back the responsibility to the product instance's DT. Step by step this DT takes decisions, where only data from its passive part and from the interactions with other DTs is taken into account. This decentralized decision-making process results in a manufacturing system that is efficient and highly adaptable, e.g., with regard to customer requests, deviations from schedules, and the integration of new manufacturing resources.

The presented platform comprises some similarities to the cloud manufacturing platform (CMfg) as both represent manufacturing resources and tasks by digital representations in platforms in order to organize manufacturing [50, 51]. Whereas a central service allocates and schedules in CMfg, this paper's platform, in the spirit of Industrie 4.0 and based on the conviction that the determination of global optima is too complex to cope with rapidly changing manufacturing environments, enables resources, tasks, and their Digital Twins to autonomously build highly dynamic networks and flexibly negotiate the scheduling among themselves.

Future research and development of the DT platform for I4.0 focuses on further optimization and enhancements like recovery mechanisms for process states in crashed business microservices. Taking into account the distributed knowledge in the decentralized platform, how can a product instance's manufacturing be resumed loss-free in the right process step?

Furthermore, every data infrastructure service should be equipped, like I4.0 business microservices and I4.0 manufacturing resources, with an AAS composing an I4.0 data infrastructure service. This would achieve one of the main goals of I4.0, one standardized interface in the form of the Asset Administration Shell (AAS). This way, DT interaction, adaptability, and autonomy would be improved even further. Every DT and I4.0 microservice would only require a free port for serving its own AAS and one endpoint of an AAS and submodel registry for, on the one hand, registering and promoting its own service and, on the other, for finding and interacting with every other I4.0 service present in the Digital Twin platform for Industrie 4.0. Finally, I4.0 services could be used globally in an Industrie 4.0 data space based on Gaia-X Federation Services.

Acknowledgments The research and development projects "Technical Infrastructure for Digital Twins" (TeDZ) and "Industrial Automation Platform" (IAP) are funded by the Ministry of Economic Affairs, Innovation, Digitalization and Energy (MWIDE) of the State of North Rhine-Westphalia within the Leading-Edge Cluster "Intelligent Technical Systems OstWestfalenLippe" (it's OWL) and managed by the Project Management Agency Jülich (PTJ). The research and development project "KI-Reallabor für die Automation und Produktion" initiated by the German initiative "Platform Industrie 4.0" is funded by the German Federal Ministry for Economic Affairs and Energy (BMWi) and managed by the VDI Technologiezentrum (VDI TZ). The authors are responsible for the content of this publication.

References

1. Panetto, H., Iung, B., Ivanov, D., Weichhart, G., & Wang, X. (2019). Challenges for the cyber-physical manufacturing enterprises of the future. *Annual Reviews in Control, 47*, 200–213.
2. Al-Gumaei, K., Müller, A., Weskamp, J. N., Longo, C. S., Pethig, F., & Windmann, S. (2019). Scalable analytics platform for machine learning in smart production systems. In *24th IEEE ETFA, 2019*. https://doi.org/10.1109/ETFA.2019.8869075
3. Asset Administration Shell Reading Guide (11/2020). https://www.plattform-i40.de/PI40/Redaktion/DE/Downloads/Publikation/Asset_Administration_Shell_Reading_Guide.html
4. 2030 Vision for Industrie 4.0—Shaping Digital Ecosystems Globally. https://www.plattform-i40.de/PI40/Navigation/EN/Industrie40/Vision/vision.html
5. Gaia-X—A Federated Data Infrastructure for Europe. https://www.data-infrastructure.eu
6. Bitkom e.V. and Fraunhofer-Institut für Arbeitswirtschaft und Organisation IAO. (2014). Industrie 4.0 – Volkswirtschaftliches Potenzial für Deutschland
7. Deuter, A., & Pethig, F. (2019). The digital twin theory—Eine neue Sicht auf ein Modewort. Industrie 4.0 Management 35.
8. Jasperneite, J., Hinrichsen, S., Niggemann, O. (2015). Plug-and-Produce für Fertigungssysteme. *Informatik-Spektrum, 38*(3), 183–190.
9. Pethig, F., Niggemann, O., Walter, A. (2017). Towards Industrie 4.0 compliant configuration of condition monitoring services. In *15th IEEE INDIN*.
10. Lang, D., Friesen, M., Ehrlich, M., Wisniewski, L., & Jasperneite, J. (2018). Pursuing the vision of industrie 4.0: Secure plug-and-produce by means of the asset administration shell and blockchain technology. In *16th IEEE INDIN*.
11. Heymann, S., Stojanovci, L., Watson, K., Nam, S., Song, B., Gschossmann, H., Schriegel, S., & Jasperneite, J. (2018). Cloud-based plug and work architecture of the IIC testbed smart factory web. In *23rd IEEE ETFA*.
12. Stock, D., Bauernhansl, T., Weyrich, M., Feurer, M., & Wutzke, R. (2018). System architectures for cyber-physical production systems enabling self-X and autonomy. In *25th IEEE ETFA*. https://doi.org/10.1109/ETFA46521.2020.9212182
13. Redeker, M., Volgmann, S., Pethig, F., & Kalhoff, J. (2020). Towards data sovereignty of asset administration shells across value added chains. In *25th IEEE ETFA*. https://doi.org/10.1109/ETFA46521.2020.9211955
14. Redeker, M., Weskamp, J. N., Rössl, B., & Pethig, F. (2021). Towards a digital twin platform for industrie 4.0. In *4th IEEE ICPS*. https://doi.org/10.1109/ICPS49255.2021.9468204
15. Redeker, M., Klarhorst, C., Göllner, D., Quirin, D., Wißbrock, P., Althoff, S., & Hesse, M. (2021). Towards an autonomous application of smart services in industry 4.0. In *26th IEEE ETFA*. https://doi.org/10.1109/ETFA45728.2021.9613369
16. DIN SPEC 91345: Reference Architecture Model Industrie 4.0 (RAMI4.0), DIN (2016). https://doi.org/10.1109/10.31030/2436156
17. The Asset Administration Shell: Implementing Digital Twins for use in Industrie 4.0—a starter kit for developers. https://www.plattform-i40.de/PI40/Redaktion/EN/Downloads/Publikation/VWSiDV2.0.html
18. International Electrotechnical Commission (IEC). (2002). IEC 61360 -1 to 4- Standard data element types with associated classification scheme for electric components.

19. Platform Industrie 4.0: Details of the Asset Administration Shell—Part 1—The exchange of information between partners in the value chain of Industrie 4.0 (V3.0), Federal Ministry for Economic Affairs and Energy (2020). https://www.plattform-i40.de/IP/Redaktion/EN/Downloads/Publikation/Details_of_the_Asset_Administration_Shell_Part2_V1.html
20. International Electrotechnical Commission (IEC). (2016). IEC 61360—Common Data Dictionary (CDD—V2.0014.0014).
21. International Electrotechnical Commission (IEC): IEC 63278-1 ED1 Asset administration shell for industrial applications—Part 1: Administration shell structure. https://industrialdigitaltwin.org/
22. OPC Foundation: OPC 30270: OPC UA for Asset Administration Shell (AAS), Release 1.00. https://reference.opcfoundation.org/v104/I4AAS/v100/docs/
23. AASX Server. https://github.com/admin-shell-io/aasx-server
24. Eclipse BaSyx. https://wiki.eclipse.org/BaSyx
25. Platform Industrie 4.0: Details of the Asset Administration Shell—Part 2—Interoperability at Runtime (V1.0), Federal Ministry for Economic Affairs and Energy (2020). https://www.plattform-i40.de/IP/Redaktion/EN/Downloads/Publikation/Details_of_the_Asset_Administration_Shell_Part2_V1.html
26. VDI/VDE 2193 Blatt 1:2020-04: Language for I4.0 Components—Structure of messages. https://www.beuth.de/de/technische-regel/vdi-vde-2193-blatt-1/318387425
27. Industrial Digital Twin Association (IDTA). https://industrialdigitaltwin.org/
28. Zillner, S., Curry, E., Metzger, A., Auer, S., & Seidl, R. (Eds.). (2017). *European big data value strategic research & innovation agenda*. Big Data Value Association. https://www.bdva.eu/sites/default/files/BDVA_SRIA_v4_Ed1.1.pdf
29. Zillner, S., Bisset, D., Milano, M., Curry, E., García Robles, A., Hahn, T., Irgens, M., Lafrenz, R., Liepert, B., O'Sullivan, B., & Smeulders, A. (Eds.). (2020). *Strategic research, innovation and deployment agenda—AI, data and robotics partnership. Third release.* Big Data Value Association, euRobotics, ELLIS, EurAI and CLAIRE. https://ai-data-robotics-partnership.eu/wp-content/uploads/2020/09/AI-Data-Robotics-Partnership-SRIDA-V3.0.pdf
30. Tanenbaum, A.S., van Steen, M. (2006). *Distributed systems: Principles and paradigms*, 2nd ed. Prentice-Hall, Inc., USA.
31. Richardson, C. (2018). *Microservices patterns: With examples in Java*. Manning Publications. https://books.google.de/books?id=UeK1swEACAAJ
32. Oxford Reference: Service Access Point. https://www.oxfordreference.com/view/10.1093/oi/authority.20110803100456475
33. VDI/VDE 2193 Blatt 2:2020-01: Language for I4.0 components—Interaction protocol for bidding procedures. https://www.beuth.de/de/technische-regel/vdi-vde-2193-blatt-2/314114399
34. ISO/IEC 20922:2016. Information technology—Message Queuing Telemetry Transport (MQTT) v3.1.1. https://www.iso.org/standard/69466.html
35. AASX Package Explorer, Server and Registry. https://github.com/admin-shell-io
36. Eugster, P. T., Felber, P. A., Guerraoui, R., & Kermarrec, A.-M. (2003). The many faces of publish/subscribe. *ACM Computing Surveys, 35*(2), 114–131.
37. Apache Kafka. https://kafka.apache.org
38. InfluxDB. https://www.influxdata.com
39. Kubernetes. https://kubernetes.io
40. Fluentd. https://www.fluentd.org
41. Elasticsearch. https://www.elastic.co/elasticsearch
42. OpenSearch. https://opensearch.org
43. GitLab. https://about.gitlab.com
44. Chen, L. (2015). Continuous delivery: Huge benefits, but challenges too. *IEEE Software, 32*, 2.
45. Weskamp, J. N., Chowdhury, A., Pethig, F., Wisniewski, L. (2020). Architecture for knowledge exploration of industrial data for integration into digital services. In *3rd IEEE ICPS*. https://doi.org/10.1109/ICPS48405.2020.9274700

46. Ghosh Chowdhury, A., Illian, M., Wisniewski, L., Jasperneite, J. (2020). An approach for data pipeline with distributed query engine for industrial applications. In *25th IEEE ETFA*.
47. Weskamp, J. N., & Tikekar, J. (2021). An Industrie 4.0 compliant and self-managing OPC UA Aggregation Server. In *26th IEEE ETFA*. http://dx.doi.org/10.1109/ETFA45728.2021.9613365
48. Collaborative data-driven business models: Collaborative Condition Monitoring—How cross-company collaboration can generate added value. https://www.plattform-i40.de/PI40/Redaktion/EN/Downloads/Publikation/collaborative-data-driven-business-models.html
49. Submodel Templates of the Asset Administration Shell—ZVEI Digital Nameplate for industrial equipment (Version 1.0). https://www.plattform-i40.de/PI40/Redaktion/DE/Downloads/Publikation/Submodel_Templates-Asset_Administration_Shell-digital_nameplate.html
50. Tao, F., Zhang, L., Venkatesh, V. C., Luo, Y., & Cheng, Y. (2011). Cloud manufacturing: A computing and service-oriented manufacturing model. *Proceedings of the Institution of Mechanical Engineers, Part B: Journal of Engineering Manufacture, 225*(10), 1969–1976. https://doi.org/10.1177/0954405411405575
51. Zhang, L., Luo, Y., Tao, F., Li, B. H., Ren, L., Zhang, X., Guo, H., Cheng, Y., Hu, A., Liu, Y. (2014). Cloud manufacturing: A new manufacturing paradigm. *Enterprise Information Systems, 8*(2), 167–187. https://doi.org/10.1080/17517575.2012.683812

A Framework for Big Data Sovereignty: The European Industrial Data Space (EIDS)

Christoph Mertens, Jesús Alonso, Oscar Lázaro, Charaka Palansuriya, Gernot Böge, Alexandros Nizamis, Vaia Rousopoulou, Dimosthenis Ioannidis, Dimitrios Tzovaras, Rizkallah Touma, Miquel Tarzán, Diego Mallada, Paulo Figueiras, Ruben Costa, Diogo Graça, Gisela Garcia, Begoña Laibarra, Aitor Celaya, Piotr Sobonski, Azzam Naeem, Alberto Mozo, Stanislav Vakaruk, J. Enrique Sierra-García, Antonio Pastor, Juan Rodriguez, Marlène Hildebrand, Tomasz Luniewski, Wojciech Zietak, Christoph Lange, Konstantinos Sipsas, and Athanasios Poulakidas

---

C. Mertens
International Data Spaces e.V. (IDSA), Berlin, Germany
e-mail: christoph.mertens@internationaldataspaces.org

J. Alonso (✉) · O. Lázaro
Asociación de Empresas Tecnológicas Innovalia, Bilbao, Spain
e-mail: jalonso@innovalia.org; olazaro@innovalia.org

C. Palansuriya
EPCC, The University of Edinburgh, Edinburgh, UK
e-mail: charaka@epcc.ed.ac.uk

G. Böge
FIWARE Foundation e.V, Berlin, Germany
e-mail: gernot.boege@fiware.org

A. Nizamis · V. Rousopoulou · D. Ioannidis · D. Tzovaras
Centre of Research and Technology Hellas (CERTH), Thessaloniki, Greece
e-mail: alnizami@iti.gr; vrousop@iti.gr; djoannid@iti.gr; Dimitrios.Tzovaras@iti.gr

R. Touma · M. Tarzán
Fundació i2CAT, Barcelona, Spain
e-mail: rizkallah.touma@i2cat.net; miquel.tarzan@i2cat.net

D. Mallada
GESTAMP, Pillaipakkam, Tamil Nadu, India
e-mail: dmallada@gestamp.com

P. Figueiras · R. Costa
UNINOVA, Guimarães, Portugal
e-mail: paf@uninova.pt; rddc@uninova.pt

D. Graça · G. Garcia
Volkswagen Autoeuropa (VWAE), Lisbon, Portugal
e-mail: Diogo.Graca@volkswagen.pt; Gisela.Garcia@volkswagen.pt

© The Author(s) 2022
E. Curry et al. (eds.), *Data Spaces*, https://doi.org/10.1007/978-3-030-98636-0_10

Abstract The path that the European Commission foresees to leverage data in the best possible way for the sake of European citizens and the digital single market clearly addresses the need for a European Data Space. This data space must follow the rules, derived from European values. The European Data Strategy rests on four pillars: (1) Governance framework for access and use; (2) Investments in Europe's data capabilities and infrastructures; (3) Competences and skills of individuals and SMEs; (4) Common European Data Spaces in nine strategic areas such as industrial manufacturing, mobility, health, and energy. The project BOOST 4.0 developed a prototype for the industrial manufacturing sector, called European Industrial Data Space (EIDS), an endeavour of 53 companies. The publication will show the developed architectural pattern as well as the developed components and introduce the required infrastructure that was developed for the EIDS. Additionally, the population of such a data space with Big Data enabled services and platforms is described and will be enriched with the perspective of the pilots that have been build based on EIDS.

B. Laibarra · A. Celaya
Software Quality Systems (SQS), Oswego, NY, USA
e-mail: blaibarra@sqs.es; acelaya@sqs.es

P. Sobonski · A. Naeem
United Technology Research Centre Ireland, Cork, Ireland
e-mail: piotr.sobonski@rtx.com; azzam.naeem@rtx.com

A. Mozo · S. Vakaruk
Universidad Politécnica de Madrid (UPM), Madrid, Spain
e-mail: a.mozo@upm.es; stanislav.vakaruk@upm.es

J. E. Sierra-García
Universidad de Burgos, JRU ASTI-UBU, Burgos, Spain
e-mail: jesierra@ubu.es; enriquesg.cext@asti.tech

A. Pastor · J. Rodriguez
Telefonica I+D, Madrid, Spain
e-mail: antonio.pastorperales@telefonica.com; juan.rodriguezmartinez@telefonica.com

M. Hildebrand
EPFL, Lausanne, Switzerland
e-mail: marlene.hildebrand@epfl.ch

T. Luniewski · W. Zietak
CAPVIDIA, Houston, TX, United States
e-mail: tl@capvidia.com; wz@capvidia.com

C. Lange
Fraunhofer Institute for Applied Information Technology FIT, RWTH Aachen University,
Aachen, Germany
e-mail: christoph.langebever@fit.fraunhofer.de

K. Sipsas · A. Poulakidas
INTRASOFT, Mumbai, Maharashtra, India
e-mail: Konstantinos.SIPSAS@intrasoftintl.com; Athanasios.Poulakidas@intrasoft-intl.com

Keywords Data Spaces · Data treasures · Data sharing · Trust · Digital sovereignty · EIDS · IDSA · Open source · Interoperability · Semantic model · QIF · FIWARE · Certification

1 Introduction

Over the last two decades, data economy has emerged as a global growing trend, with heterogeneity of data sources becoming available. Economic development has evolved from a public open data or science data smart usage to the mass adoption and exploration of the value of industrial Big Data. And now, a new paradigm shift looms, evolving toward the economy of data sharing or common data.

Industry is now starting to learn how to extract value out of its own industrial data to gain industrial competitiveness; companies are also starting to realize that it is extremely unlikely that a single platform, let alone a single company, will drive the industrial data business. Hence, the development of new services and products relying on access to common data in the industrial environment calls for the transformation of various aspects. One of the central challenges in the production environment is the handling of the data that is generated, whether it comes from sensors of machines, in planning processes, or in product development. The news that this data is very valuable has now reached the broad base of society. The consequence, however, is not that the data is purposefully turned into money, but that this data treasure is locked away in many cases.

With the Data Governance Act and AI regulation, Europe is already laying the foundations for a European common data space. However, this policy framework needs to be matched with transformation at the industrial level of data sharing business culture and convergence of data infrastructures and soft data space services. It must be acknowledged that, on one hand, the common data explosion will arrive when industry exhausts its ability to extract value out of its own industrial data. On the other hand, it must also be acknowledged that it is very unlikely that the B2B common data economy will be based on access to the raw data entities themselves.

On the contrary, data ecosystems and future value chains will more likely be developed on sovereign access to well-curated and well-controlled data endpoints of high quality. Hence, the most urgent need is to develop not only a data culture but a data sharing or common data one.

Therefore, the development of data ecosystems should be envisioned as an evolutionary rather than a revolutionary process that will support new business models and disruptive value chains. This might be a chicken-egg problem. Data sharing and the data economy bring challenges that must be faced, like trust and openness, interoperability and cost, continuity, and controlled free flow of data. These are challenges, yes, but what is most important is that all of these lead to necessities.

In order to avoid a data divide and provide truly free access to a European common sovereign data space, there's a need for both a comprehensive digital infrastructure, not only a physical one but a soft one, related to all the services for sovereign data sharing and controlled data usage among different organizations. This should be accompanied by specific industrial agreements that will define the governance rules of future value chains. A well-balanced development of such dimensions is the best guarantee for the growth of the emerging and embryonic Data Spaces that the European industry is already putting in place to keep the leadership of the industrial (B2B) data and intelligence economy. The industry is already collaborating in finding an agreement on the design principles that should drive the development of such Data Spaces and ecosystems. One of the first large-scale implementations of an industrial embryonic data space has been implemented in the Boost 4.0 project [1]: the European Industrial Data Space (EIDS).

The EIDS is based on the International Data Spaces (IDS) Reference Architecture Model developed by the International Data Spaces Association (IDSA), a nonprofit association with more than 140 members from all over the world. The association aims to give back full control over data to companies, by coming up with the so-called IDS standard.

The main goal of the IDS standard is to provide trust between the different participants in the data space. To achieve this, it relies both in standardized technical components (connectors, applications, etc.) and in a strong certification program both for these components and for the entities that participate in the ecosystem (specially the operational environments). This way, it can be ensured that no "backdoors" are built into the software. Moreover, the security mechanisms are checked, as well, and they are made visible to other participants as a "trust level." Based on this information, everyone can decide whether the specified trust level of a potential partner is sufficient for their own use case or whether other partners need to be searched for.

To not create a singular and isolated perspective on the Data Spaces for manufacturing and Big Data, Boost 4.0 created a reference architecture (RA) [2]. The Boost 4.0 (RA) facilitates a common understanding within the Boost 4.0 project partners and supports the pilots in identifying Big Data assets required in order to fulfill their requirements. This model also brings together aspects from existing reference architectures and models targeting either Big Data (BDVRM [3], NBDRA) or the manufacturing domain (RAMI4.0, IIRA) and thus a clear alignment has been achieved. Also, the specific IDS standard was aligned with this RA and therefore the RA builds the foundation for the EIDS and gives the chance to put EIDS into perspective with many other reference architectures.

This chapter explains the basic design principles that were followed by Boost 4.0 to build the EIDS and presents some of the technical developments that enable the deployment of the EIDS. On the core of the EIDS there's the infrastructure that holds the entire data space, which is the main enabler for the data economy. On the next layer, there are the data space commonalities, which are the domain-specific specifications and structures. The outmost layer represents the raison d'être of the EIDS: this is where the data providers and the data users are represented, in the form of applications/platforms and specific use cases. Last, but not least, the EIDS certification framework is the main source of trust, as it ensures the compliance of components and applications with the essence of the EIDS.

2   The European Industrial Data Space: Design Principles

The European Industrial Data Space builds its solid foundations on the IDS Infrastructure described in the IDS Reference Architecture Model. The value proposition of IDS can be summarized with three main pillars:

1. Unlimited **interoperability**, to enable connectivity between all kinds of different data endpoints.
2. **Trust**, to ensure a transparent and secure data exchange.
3. **Governance** for the data economy, which takes the form of usage control mechanisms and policy enforcement.

These three core pillars are put into practice in the so-called "Data Spaces" concept. This term refers to an approach to data management having the following characteristics:

- A federated data architecture. In fact, data do not need to be physically integrated. This is a crucial difference with central platforms, client service solutions, or cloud.
- No common schema. The integration will be at the semantic level and this does not affect domain-specific vocabularies.
- Data is an economic asset. In Data Spaces data is used as an economics asset whose value increases while you share the data.
- Ecosystem of Data Spaces. The power of Data Spaces lies in their scalability and nesting opportunities. There can be Data Spaces in Data Spaces, which are referred to as "ecosystem of Data Spaces," and this leads to large pools of cross-domain data.

The concept of Data Spaces combined with the pillars provided by IDSA makes sure to exploit the potential of the former, avoiding the lack of shared rules which is typical of the federated approach. Therefore, IDS is a crucial element to ensure data sovereignty, traceability, and trust among participants of a data space. The data space approach is foreseen in the European Strategy for Data of the European Commission, which suggests rolling out common European Data Spaces in nine strategic economic sectors. The manufacturing sector is one of them.

Boost 4.0 positions itself as a realization of such an embryonic data space built over IDS, to unlock the value of data treasures and Big Data services among the partners in its consortium, as displayed in Fig. 1. To achieve this, Boost 4.0 has deployed the required IDS infrastructure, has adopted the semantic model proposed by the IDS RAM which has been combined with different domain-specific standard vocabularies, and developed the certification framework that validates the implementation of the connectors that enable participation in the EIDS.

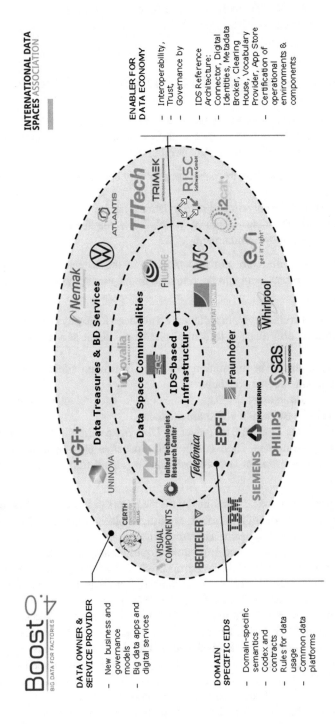

Fig. 1 Onion model for the European Industrial Data Space (EIDS)

Once the foundations of the EIDS are laid, it is necessary to adapt digital manufacturing platforms and to develop applications to provide an offering of services that attract data owners and other participants to the ecosystem.

3   EIDS-Based Infrastructure

The EIDS is based on the IDS Reference Architecture Model (RAM) V3.0 [4]. The IDS RAM defines how existing technologies must be orchestrated to create Data Spaces that deliver data sovereignty via a trusted environment to its participants. Among its five layers, the business layer defines a set of essential and optional roles for IDS-based Data Spaces. The essential components are the ones ensuring interoperability and a trustworthy data exchange thanks to clearly identifiable participants:

- The **connector** is defined as a dedicated communication server for sending and receiving data in compliance with the general connector specification (DIN SPEC 27070 [5]); different types of connectors can be distinguished (base connector vs. trusted connector, or internal connector vs. external connector).
- **DAPS** (Dynamic Attribute Provisioning Service), which issues dynamic attribute tokens (DATs) to verify dynamic attributes of participants or connectors.
- The **certificate authority** (CA), a trusted entity issuing digital certificates (X.509 certificates), may host services to validate the certificates issued.

The optional roles of IDS are the ones that make the data space more effective for discovering data endpoints, app search, and more:

- **Vocabulary Provider**. Data integration within IDS occurs at semantic level and this does not affect domain-specific vocabularies. IDS foresees vocabulary providers that help participants to discover domain-specific ontologies.
- **Metadata Broker**. This component is an intermediary managing a metadata repository that provides information about the data sources available in the IDS; multiple broker service providers may be around at the same time, maintaining references to different, domain-specific subsets of data endpoints.
- **App Store**. It is defined as a secure platform for distributing data apps; it features different search options (e.g., by functional or nonfunctional properties, pricing models, certification status, community ratings, etc.).
- **Clearing House**. This is an intermediary providing clearing and settlement services for all financial and data exchange transactions within the IDS.

Since the IDS RAM is technology agnostic for each data space that builds upon it, the challenge is to define its very own technology stack. The following paragraphs introduce some examples of technology stacks for some of the components that the EIDS uses.

3.1 HPC as Foundation for Resource-Intensive Big Data Applications in the EIDS

The Cirrus HPC service hosted at EPCC, the University of Edinburgh (UEDIN), provides both a high-performance computing (HPC) infrastructure and access to large storage systems. These infrastructures are suitable for Big Data analytic, machine learning, and traditional compute-intensive applications (e.g., CFD applications). The EIDS can access this infrastructure within the Boost 4.0 context.

The Cirrus [6] facility is based around an SGI ICE XA system. There are 280 standard compute nodes and 2 GPU compute nodes. Each standard compute node has 256 GiB of memory and contains two 2.1 GHz, 18-core Intel Xeon (Broadwell) processors. Each GPU compute node has 384 GiB of memory and contains two 2.4 GHx, 20-core Intel Xeon (Skylake) processors and four NVIDIA Tesla V100SXM2-16GB (Volta) GPU accelerators connected to the host processors and each other via PCIe. All nodes are connected using a single InfiniBand fabric and access the shared, 406 TiB Lustre file system. In addition to the Lustre file system that is included in Cirrus, it also has access to a high-capacity (i.e., in petabyte scale) object store system. This object store is similar to the Amazon S3 service and used to address medium- to longer-term data storage requirements.

In addition, for EIDS, a generic virtual machine (VM)-based infrastructure is deployed at EPCC to accommodate various infrastructure requirements where the above specialized infrastructure may not be suitable. Presently 10 VMs are available to Boost 4.0: each with 4 cores, 8 GB RAM, and 100 GB storage and runs Ubuntu version 18.04.

Furthermore, a certificate authority (CA) provided by the Hyperledger Fabric is deployed on the above generic EIDS VM infrastructure as described above. For more details about the Hyperledger Fabric CA, see the official documentation [7].

Additionally, a blockchain infrastructure based on the Hyperledger Fabric (v1.4) is deployed on the above EIDS VM infrastructure. Also, an Ethereum-based blockchain is deployed on this infrastructure. EIDS connectors could be linked, for example, to access these blockchain infrastructures.

3.2 Open-Source-Based FIWARE's EIDS Connector

During the runtime of Boost 4.0 FIWARE, in cooperation with the community members, Universidad Politécnica de Madrid (UPM) and Martel Innovate developed the EIDS CIM REST Connector. It can be implemented with the FIWARE Context Broker providing the NGSI interface in its proven version 2 or the latest ETSI GS CIM standard NGSI-LD embracing the concepts of Linked Data and the Semantic Web.

Both allow users to provide, consume, and subscribe to context information in multiple scenarios and involve multiple stakeholders. They enable close to real-time access to and exchange of information coming from all kinds of data sources. Combined with FIWARE Smart Data Models, this approach offers easy and seamless interoperability within the IDS, while reusing the existing and well-adapted FIWARE standards and technology. Beyond that, also other existing REST(ful) APIs can be connected to the IDS.

The development of the FIWARE EIDS CIM REST Connector has been guided by several design goals. SME, developers, and interested parties shall be empowered to start quickly and easily into the world of FIWARE and IDS, with minimal upfront efforts but the full potential to scale up and adapt quickly. The technology stack shall be open, well-adapted, and cloud-ready. Service providers and diverse business models shall be supported with multi-cluster, multi-connector, multi-tenant, and multi-API options available from the very beginning to facilitate SME on-boarding, letting the EIDS grow. Four existing open-source projects were chosen to fulfill the ambitious design goals:

1. **Kubernetes** for automating the deployment, scaling, and management of containerized applications. As the current de facto industry standard, it groups containers that divide an application into logical units for easy management and discovery. Kubernetes can scale on demand and offers a vibrant community and an existing market for professional services.
2. **Rancher** for teams adopting Kubernetes management. It addresses the operational and security challenges of managing multiple Kubernetes clusters by offering extended security policies, user access management, Istio integration, and multi-provider and multi-cluster support. Fully featured tools for monitoring, logging, alerting, and tracing are provided right out of the box.
3. **Istio** as a service mesh to route and manage the traffic within the connector. It layers transparently onto existing distributed applications and microservices that can integrate into any logging platform or telemetry or policy system. Istio's diverse feature set lets you successfully, and efficiently, run a distributed microservice architecture and provide a uniform way to secure, connect, and monitor microservices, from simple to highly complex and restrictive scenarios.
4. **Ballerina** as an open-source programming language and platform for cloud-era application programmers. Ballerina includes the programming language and a full platform, which consists of various components of the surrounding ecosystem to offer a rich set of tools for integrated cloud development from coding to testing and deploying directly to the cloud.

As a result, the FIWARE EIDS CIM REST Connector offers a universal, unique, transparent, and fully integrated base-level approach to bring NGSI support to the IDS without the need for changes in application codes.

The generic approach of the FIWARE software components enables use cases ranging from Big Data, AI, or complex event processing to extended data analytics, data aggregation, and data anonymization, all combined with fine-grained access control down to the attribute level. Actionable insights can be conditionally propagated and notified to receiving devices and systems to leverage the highest value of near real-time data as best as possible. And last but not least, an innovative approach for providing access and usage control to such publish/subscribe scenarios has been developed, incorporating the core trust concepts of the UCON model, the ORDL description language, and the XACML 3.0 specification.

4   Data Space Commonalities: Semantics as Prerequisite for Data Economy

Nowadays, vast amounts of data are produced daily by machines on assembly lines. These usually contain very useful information regarding manufacturing processes, and performing analysis might help offer precious insights to improve them. Unfortunately, when trying to achieve this goal, one is often confronted with the issue of data quality. Currently, in most companies, data exist in silos and data schemas underlying these systems are not harmonized. In concrete terms, this means that data comes in different formats, such as SQL databases, XML, Excel sheet, or CSV files. This implies that from one database or data file to another, the schemas can drastically change, since each service can decide on its own data schema and will select the one that better fits its needs. This means that from one service to another, the same piece of information can be attributed with different field names.

Another issue is the clarity of data schemas. To be comprehensible, data needs to follow a schema which helps associate the values to their meaning. Unfortunately, data schemas are not always easily understandable. Sometimes, the names of the fields in a data schema do not provide the user with helpful meaning. This is particularly prevalent in the case of legacy data, where sometimes fields had to follow a certain naming convention that did not coincide with natural language, making them particularly difficult to interpret in retrospect. Another issue that can happen with legacy data is when datasets had a fixed size, forcing users to reuse an existing field to store unrelated information, losing the meaning of the data in the process.

This creates a need within companies, but also within manufacturing domains at large, for a harmonized, universally understandable data schema. Ontologies can play a major role to help in that regard. The purpose of ontologies is to provide a shared common vocabulary among a domain context.

Thanks to ontologies, data can be marked up with meaningful metadata, which help to ensure that data is understood the same way across different parties. The different actors involved simply need to know the same ontology to understand the data. This leads to great time improvements, especially since concepts can easily be reused from one ontology to another. It also leads to increased interoperability, since concepts asserted in the ontology remain consistent in different applications. Whenever a piece of software uses the same ontology, the concepts and their relations cannot be misinterpreted.

This last point is another strong point in favor of ontologies: they help make data machine interpretable. Thanks to metadata tags, a piece of code can easily read the data, understand what it stands for, and treat it accordingly.

4.1   The IDS Information Model

Exchanging data and operating the infrastructure of an ecosystem that supports such exchange, both require a common language. The IDS Information Model [8, 9] is the common language used for:

- Descriptions of digital resources offered for exchange.
- The self-description of infrastructural components, such as connectors, providing data or metadata brokers enabling potential consumers to find them.
- The headers of the messages that such components send to each other.

The IDS Information Model is conceptually specified in Sect. 3.4 "Information Layer" of the IDS RAM as human-readable text and UML diagrams. Its declarative representation as an ontology [10] based on the World Wide Web Consortium's family of Semantic Web standards around the RDF Resource Description Framework provides an unambiguous machine-comprehensible implementation for the purpose of validation, querying, and reasoning. Finally, a programmatic representation addresses the purposes of integration into services, tool support, and automated processing [11].

The IDS Information Model addresses all concerns of sharing digital resources, i.e.,

- Their content and its format and structure.
- Concepts addressed by a resource.
- The community of trust, whose participants exchange digital resources.
- Digital resources as a commodity.
- The communication of digital resources in a data ecosystem.
- The context of data in such a resource.

As shown in Fig. 2, the conceptualization and implementation of most of the concerns addressed by the IDS Information Model builds on established standards, key pillars being an extension of the W3C Data Catalog Vocabulary (DCAT) [12] and a usage policy language based on the W3C Open Digital Rights Language (ODRL) [13].

Depending on the requirements of each concern, the Information Model's declarative representation as an ontology literally reuses existing standards, or it adapts and extends them to the specific requirements of data ecosystems.

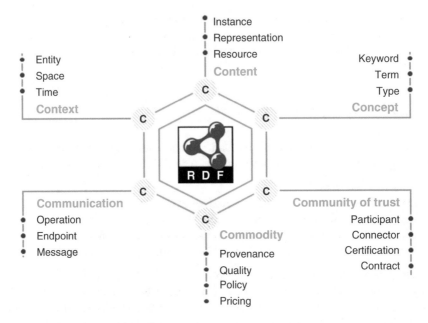

Fig. 2 The Concern Hexagon of the IDS Information Model in detail, with references to standards built on IDS Reference Architecture Model

4.2   Domain-Specific Ontologies in EIDS: QIF

The IDS Information Model takes a domain-agnostic perspective on data. To reasonably exchange data in specific domains such as manufacturing in the context of the EIDS, a connection with domain ontologies is required. While the IDS mandates semantic descriptions of data resources by metadata in terms of the IDS Information Model for interoperability, it does not enforce but strongly encourages the use of standardized, interoperable representations for the actual data as well. In other words, data should use a standardized domain ontology as their schema. The IDS Information Model bridges the domain-agnostic metadata level and the domain-specific domain knowledge with regard to multiple aspects of the content and concept concerns. It is recommended to use structured classification schemes, e.g., taxonomies, instead of just string-based keywords to characterize the concepts covered by a digital resource. The content of a digital resource in manufacturing may, for example, be sensor measurements expressed using the W3C Sensor, Observation, Sample, and Actuator/Semantic Sensor Network (SOSA/SSN) ontology [14]. The bridge to the metadata level is established by using, for example, the W3C Vocabulary of Interlinked Datasets (VoID) [15] to express that the digital resource mainly contains instances of the class *Sensor*, or using the W3C Data Cube Vocabulary [16] to express that the digital resource consists of a three-dimensional matrix with temperature measurements in the dimensions: (1) time, (2) geo-coordinates, and (3) sensor used.

One example, used in Boost 4.0, of a domain-specific standard ontology is the Quality Information Framework (QIF). QIF is a feature-based ontology of manufacturing metadata, built on XML technology, with the foundational requirement of maintaining traceability of all metadata to the "single source of truth"—the product and all its components as defined in CAD/MBD. It is an ISO 23952:2020/ANSI standard which includes support for a complete semantic derivative model-based definition (MBD) model, measurement planning information, and measurement result information. This characterization allows users to establish a Digital Twin by capturing the duality of aspects of manufacturing data: the as-designed product and the as-manufactured product – including the mappings between the two aspects.

QIF is a highly organized grammar and vocabulary for software systems to communicate manufacturing data structures. With software interoperability as the goal, vendors and end-users were available to verify (through both pilot studies and production-system deployments) the robustness of QIF and its truly semantic structure. Another important design objective is data traceability to the authority product definition model. Each piece of metadata in a QIF dataset is mapped to the owning GD&T (geometric dimensioning and tolerancing), feature, and model surface in the authority CAD model. This high-resolution mapping to the "single source of truth" ensures that any other data derived from the model can be traced through a graph to the mass of QIF data. This is why QIF matters beyond just metrology: it provides the mapping to metrology data for the entire model-based enterprise.

The software vocabulary and grammar of QIF is defined by the QIF XML Schema Definition Language (XSDL) schemas. These schemas are at the core of what defines QIF. At a high level, there are three primary pillars to QIF's approach to these data quality: XSD validation, XSLT integrity checks, and digital signatures. XML Schema Definition (XSD) validation is a test where a QIF instance file is checked for validity to the QIF digital "language" defined in the QIF schemas. Extensible Stylesheet Language Transformations (XSLT) is a Turing-complete language for processing XML files. Included with the QIF standard is a set of XSLT scripts which can carry out integrity checks on QIF instance files that are not possible with a simple XSD validation test. QIF has the ability to control the provenance of the data via digital signatures. This infrastructure helps to ensure that QIF instance files can be imbued with the trust necessary to use it throughout the entire product life cycle.

5   Data Treasures and Big Data Services in EIDS: Big Data Applications and Use Cases

This section describes in high level the types of services, applications, and platforms that are available in EIDS ecosystem and some examples that have been deployed in the Boost 4.0 project, each of which focuses on the integration with one of the design principles, although obviously they mostly integrate several of them.

The Boost 4.0 project provides plenty of services and tools for the application of Big Data technologies at different stages of the factory life cycle and the supply chain (cf. Table 1). The project contributes on the state of the art in cognitive

Table 1 Big Data solutions that are made available in the European Industrial Data Space (EIDS)

Platform/app	Key features
RISC Data Analysis Platform for the Optimization of Flexible Machines Manufacturing	Combination of BD technologies and semantics to process and store data. It is based on cluster computing and data analytics services
ESI Hybrid Twin for Zero Defect Production	Cloud platform dedicated to simulation solutions. Provides 3D remote visualizations and a vertical application framework
Interactive Analytics for Smart injection Molding	Live monitoring of the machine's behavior. Graphical interface for mapping data with concepts of a generic data model. Outlier detection
ESI Platform for Cognitive Value Chain	Big Data and visual analytics. Use of ontology for the extraction of reliable training datasets for predictive maintenance
Visual Components 4.0 and UNINOVA Analytics for Inventory Optimization and Dynamic Production Planning	BD infrastructure for data management, storage, and processing. Visualization services based on Grafana. Data aggregation into a 3D Digital Twin by exploiting the connection with Visual Components Simulation Software
Siemens MindSphere Extensions for third Parties	Cloud-based IoT platform for BD management and analysis with application in various domains
IBM Blockchain Application for Certification of Origin	A blockchain-based application for enabling traceability and full visibility of transaction data related to the production of fabrics
CERTH Cognitive Analytics Platform for Industry 4.0	Cloud platform for BD analytics and advanced visualization. Real-time model retraining and ensemble methods for predictive maintenance. EIDS compatible
SAS Analytics Platform for Production Planning	Powerful analytics platform with plenty of real-world applications. Support of extensions through web services
ATLANTIS DSS and Fraunhofer IEM Analytics for Predictive Maintenance	Complete solution that provides data analytics services alongside with decision support mechanisms. EIDS compatible
TRIMEK M3 Platform for Zero Defect Manufacturing	Metrology platform that provides BD analysis and visualizations for scanning processes
I2CAT IoT Real-Time Location System	IoT location monitoring system with high precision (1–2 cm). EIDS compatible

manufacturing using the latest available techniques and methodologies in Big Data management, processing, modeling, analysis, and visualization. More precisely, Big Data services, tools, and platforms related to the detection of deterioration rate of production machines and their root cause are available [17]. Furthermore, predictive and cognitive modeling methodologies and early anomaly detection algorithms are available as well through the EIDS infrastructure [18]. The list of provided analytics services is completed by the implemented planning optimization algorithms and advanced data visualization and visual analytics tools [19]. In addition to these solutions, services and apps related to the connection of Big Data analytics platforms with data sources from shopfloor and other tools are also provided. These services/apps are EIDS connector-compliant application for specific analytics platforms that participated in the project and they are major players in analytics services in global market at the same time. Table 1 presents an overview of the solutions for which the compatibility was determined during the project runtime. A selection of these solutions is described within this section.

5.1 Types of Connection for Big Data Analytics Services and Platforms in the EIDS

There are two main ways that Boost 4.0 analytics services or platforms can be made available through the EIDS enabled ecosystem as it is depicted in Fig. 3.

1. An analytics service or platform can be connected to the EIDS by using its own connector. This type of connection fits better for Big Data platform integration with EIDS as these platforms have a lot of dependencies and IPR restrictions so it is difficult to be packaged and deployed as apps in order to be available for downloading and execution in an IDS connector.

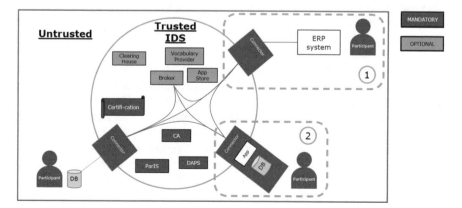

Fig. 3 Connection of EIDS ecosystem with Boost 4.0 Big Data analytics apps/platforms

2. Analytics method/service/app can be deployed by using microservice concepts (and Docker containers) and be compiled internally into a connector. This solution fits better for stand-alone applications and algorithms as it is easier to be compiled and executed in a connector.

5.2 Example of Integration of an Open-Source-Based Connector Within a Big Data Analytics Platform Connected: The CERTH Cognitive Analytics Platform

5.2.1 Manufacturing Scenario

The fast growth of data creation and gathering from a wide range of sources in factories and supply chains led to significant challenges in data analytics solutions, data collection, and handling. Big Data availability boosts predictive maintenance solution for smart factories. However, the data availability itself cannot create a self-learning factory that is able to continuously learn and act. This Big Data platform example presents a solution that aims to solve both problems of self-learning and trusted data connection and transfer.

5.2.2 Digital Business Process

The Cognitive Analytics platform for Industry 4.0 is one of the EIDS Big Data applications of Boost 4.0. The Cognitive Analytics platform exploits the Big Data of factories and feeds them to machine learning algorithms aiming to enhance industrial automation through real-time fault diagnosis and production planning optimization. It is an end-to-end user platform supporting predictive maintenance functionalities, and it is delivered as a Big Data application compatible with the EIDS ecosystem of Boost 4.0 as it is able to provide and consume data by using the IDS trusted connector. Figure 4 depicts the functionalities and achievements of the Cognitive Analytics platform.

Two core functionalities are provided within the platform: anomaly detection for predictive maintenance and production scheduling optimization. The anomaly detection feature of the platform receives multivariate time series data coming from machines. Historical data are stored in the platform's database and live streaming data are coming through the IDS connector. The platform provides online training of historical data, live monitoring of real-time data, and automatic cognitive retraining which keeps the system's performance at high levels. The EIDS application of production scheduling optimization is demonstrated on a real-case scenario of a fabric production's weaving process. The main goals of the scheduling optimization of the weaving process are finishing orders by their deadline, cost-efficient production, and prioritization of orders. Based on advanced visualization

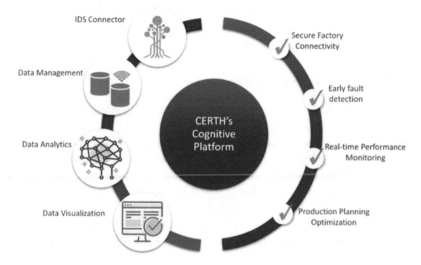

Fig. 4 The functionalities of the Cognitive Analytics platform in conjunction with its high-level achievements

techniques, a monitoring user interface was developed for this application showing the status and calculating the arrangement of orders to machines in real time.

5.2.3 EIDS Integration Approach

The open-source-based IDS connector accomplishes the integration of real-time machine data with the Cognitive Analytics. Secure Big Data exchange is achieved with the configuration of IDS trusted connector ecosystem. The system encompasses two IDS trusted connectors. The first connector is deployed on the factory site. The IDS connector receives the machines' data through an MQTT broker which is connected with the factory's cloud infrastructure. The factory cloud repository is the data provider of IDS architecture. The second IDS connector is placed in the data consumer site, specifically the Cognitive Analytics framework. A MQTT broker is also used to enable data exchange with the data provider. The architecture accomplishes secure and trusted communication between all system components, especially in the case of sensitive and private data of factories.

5.3   Example of Integration of the Open-Source-Based Connector with Manufacturing Supermarkets 4.0: Volkswagen Autoeuropa

5.3.1   Manufacturing Scenarios

Data collection, harmonization, and interoperability are difficult to achieve in all datasets, but in Big Datasets it is even worse, due to the sheer volume of the data to be transformed. Volkswagen Autoeuropa is an automotive manufacturing industrial plant located in Portugal (Palmela) since 1995 and part of Volkswagen Group. Currently, the logistics process is heavily reliable on manual processes and each step of the process creates data that is stored in a silo-based approach. In order to develop a true Digital Twin of the logistics process, there is the need to integrate all data silos and to make them interoperable between themselves and with other, external data sources, so as to enable real-world data provision to the Digital Twin.

5.3.2   Digital Business Process

The solution for this issue is based on a dedicated IDS-supported Big Data application that enables data harmonization and integration across the different data silos, by collecting and harmonizing the data from each silo into a common database system. The proposed Big Data app addresses the following technical requirements: (1) able to deal with raw data in many formats and sizes; (2) assure data quality; (3) efficient Big Data transformation and storage; (4) being able to address interoperability at the data level, enabling the development of additional added value services for users; (5) inclusion of custom schemas, in order to transform and harmonize data into standardized or proprietary schemas; and (6) a robust and efficient distributed storage system that is scalable in order to process data from data sources.

5.3.3   EIDS Integration Approach

The main open-source tools used for developing the proposed architecture were (1) Apache Spark, used for large-scale data processing, which includes the tasks of data cleaning and transformation; (2) Redis, as a NoSQL in-memory approach, for storing and managing raw data; and (3) PostgreSQL as the final database system that stores harmonized data (PostgreSQL could be replaced by any other database system). An IDS trusted connector is used to access raw, unharmonized data within the IDS ecosystem, and another IDS connector is used to publish harmonized data back to the IDS ecosystem. The fact that the IDS ecosystem is compliant with the containerization approach adopted for the Big Data app, in this case using

Docker Swarm orchestration, is a big advantage in terms of integration with the IDS ecosystem and with the IDS App Store.

5.4 Example of Integration of the Domain-Specific Ontologies with Predictive Maintenance Processes: OTIS

5.4.1 Manufacturing Scenarios

OTIS manufacturing system for elevator panel production line generates multiple data silos coming from MES, machines, and ERP systems. Dispersed information is difficult to integrate, thus rendering process to run in local optima. Aggregated data will have significant impact on overall manufacturing process improvements using data integration, analytics, and modeling. Inside the process team looked at optimization of the following adverse effects observed in the process:

- Maintenance cost reduction—due to various mechanical breakdowns or incorrect maintenance during production process. This results in production stop and higher machine maintenance cost to resume production processes.
- OEE (Overall Equipment Effectiveness)—increase equipment operation time vs. maintenance time.
- Discover hidden causes of production stops—combining distributed data silos and performing dataset mining.

5.4.2 Digital Business Process

Envisioned and realized data aggregation and integration using Boost 4.0 technologies enables production process to run more optimally. The solution consists of two parts: (1) Bayesian causal model that describes details of manufacturing processes derived from data mining and analysis and (2) information aggregation and sharing with supply chain via FIWARE IDS connector to enable global-level production optimization.

5.4.3 EIDS Integration Approach

The replication pilot final version is using two technologies to aggregate data coming from production sensors. On local level MQTT broker is used to aggregate data from machine sensors and systems. On Boost 4.0 level pilot used FIWARE IDS connector that integrates with MQTT broker via plugin to share production information with Boost 4.0 consortium. The usage of domain-specific vocabularies is extremely important in this example, as it deals with diverse sources, with different machine providers and sensors of all kinds.

5.5 Example of Integration of EIDS-Based Infrastructure with Warehouse Management Processes—Gestamp

5.5.1 Manufacturing Scenarios

Gestamp's participation in Industry 4.0 initiatives aims to create more efficient, consistent, and reliable manufacturing plants by adding intelligence to the production processes and getting the right information to the right people. These efforts are often hindered by the unavailability of precise, fine-grained, raw data about these production processes. Moreover, the skills required to apply advanced predictive analytics on any available data might not be present inside the company, which entails the need to securely share the data with expert third parties.

5.5.2 Digital Business Process

These issues have been addressed by providing a state-of-the-art indoor real-time locating system (RTLS). The system deploys IoT sensors on the plant shop floors to gather raw data directly from key mobile assets involved in the manufacturing processes, such as coils, dies, semi-finished and finished products, containers, fork-lifts, and cranes. Seamlessly integrating these IoT sensors with the IDS ecosystem, the solution grants access to the gathered data from outside of Gestamp's in-house network only through secure IDS connectors, thus facilitating data sharing while simultaneously adhering to Gestamp's strict security restrictions.

This previously untapped data can then be exploited to increase the logistical efficiency of the production processes by:

- Improving operational effectiveness.
- Increasing flexibility of production.
- Allowing more dynamic allocation of resources.
- Reducing changeover time.
- Refining warehouse and storage area management.

5.5.3 EIDS Integration Approach

The RTLS consists of ultra-wideband (UWB) tags and anchors, radio frequency identification (RFID) tags, and the proprietary *i2Tracking* stack. Other technologies used in the system include MQTT message brokers, Python-based data processors, and MongoDB and SQL databases. The design and technologies of the system and the integration in the EIDS infrastructure guarantee that the collected real-time data is made available in the IDS space with minimal overhead, which allows for more robust and precise analysis of the data. The solution also includes sample IDS consumer applications that show how the provided data can be ingested and utilized by third-party consumers.

5.6 Example of Integration of EIDS-Based Infrastructure with Logistics Processes: ASTI

5.6.1 Manufacturing Scenarios

Due to the special features of 5G networks such as high availability, ultra-low latency, and high bandwidth, Industry 4.0 proposes the use of this technology to support the intra-factory communications in replacement of the current communication practices mainly based on WLAN (IEEE 802.11 family). 5G networks, in addition to improved transmission capabilities, include the allocation of the computational resources closer to the factories for reducing latencies and response times.

Furthermore, the use of Artificial Intelligence and machine and deep learning techniques is substantially boosting the possibilities for prediction of complex events that help to take smart decisions to improve the industrial and logistic processes.

5.6.2 Digital Business Process

In this context, an interesting use case that combines Industry 4.0, 5G networks, an IDS trusted connector, and Artificial Intelligence and deep learning techniques is proposed. By this combination it is possible to predict the malfunctioning of an automated guided vehicle (AGV) connected through 5G access with its PLC controller deployed and virtualized in a multi-access edge computing (MEC) infrastructure, by exclusively using network traffic information and without needing to deploy any meter in the end-user equipment (AGV and PLC controller).

5.6.3 EIDS Integration Approach

Intensive experiments with a 5G real network and an industrial AGV in the 5TONIC [20] environment validate and prove the effectiveness of this solution. By using deep neural networks, and only analyzing the network parameters of the communication between the AGV and the PLC controller, several time series are built based on 1-D convolutional neural network (CNN) models that are able to predict in real time that the AGV is going to lose its trajectory 15 s ahead, which allows taking preemptive actuations. An IDS trusted connector acts as a bridge to transmit the CNN prediction outside the MEC infrastructure to an external dashboard based on the Elasticsearch-Logstash-Kibana (ELK) stack.

6 Certification as Base for Trust in the EIDS

Data security and data sovereignty are the fundamental value propositions of the EIDS. Any organization or individual seeking permission to access the EIDS must certify the core components, like connectors, to securely exchange data with any other party which is part of the data space.

The EIDS components are based on the International Data Space Reference Architecture Model V3.0, which also defines a certification criteria catalogue. Both Data Spaces, IDS and EIDS, are referring to the same criteria catalogues. The catalogue is split into three thematic sections, i.e., IDS-specific requirements, functional requirements that are taken from ISA/IEC 62443-4-2, and best practice requirements for secure software development.

The EIDS core components must provide the required functionality and an appropriate level of security. Therefore, the IDS certification scheme defines three security profiles for the core components:

- **Base Security Profile**: includes basic security requirements: limited isolation of software components, secure communication including encryption and integrity protection, mutual authentication between components, as well as basic access control and logging. However, neither the protection of security-related data (key material, certificates) nor trust verification is required. Persistent data is not encrypted and integrity protection for containers is not provided. This security profile is therefore meant for communication inside of a single security domain.
- **Trust Security Profile**: includes strict isolation of software components (apps/services), secure storage of cryptographic keys in an isolated environment, secure communication including encryption, authentication and integrity protection, access and resource control, usage control, and trusted update mechanisms. All data stored on persistent media or transmitted via networks must be encrypted.
- **Trust + Security Profile**: requires hardware-based trust anchors (in the form of a TPM or a hardware-backed isolation environment) and supports remote integrity verification (i.e., remote attestation). All key material is stored in dedicated hardware-isolated areas.

Within the Boost 4.0 project, the Spanish company SQS has developed the following infrastructures to test IDS and EIDS components.

6.1 IDS Evaluation Facility

SQS has defined a test laboratory (Lab. Q-IDSA) that integrates infrastructures already available in SQS quality as a service (QaaS) offer with new developments and processes required to validate IDS components. It will be accredited by IDSA and has also the scope of being ISO17025 accredited, which will make the lab a test center.

To carry on the certification process, SQS has defined a set of activities that imply technical assessment, where functional, interoperability, and security testing is performed, and documentation and processes are reviewed. As such, the adequacy and completeness of the installation and operational management documents are judged, and the adequacy of the security procedures that the developer uses during the development and maintenance of the system is determined.

A detailed description of the evaluation process of IDS-based components can be found in the position paper "IDS Certification Explained" [21].

6.2 Integration Camp

SQS has built an architecture with real IDSA components with the goal of having a full IDSA environment. The architecture was first built with minimum components needed to test the interoperability of connectors and base of IDSA environment, and it is in constant evolution, including more components (i.e., DAPS, Broker), building an architecture where every IDS component can be tested.

This architecture is opened for everyone, in a monthly event (Fig. 5), as a remotely accessible infrastructure where participants can test if their components are ready to work in a real IDS environment. The evaluation facility is the ideal place for those who want to prepare their IDS connectors and other IDS components for certification.

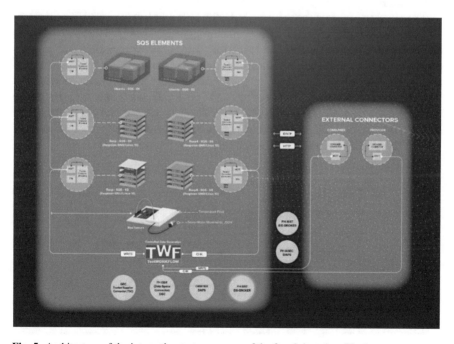

Fig. 5 Architecture of the integration test camp—as of the fourth iteration (Nov)

7 Data Space for the Future of Data Economy

This chapter gave a high-level overview on how Data Spaces are designed in the context of the IDS Reference Architecture and standards and on how the Boost 4.0 project used it in order to come up with its very own embryonic data space for the manufacturing domain—the European Industrial Data Space.

The fundamental design principle for Data Spaces must fulfill three core functionalities, all of which were encountered in the IDS-based EIDS ecosystem:

1. **Interoperability.**
2. **Trust.**
3. **Governance.**

Figure 6 puts these three pillars into context with generic reference models, domain-specific aspects, and business case-specific enablers. This figure shows all topics which Boost 4.0 dealt with, like domain-specific standards and formats, metadata, exchange of data, identification and authentication, authorization, certification and monitoring, and governance. Furthermore, it shows the topics that are still to be solved yet and must be encountered in future projects. Among those topics, like authorization and usage control, legal and operational agreements will play a significant role.

The creation of a data space for manufacturing is a milestone which is of tremendous importance for the future of data economy, since it brings data sovereignty to those who own the data treasure and therefore helps to break data silos and leverage its value. The fact that the manufacturing domain is one of four core domains that have been declared as the focus areas by the European Commission in the Open DEI [22] project (besides health, energy, and agri-food) shows how meaningful the development and exploration of the EIDS is. Besides, there are further first projects and initiatives that are aiming at the goal to come up with Data Spaces, either for a specific domain or with a broader scope:

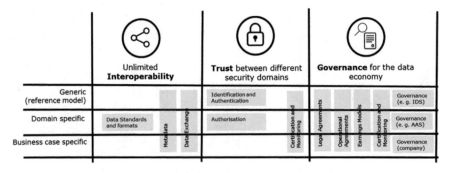

Fig. 6 Three pillars for Data Spaces and their enablers

- Gaia-X [23] that is a federated data infrastructure for Europe based on IDS in the field of data sovereignty.
- The IDS Launch Coalition [24] that focuses on creating an offering of.
  IDS-based products for the market
- The German "Datenraum Mobilität" [25] (eng. Mobility Data Space), also based on IDS and focusing on the mobility domain.

A growing amount of use cases that do not only connect two data endpoints but rather show the potential of Data Spaces, like the EIDS, by supporting many-to-many connections is on its way to change our perception of how data will be treated and traded in the future.

Acknowledgments This research work has been performed in the framework of the Boost 4.0 Big Data lighthouse initiative, a project that has received funding from the European Union's Horizon 2020 research and innovation program under grant agreement no. 780732. This data-driven digital transformation research is also endorsed by the Digital Factory Alliance (DFA) (www.digitalfactoryalliance.eu).

References

1. Boost 4.0. https://boost40.eu
2. Boost 4.0. *Public Deliverables*. http://boost40.eu/deliverables/
3. Curry, E., Metzger, A., Berre, A. J., Monzón, A., & Boggio-Marzet, A. (2021). A reference model for big data technologies. In E. Curry, A. Metzger, S. Zillner, J.-C. Pazzaglia, & A. García Robles (Eds.), *The elements of big data value* (pp. 127–151). Springer International Publishing.
4. International Data Spaces Association. (2019). *Reference architecture model*. Available at: https://www.internationaldataspaces.org/wp-content/uploads/2019/03/IDS-ReferenceArchitecture-Model-3.0.pdf
5. International Data Spaces Association. (2020). *IDS is officially a standard: DIN SPEC 27070 is published*, February 21 2020. https://internationaldataspaces.org/ids-is-officially-a-standard-din-spec-27070-is-published/
6. *Cirrus Documentation*. https://cirrus.readthedocs.io/en/master/index.html
7. Hyperledger Fabric CA Documentation. https://hyperledger-fabric-ca.readthedocs.io/en/release-1.4/
8. IDSA Tech Talk on 3 September 2020; cf. https://www.internationaldataspaces.org/recordedidsa-live-sessions/#techtalks
9. Bader et al. (2020). The international data spaces information model – an ontology for sovereign exchange of digital content. *International Semantic Web Conference*. Springer.
10. Available at https://w3id.org/idsa/core, source code maintained under the Apache 2.0 license at https://github.com/International-Data-Spaces-Association/InformationModel/
11. *Java implementation*. Available at https://maven.iais.fraunhofer.de/artifactory/eis-ids-public/de/fraunhofer/iais/eis/ids/infomodel/
12. *W3C Data Catalog Vocabulary DCAT*: https://www.w3.org/TR/vocab-dcat-2/
13. *W3C Open Digital Rights Language ODRL*: https://www.w3.org/TR/odrl-model/
14. *W3C Sensor, Observation, Sample and Actuator/Semantic Sensor Network (SOSA/SSN) ontology*: https://www.w3.org/TR/vocab-ssn/
15. *W3C Vocabulary of Interlinked Datasets VoID*: https://www.w3.org/TR/void/
16. *W3C Data Cube Vocabulary*: https://www.w3.org/TR/vocab-data-cube/

17. Lázaro, O., Alonso, J., Figueiras, P., Costa, R., Graça, D., Garcia, G., et al. (2022). Big data driven industry 4.0 service engineering large scale trials: The Boost 4.0 experience in technologies and applications for big data value. In E. Curry, S. Auer, A. J. Berre, A. Metzger, M. S. Perez, & S. Zillner (Eds.), *Technologies and applications for big data value* (pp. 373–397). Springer.

18. Lázaro, O., Alonso, J., Holom, R. M., Rafetseder, K., Kritzinger, S., Ubis, F., et al. (2022). Model based engineering and semantic interoperability for trusted digital twins big data connection across the product lifecycle in technologies and applications for big data value. In E. Curry, S. Auer, A. J. Berre, A. Metzger, M. S. Perez, & S. Zillner (Eds.), *Technologies and applications for big data value* (pp. 399–429). Springer.

19. Lázaro, O., Alonso, J., Ohlsson, P., Tijsma, B., Lekse, D., Volckaert, B., et al. (2022). Next-generation big data-driven factory 4.0 operations and optimization: The Boost 4.0 experience in technologies and applications for big data value. In E. Curry, S. Auer, A. J. Berre, A. Metzger, M. S. Perez, & S. Zillner (Eds.), *Technologies and applications for big data value* (pp. 345–371). Springer.

20. *5Tonic, an Open Research and Innovation Laboratory Focusing on 5G Technologies*: https://www.5tonic.org/

21. *IDSA position paper Certification Explained v 1.0*: https://www.internationaldataspaces.org/wp-content/uploads/2020/09/IDSA-Position-Paper-IDS-CertificationExplained.pdf

22. *OPEN DEI*: https://www.opendei.eu/

23. *Gaia-X*: https://www.data-infrastructure.eu/

24. *IDSA Launching Coalition*: https://www.internationaldataspaces.org/the-launching-coalition-start-of-a-new-journey-on-sovereign-data-sharing/

25. Delhaes, D. (2020-10-28). *Merkel drängt Autokonzerne: BMW, Daimler und VW sollen Daten-schatz teilen [Merkel urges car companies: BMW, Daimler and VW should share a wealth of data]*. Available at: https://www.handelsblatt.com/politik/deutschland/autogipfelmerkel-draengt-autokonzerne-bmw-daimler-und-vw-sollen-datenschatz-teilen/26308418.html

Deploying a Scalable Big Data Platform to Enable a Food Safety Data Space

Mihalis Papakonstantinou, Manos Karvounis, Giannis Stoitsis, and Nikos Manouselis

Abstract The main goal of this chapter is to share the technical details and best practices for setting up a scalable Big Data platform that addresses the data challenges of the food industry. The amount of data that is generated in our food supply chain is rapidly increasing. The data is published by hundreds of organizations on a daily basis, in many different languages and formats making its aggregation, processing, and exchange a challenge. The efficient linking and mining of the global food data can enable the generation of insights and predictions that can help food safety experts to make critical decisions. All the food companies as well as national authorities and agencies may highly benefit from the data services of such a data platform. The chapter focuses on the architecture and software stack that was used to set up a data platform for a specific business use case. We describe how the platform was designed following data and technology standards to ensure the interoperability between systems and the interconnection of data. We share best practices on the deployment of data platforms such as identification of records, orchestrating pipelines, automating the aggregation workflow, and monitoring of a Big Data platform. The platform was developed in the context of the H2020 BigDataGrapes project, was awarded by communities such as Elasticsearch, and is further developed in H2020 The Food Safety Market project in order to enable the setup of a data space for the food safety sector.

Keywords Big Data · Artificial Intelligence · Food safety · Data platforms · Food recall prevention · Food risk prediction · Data space

M. Papakonstantinou · M. Karvounis · G. Stoitsis (✉) · N. Manouselis
Agroknow, Maroussi, Greece
e-mail: Mihalis.papakonstadinou@agroknow.com; manos.karvounis@agroknow.com; stoitsis@agroknow.com; nikosm@agroknow.com

E. Curry et al. (eds.), *Data Spaces*, https://doi.org/10.1007/978-3-030-98636-0_11

227

1 Introduction

The food system is rapidly changing, becoming more and more digitized. Data is being generated in all entities in the food supply chain. To better understand emerging risks and protect the consumers, it is essential to be able to combine, process, and extract meaning from as much data as possible.

Food safety and certification requirements throughout the supply chain have become stricter, and authorities are continuously in activity to develop and promote more selective methods of monitoring and controlling our food. That has led to a major increase in public data from official sources of food incidents across many countries that focus on a broad range of products and ingredients. A food incident can relate to an issue that could or is expected to impact multiple government jurisdictions. A food recall is an action taken by a food business to remove unsafe food from distribution, sale, and consumption. All food businesses must be able to quickly remove food from the marketplace to protect public health and safety.

When a food safety incident occurs somewhere in a global supply chain, simply gaining access to that top-level information is not enough to truly understand the repercussions. What about the product brand that was recalled? Which was the company behind the recall? And if you dive even deeper you also need to make sure you understand the company's involvement in the incident. Was it the manufacturer or the packer? Or was it possibly the importer or the trader? In other words, accessing comprehensive, reliable data in real time—and being able to properly analyze and harness that data—is critical for ensuring food safety in the increasingly complex, dynamic, and international supply chains.

Further to the food safety issues, more challenges could concern the monitoring of food fraud incidents around the world in almost real-time, large-scale data analysis in order to reveal patterns; predictions on whether someone in the supply chain has substituted, misbranded, counterfeited, stolen, or enhanced food in an unapproved way; and finally detection of increased probability of food fraud. We could further analyze all these challenges and find ourselves with even more, such as the identification of correlations between fraud incidents with price changes, the probability of fraud increase for suppliers based in countries with high corruption scores, and how the weather phenomenon is linked to an increase of food adulteration incidents in an area, amidst many more. To face the above challenges, Big Data platform that collects and processes many different data types is necessary in order to assess and predict risks [1].

During the last 15 years, very important fraud issues like the "2013 horse meat scandal" [2] and the "2008 Chinese milk scandal" (Wen et al 2016) have greatly affected the food industry and public health. One of the alternatives for this issue consists of increasing production, but to accomplish this, it is necessary that innovative options be applied to enhance the safety of the food supply chain [3]. For this reason, it is quite important to have the right infrastructure in order to manage data of the food safety sector and provide useful analytics to food safety experts.

There are several systems that are collecting and processing food safety information in order to support decision making in the food industry. This includes systems operated by public organizations like Rapid Alert System for Food and Feed [4] and commercial systems like HorizonScan (FERA [5]), gComply (Decernis [6]), and DigiComply (SGS [7]). These systems are based on a data collection and processing platform and are currently working on a production setup, serving thousands of users and providing access to hundreds of thousands of records. However, each of these systems is focusing on one specific data type and they are not combining and linking different data types.

In this chapter we present the design principles and the deployment details of a Big Data platform for the food safety sector that combines several different data types and can help food safety experts to make data-informed decisions. The data platform is designed to handle voluminous data that cannot be analyzed using the traditional data management techniques and warehousing. The proposed methodology and architecture was developed with an aim of increasing the scalability, availability, and performance of data platforms that need to handle processes such as data cleaning, data transformation, data unification, data enrichment, and data intelligence. The novelty aspects of our work consist of (a) the introduction of an architecture and a methodology that allows the processing and linking of highly heterogeneous data at a large scale and (b) an operational version of the first Big Data platform that links millions of food safety data records, and it provides food safety analytics and predictions that prove the scalability of the proposed approach. The methodology for data ingestion and processing and the architecture of the platform can be applied in other sectors in which processing of incidents announced by open data sources is very critical, such as pet food, cosmetics, medical devices, and pharmaceuticals. The chapter relates to the technical priorities: 3.2 Data Processing Architectures and 3.3 Data Analytics of the European Big Data Value Strategic Research and Innovation Agenda [8]. It addresses the horizontal concern of data processing architectures of the BDV Technical Reference Model and the vertical concerns of Big Data types and semantics. In addition, this work aims to maximize the contribution with the future European activity in AI and data by focusing on how a vertical organization like the food safety sector can be transformed, creating new opportunities [9].

2   Big Data Platform Architecture

The development of the Big Data platform for the food safety sector focused at targeting the needs of the food safety industry using Big Data processing, text mining, semantics, and Artificial Intelligence technologies. The platform is responsible for collecting, processing, indexing, and publishing heterogeneous food and agriculture data from a large variety of data sources [10]. The platform was designed using microservice architecture [11], with different technology components handling different aspects of the data lifecycle. All of the components are interconnected using well-defined connectors and application programming interface (API) endpoints, each responsible for storing and processing different types of data (Fig. 1).

More specifically, the platform includes the following components:

- The **data sources ingestion** component, which connects to numerous data sources, extracts the data, and detects the changed data.
- The **data transformation** component, which performs data transformation to an appropriate format designed for performance optimization.
- The **storage** components, which feature various storage engine technologies that are used in numerous places throughout our architecture and are responsible for the physical archiving of data collections.
- The d**ata enrichment component** implemented using **machine learning (ML) and natural language processing (NLP)** technologies, which is responsible for hosting individual text mining, machine learning, and data correlation scripts that can be used in a variety of contexts as standalone pieces of code or as web services through the so-called intelligence APIs.
- The **data processing** components, which include a machine-readable interfaces (APIs) to the different types of data collected in the platform that is used in numerous places throughout our architecture. This part of the architecture is responsible for making data discoverable, but also for submitting new data assets back to the platform.
- The **monitoring** component, which is responsible for monitoring the data platform's smooth data digestion and performance.
- The **orchestration** component, which is responsible for the overall orchestration of all the processes that run in the Big Data platform.

3   Data Modeling

To address the dynamic requirements, the data model was designed in a way that allows the addition of new data types and new properties. To that direction we used a Functional Requirements for Bibliographic Records (FRBR)-inspired logic of metadata organization, so that there are basic and common metadata attributes describing each data asset, but also space for more customized attributes per data type (International Federation of Library Associations and Institutions 1998). The main goal was the adoption of a flexible and smart data model that is able to address dynamic requirements and support many different data types.

More specifically, the two advantages of such smart data model in a data platform are that the platform (a) can accommodate many different data types and entities and (b) can support different instances of the same data object that may be published by different data sources and in different formats which is very important when you need to deal with information duplication. The introduction of a common FRBR-based metadata schema to the platform made the process of integrating new data types much easier, in a way that also conveys a basic thematic and temporal view on

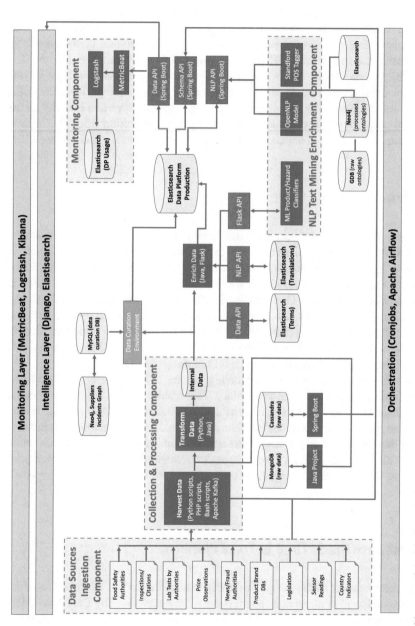

Fig. 1 Architecture of the Big Data platform

the underlying data assets. This data model was implemented using JSON, the open standard file format and data interchange format.

Each record in the data platform is structured according to a FRBR hierarchy (i.e., work, expression, manifestation, and item) where each FRBR level is described using a limited set of structural metadata elements complemented by faceted descriptions that can vary depending on the type of resource being described, its context, and the relevant FRBR level (example in the mind map below). The hierarchical data model that we developed includes a common part for all the different entities and data types with generic properties such as identifier, title, description, and date. The second part is specific to the data type and includes all the required properties that will enable data enrichment and linking (Fig. 2). The data model supports the use of standard and semantic vocabularies for several properties.

4 Data Standards Used

To enable data interoperability and data linking, we have used in our data model standard and well-adopted semantic vocabularies for several properties such as location, ingredients, materials, hazards, and suppliers. More specifically for the products, ingredients, and materials, we have used the FoodEx2 ontology [12] and the Food and Agricultural Organization (FAO) Codex commodity categories [13]. For the case of hazards we used the classification that is adopted by the European Food Safety Authority in systems like the Rapid Alert System for Food and Feed. In addition to that, we have foreseen the integration of the Global Standards 1 (GS1) for suppliers, product brands, and traceability information [14].

Data linking is achieved by annotating all the different data types with the terms from the same semantic vocabulary. This means that the recalls for specific ingredients such as cocoa are automatically linked to lab test results, border rejections, price data, and trade data. The use of hierarchical semantic vocabularies allows the automated expansion of the queries and analytics to parent and child terms. This approach enables the development of interactive charts for the analytics that allow the drill-down from generic categories of products and hazards to specific categories and instances.

5 Software Stack Used in Data Platform

An important aspect of the design of the Big Data platform was the selection of the software tools and systems that can meet the requirements for Big Data processing. In this section we present the software stack used in the food safety data platform and we analyze the rationale for the selection of each component.

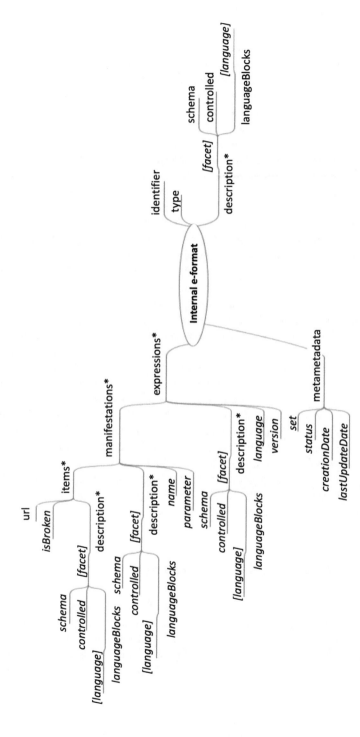

Fig. 2 The smart data model mind map for the internal data schema

5.1 Data Ingestion Components

Data ingestion is the first step in the Big Data platform for building data pipelines and also one of the toughest tasks in Big Data processing. Big Data ingestion involves connecting to several data sources, extracting the data, and detecting the changed data. It's about moving data from where it originated into a system where it can be stored, processed, and analyzed. Furthermore, these several sources exist in different formats such as images, OLTP data from RDBMS, CSV and JSON files, etc. Therefore, a common challenge faced at this first phase is to ingest data at a reasonable speed and further process it efficiently so that data can be properly analyzed to improve business decisions.

The data ingestion layer includes a set of crawling and scraping scripts for collecting information from the data sources. For each data source a different script was developed. These scripts vary in form. We utilize Scrapy, Python scripts, a custom Java project, as well as bash scripts to help in the collection of the tracked data sources. Regardless of their type, these collections of scripts take as input a starting URL and some rules and store the matching documents in the file system as output. Cron jobs are used to check every few mins whether new notifications have been published on the web site of the agency/authority. For storing the fetched data, a NoSQL database, namely, MongoDB, is used.

Apache Kafka is used to collect data streams such as environmental data and weather data from sensors. Apache Kafka is a messaging framework that is distributed in nature and runs as a cluster in multiple servers across multiple datacenters. Moreover, Kafka allows the real-time subscription and data publishing of large numbers of systems or applications. This allows streamlined development and continuous integration facilitating the development of applications that handle either batch or stream data. An important factor in data ingestion technology, especially when handling data streams, is the fault tolerance capability of the chosen technology. Kafka ensures the minimization of data loss through the implementation of the leader/follower concurrency architectural pattern. This approach allows a Kafka cluster to provide advanced fault-tolerant capability, which is a mandatory requirement for streaming data applications.

5.2 Collection and Processing Components

In the Big Data platform, each data source, depending on its format, is collected in a different way and might have a different form as mentioned in the above ingestion component. The raw data is harvested and then transformed using Python and Java code into the internal data model (schema) that was analyzed in Sect. 3. The main goal is for all the collected different types of data to have the same data structure.

5.3   Storage Components

The storage layer deals with the long-term storage and management of data handled by the platform. Its purpose is to consistently and reliably make the data available to the processing layer. The layer incorporates schemaless persistence technologies that do not pose processing overheads either when storing the data or retrieving them. Therefore, the storing and retrieving complexity is minimized. The software components that were used for the storage of the Big Data are analyzed below.

MongoDB is a distributed database which treats and stores data as JSON documents. Thus, data can have different fields and the data structure is essentially alive since it can be changed over time. Also, MongoDB provides ad hoc queries, supporting field query, range query, and regular expression searches. Moreover, MongoDB has fault-tolerant and load balancing capabilities by providing replication and sharing of the main database. In the data platform it is used to store the data fetched by the crawlers of the data sources.

Elasticsearch is a distributed database, providing a full-text search engine based on Lucene. The distributed nature of Elasticsearch allows near real-time search in all kinds of documents. The indices of Elasticsearch can be divided into shards, hence supporting automatic rebalancing and routing. Moreover, the indices can be replicated to support efficient fault tolerance. Furthermore, Elasticsearch encapsulates out-of-the-box methods for establishing connections with messaging systems like Kafka, which makes integration easier and allows the faster development of real-time applications. In our case Elasticsearch is used in many places throughout the stack. More specifically, we use it for text mining purposes by taking advantage of its analyzer capabilities, for the storage and aggregation of all the production-ready data, and for storing application performance metrics.

MySQL is a relational database management system (RDBMS), which provides a robust implementation of the SQL standard. The data platform integrated software stack also provides the phpMyAdmin user interface to monitor and query the MySQL RDBMS through a web user interface. In the data platform the MySQL database is used in the data curation environment to manage and manually enrich the records by food safety domain experts.

GraphDB is an RDF triplestore compliant with the core Semantic Web W3C specifications (RDF, RDFS, OWL). It acts as a SAIL over the RDF4J framework, thus providing functionalities for all critical semantic graph operations (storing, indexing, reasoning, querying, etc.). The query language used is the implementation of the SPARQL 1.1 specifications, while connectors with Elasticsearch and Lucene are incorporated in the system. In the Big Data platform it was used for storing the geonames ontology, which is queried through an endpoint of the internal API used for the enrichment to identify countries based on country, city, or region names, in the languages supported by geonames.

Neo4j is a native graph storage framework, following the property graph model for representing and storing data, i.e., the representation model conceptualizes information as nodes, edges, or properties. Accessing and querying the underlying data is achieved via the usage of the open-sourced Cypher query language, originally developed exclusively for Neo4j. In the food safety Big Data platform, Neo4j was used for storing the processed semantic vocabularies for products and hazards.

Apache Cassandra is a NoSQL storage engine designed to handle large amounts of write requests. Being a NoSQL engine it can easily handle model updates. It is designed to be easily configurable and deployed in a multi-node, distributed manner. In the food safety data platform Apache Cassandra is used to store numerical data such as country indicators and sensor readings.

5.4   Data Processing Components

These components are used throughout our data platform in order for our data to travel and communicate with other components.

Flask is a micro web framework written in Python. It is classified as a microframework because it does not require particular tools or libraries. It has no database abstraction layer, form validation, or any other components where pre-existing third-party libraries provide common functions. However, Flask supports extensions that can add application features as if they were implemented in Flask itself. Extensions exist for object-relational mappers, form validation, upload handling, various open authentication technologies, and several common framework-related tools. Extensions are updated far more frequently than the core Flask program. As part of the data platform, the Flask framework is used as a wrapper access layer on top of the machine and deep learning models trained for the classification of food recalls.

Django is a high-level Python web framework that encourages rapid development and clean, pragmatic design. Built by experienced developers, it takes care of much of the hassle of web development, so you can focus on writing your app without needing to reinvent the wheel. It's free and open source. As part of the data platform, Django is used for the development of incident prediction and risk assessment APIs. On the one hand, food incidents data is used to train a deep learning prediction model to predict food incidents in the future. On the other hand, the main goal of the risk assessment module is to help food safety experts to identify the ingredients with unacceptable hazard risk.

5.5   Data Enrichment Components

Data enrichment in the Big Data platform for the food safety sector is achieved by applying machine curation processes and human curation processes.

In the case of machine curation, the data enrichment components are used to autotag the data with hazards, products, and country terms. An internal API endpoint, a custom Java component for the processing of the notifications, and a set of Python scripts are used for the enrichment of the collected data.

The internal API has endpoints used for the textual processing of the notifications and identification of possible product and hazard terms. Using these endpoints we can (a) check the existence of a specific term against product and hazard vocabularies; (b) search for possible terms into text using fuzzy search, stemming, and N-grams; (c) get a stemmed version of text, without stopwords or without specific types of words (e.g., verbs); and (d) identify products in given brand names, using machine learning techniques trying to predict the product term based on the information already stored and frequently updated by human curation system.

The Java component is used to process and enrich the notifications collected by data collection component. It uses the internal API along with other endpoints to try to annotate each notification with the hazard and product terms it may involve. This component operates in any combination of the following ways:

- Controlled, in which only specific fields of the notification are used as possible input to the enrichment methods
- Smart, in which specific words and phrases usually associated with hazard terms (e.g., contains, due to, contaminated with) are used to make focused enrichment
- Country conversion, in which country names are converted into their respective ISO codes and vice versa
- Product enrichment, in which the complete text of the notification is sent to external services (OpenCalais) and the brand name to the internal API for the annotation with possible product terms
- Translate, in which the chosen text possibly containing hazard and product terms is sent to an external service, to translate it into English so it can be used by the enrichment methods

The set of Python scripts is used in two ways:

- The classification of product brands with their respective products. This extracts the information present on the curation system along with the tagged vocabulary terms, and using a set of Python packages used for machine learning (scikit-learn, NLTK) generates a predictive model used by the internal API for the annotation of text.
- The automatic approval of machine annotated hazard and product terms from the textual processing of each notification. Employed by the curation system, this script takes into account the machine-generated term, along with the reason behind the annotation, and using a set of Python libraries (pyjarowinkler, Porter Stemmer), it calculates the string distance between the two and automatically approves or not the generated term.

Furthermore, enrichment is employed for the identification of the country of origin for a recall, news item, outbreak, etc. This is achieved using the geonames ontology, imported into a GraphDB database which is queried through an endpoint of the internal API used for the enrichment. This endpoint identifies countries based on country, city, or region names, in the languages supported by geonames.

A machine learning API has been created for the enrichment of data platform's entities using machine learning techniques. This API uses two different models based on the annotation that will be attempted:

- One is using the title and textual description of the recall and is trained to identify the hazard which caused the recall.
- The other one is also using the title and description of the recall and identifies the products.

For both of them the SGDClassifier was used, along with a TFIDF vectorizer. This API has endpoints for the generation of the model with a new train dataset and for the identification of hazard and product terms and is built using Flask framework for Python.

For the human curation part the Drupal 7.0 is used as the data curation system of the data platform. The curation system includes:

- Feed importers to import the information that is collected and enriched
- Data curation workflow with specific roles and access rights
- Drupal rules for enriching and publishing information
- Drupal exporter to publish the curated and reviewed information

5.6   Monitoring Components

This component provides insights and visibility into the health and status of data platform's data clusters by tracking specific metrics in real time and sending alerts or notifications when readings exceed or fall below the set thresholds. Data collected from monitoring our data clusters can be stored, analyzed, and displayed in business intelligence and analytics dashboards and reports. The following software tools were used to monitor the health of the platform.

- **Logstash** is an open-source data collection engine with real-time pipelining capabilities. Data flows through a Logstash pipeline in three stages: the input stage, the filter stage, and the output stage. Logstash can dynamically unify data from disparate sources and normalize the data into destinations of your choice. Clean and democratize all your data for diverse advanced downstream analytics and visualization use cases.
- **Kibana** is an open-source data visualization and exploration tool used for log and time-series analytics, application monitoring, and operational intelligence use cases. It offers powerful and easy-to-use features such as histograms, line graphs, pie charts, heat maps, and built-in geospatial support. Also, it provides tight integration with Elasticsearch, which makes Kibana the default choice for visualizing the data stored in Elasticsearch.
- Finally, we employ **Metricbeat**, i.e., the proposed tool for monitoring the scalability of Big Data platform, over the Elastic Stack to monitor and report our chosen metrics.

5.7   Intelligence Layer

The data platform includes an intelligence layer that is responsible for implementing the risk assessment and risk prediction algorithms using machine learning and deep learning methods [15].

Incident prediction module: Food incident data of the last 30 years is used to train a deep learning prediction model to predict food incidents in the future. The incidents' dataset becomes available through the data API of the data platform. To that direction, a request to the data API is sent with the product (ingredient or material) for which we want to predict the incidents of the next 12 months. We can build prediction models for specific hazards and specific regions by filtering the result set with the country of interest and the hazard.

For the implementation of the prediction models that are based on the historical food safety incidents, Prophet is used [16]. Prophet is a procedure for forecasting time-series data based on an additive model where nonlinear trends are fit with yearly, weekly, and daily seasonality, plus holiday effects. It works best with time series that have strong seasonal effects and several seasons of historical data. Prophet is robust to missing data and shifts in the trend and typically handles outliers well.

Risk assessment module: For the implementation of risk estimation, we used a mathematical model that is based on the frequency of the incidents and the severity of the identified hazards in the incidents. The risk is estimated for all the possible product (ingredient, raw material) hazard pairs. Considering that the data platform includes more than 12,000 products (ingredients and materials) and more than 4300 hazards, the risk estimation should be conducted for a large number of pairs.

The risk assessment module can be used in the following ways:

- A request to the risk API to calculate the risk for a specific ingredient, origin, and hazard. This option can be used every time that we need a risk estimation for a specific time point.
- A request to risk API to generate a batch of risk time series. This option can be used to estimate the risk for a specific time period, e.g., evolution of the risk during the last 10 years.

6   Operational Instance of the Data Platform

An operational version of the data platform for the food safety records was deployed using the analyzed architecture and Big Data and Artificial Intelligence technologies. The main source of the datasets in the Big Data platform is the open data published by the national authorities, international systems, and food safety portals. The data platform includes records of news, food recalls, border rejections, regulation, fraud cases, country indicators, price data, sensor data, supplier data, product brands, and inspections. More specifically, as of March 2021, the operational version of the Big Data platform includes the following data types:

- **Food Incidents**: 412,308 food recall warnings on a big variety of products with a growth rate of 6.3% per year. The data is announced on a daily basis by each food safety authority worldwide. The data has a variety of types (food recalls happening at market level, border rejections, information for attention) and with specific attributes following a different schema for each provider.
- **Food Inspections**: 227,603 inspections with a growth rate of 1.46% per year. The data is announced on a 2–5-month basis. The data has a variety of types with specific attributes following a different schema for each provider.
- **Lab Tests**: 102,187,114 laboratory analysis results that come from 34 national monitoring programs with a growth rate of 14.68% per year. The dataset is announced on a yearly basis by each food safety authority worldwide. The data are of a specific type with specific attributes and in an xls format following a different schema, depending on the provider.
- **Suppliers**: 702,578 different company suppliers with a growth rate of 1.46% per year. The data is announced on a daily basis.
- **Price**: 415,670 prices with a growth rate of 14.9% per year. The data is announced on a daily basis.
- **Country Indicator**: 38,061,648 country indicators with a growth rate of 5.14% per year. The data is announced on a yearly basis.
- **News:** 68,971 news records with a growth rate of 6.57% per year. The data is announced on a daily basis.
- **Maximum Residue Level Limit:** 141,594 records with a growth rate of 0.7% per year. The data is announced on a yearly basis.
- **Product Brand**: 28,081 brand records with a growth rate of 15.6% per year. The data is announced on a daily basis.
- **Sensor data:** 12,437,743 sensor readings from weather stations that are installed on the farms and temperature and humidity sensors that are installed on the processing units. The growth rate of the sensor data is 5.6% per year.

All of the above amount to 142,245,567 data records.

7 Identifying Records in the Big Data Platform

A very important aspect that you need to take into consideration when building a (Big) Data platform instance is on how to assign IDs to records in the data platform. Several practices may be applied to tackle this problem:

- Applying a hash function over crawled/scraped urls
- Some kind of internal identification process
- Attempting to identify/extract each source's unique identification method

7.1 Hash Function over Crawled Urls

This is a somewhat safe approach; urls are unique throughout the web so chances are a hash function on top can prove to be successful. It however does not come without any drawbacks. What if there are updates to the content crawled? It is not uncommon for urls of websites to be generated based on the title of the source. It is the piece of text containing the most important information on the generated content and the most SEO-friendly one. So what about updates to the titles? This can lead to updates to the url as well. So even though that is a rather straightforward choice, special care should be taken to such updates in order to avoid duplicates.

7.2 Internal Identification Process

This can be implemented either by deploying an API endpoint responsible for assigning an ID to each resource collected or a simple method/function/bash script.

The above suggested method has some very important pros, the most important of them being its blackbox way of working. Once it has been perfected, you no longer have to worry about duplicates in your platform or assigning the same ID to two different resources.

However, they have some cons as well. First and foremost, time should be spent perfecting such a mechanism. Due to the importance of ID assignment in data-related projects and platforms, one should definitely allow many hours (or story points) to such a project/task since it will be the backbone of pretty much everything you build. Another drawback we should point out is the rationale behind the identification process. Basing it uniquely on the collected content can lead to duplicates as described in the previous case.

7.3 Remote Source Identification

This approach is considered as the most challenging choice available. Although one may think of this as trivial if the data collected is in an xls or csv format where identification is rather straightforward, what if a content management system (CMS) is employed? Knowledge of it should be present if one wants to successfully assign a unique ID able to avoid duplicates. For instance, Drupal assigns a unique id to each piece of content (nid) always present in meta-tags and by default in CSS classes of article tags. However, if employed correctly one should never worry about their ID assignment or almost never. Care should be taken only when some major migration takes place on the remote source's side, a rather infrequent case.

In the Big Data platform we applied a hybrid approach. More specifically, all of the aforementioned approaches are utilized in a priority manner:

- First, we attempt to identify the ID given to the data record by the remote source.
- If this method fails, we employ our internal ID assignment that hashes a concatenation of the important and never-changing properties of the data type. For example, such a vector of properties for a food recall is the date, the reason behind the recall, and the product involved in it.
- Finally, if we are not able to safely extract these properties from the official source of information, we employ the hashing over the collected url.

Regardless of the technique utilized for our collected records, we also include our internal ID uniquely given to each of the sources we collect our data from.

8 Orchestrating the Big Data Platform

As already analyzed, the data platform collects, translates, and enriches global food safety data. A number of workflows are involved in the process. Tasks triggering one another, signifying the collection, processing, and enrichment of each of the close to 200M data points that are present in the infrastructure.

There is a very important challenge that we had to take into account in designing and implementing our Big Data platform stack. Initially we utilized cron jobs for these workflows. Every data source we track has its dedicated directory in the backend and processing servers, and within each of these directories a run.sh script is used. This is the script that manages all the action. Every single task in each workflow triggered is managed by such a script, calling other scripts created with the responsibility to handle each task. And this run.sh script is triggered by crontab.

Depending on the source, translation endpoint triggering scripts may be present. Text mining or text classification workflows may take place with their respective scripts. All initiate calls to the respective projects and endpoints.

The key points in the orchestrating process are:

1. We need to dive into the data and identify the rate at which new records are published in order to configure the extract, transform, load (ETL) workflows to be triggered only when chances of new data are present.
2. Only new data needs to be taken into account. In our stack, each of the data records collected comes with a collection timestamp and a flag signifying whether the record has been processed or not.
3. Implementing and deploying a workflow capable of executing regardless of the stress levels of a server is really challenging. A good choice at this point is splitting the workflow into atomic operations; this ensures that even though a task or a workflow may not be complete, no data loss will be observed since each new workflow triggered will always check for the previous workflows' leftovers.

In the Big Data platform, we are using Apache Airflow for the ETL pipelines. Apache Airflow is one of the best workflow management systems (WMS) that provides data engineers with a friendly platform to automate, monitor, and maintain their complex data pipelines.

Just to give a quick overview in terms of numbers, in the current infrastructure of the Big Data platform:

- **113 ETL** workflow cron jobs are present.
- On average workflows are triggered once every **10 min**.
- **9 dedicated servers** are involved in this part of the infrastructure.
- **11** workflow jobs have been switched to Apache Airflow DAGs.
- **1 Elastic Stack** instance (involving Elasticsearch, Kibana and Metricbeat) is employed to keep track of the health of our infrastructure.

In terms of performance, the approach described in this section has proven to be a very robust one. Since Apache Airflow is deployed as a system service in the dedicated server tasked with these workflows, monitoring its uptime is a very straightforward task. Furthermore, it provides an out-of-the-box logging mechanism to easily identify possible bottlenecks and CPU/memory-hog steps in each workflow. Finally, since internally it depends on the cron daemon of the underlying infrastructure (present in all *nix distributions), we can be certain that every workflow will be triggered at the time requested.

9   Monitoring the Big Data Platform

Using the Big Data software stack you may build a very robust data platform instance, capable of handling huge amounts of data, harmonizing, linking together, and enhancing in any ML/DL or any other possible way available. However, what if:

1. A node in your Elasticsearch instance stops working?
2. Your MongoDB deployment consumes too much memory?
3. The awesome model you have designed and implemented takes up too much CPU/GPU?
4. The Apache Airflow webserver and/or scheduler stops working?

In the data platform stack, we have employed two levels of uptime monitoring: one for the uptime of the components of the stack and another for their performance. We will further analyze both in the rest of this section. The integration of these monitoring approaches has led to a downtime of under 1 h over the past 5 years.

9.1   Ensure Uptime for Every Component Deployed to the Stack

As already presented earlier in the chapter, the Big Data platform infrastructure is a microservice one. This means that every single piece of component is wrapped up with a bunch of API endpoints, accessible through various ports in our servers. In our infrastructure, every newly deployed project, API service, and storage engine in the infrastructure come with a checkup script. This means that each and every time a new framework, tool, storage engine, and project are deployed, a script is also set up as a cron job.

This accompanying script is actually quite simple and is automatically generated with a command line tool we have implemented. All it does is that it uses *nmap* to check whether a port is open and accepting traffic or not. If it is not an automatic attempt to restart, the respective service is made and an email is sent with the last 50 lines of the respective service's logfile. To handle out-of-memory issues, the system's cache is also cleared. To ensure uptime of the stack, these scripts are added as cron jobs and are executed every 2 min.

9.2 Problems That Are Not Related to the Infrastructure

Many problems may come from the implementation and wrap-up code in a Big Data platform. One has to ensure that such cases are also monitored. Elastic has made available a tool to that end. APM is released by Elastic, fully maintained and easily integratable. In our data platform, we have connected APM to our dedicated Elasticsearch instance for monitoring purposes and the respective Kibana instance for visualization purposes. Using this Kibana instance we analyze and visualize the performance of each of the components of the stack and can go deeper into our analysis, identifying endpoints that take too long to respond and consuming too much CPU or memory, or even identifying hacking attempts leading to many UNAUTHORIZED HTTP responses and where they were made from.

10 Performance and Scalability of the Data Platform

In order to evaluate the performance and the scalability of the operational data platform, we performed a rigorous testing experimentation for three critical steps, namely, data ingestion, incident prediction, and risk assessment. We focused our experimentation on the performance of the Big Data platform stack by tracking system indicators such as (a) completion time, both on a step-by-step level and on the whole end-to-end data flow, (b) CPU (central processing unit) usage (we will track the CPU usage by each of the components as they are triggered by the flow), (c) memory usage of each of the components and technologies, and (d) network usage in terms of bytes (we employ this metric, since the whole stack is based on a microservice architecture).

More specifically, we performed the experimentation using 3 real scenarios for the dataset upload, 20 different use cases for recall prediction experiment, and risk assessment experiments to better showcase the scalability potential of the deployed

Table 1 Different data types processed by the Big Data platform and their growth rate

Data type	Number of records (as of March 2021)	Annual growth (%)
Food safety incidents	412,308	6.3
Inspections	227,603	1.46
Laboratory testing results	102,187,114	14.68
Suppliers	702,578	1.46
Prices	415,670	14.9
Country indicator	38,061,648	5.14
News items	68,971	6.57
Maximum residue level limit	141,594	0.7
Product brands	28,081	15.6
Sensor data	12,437,743	5.6

Big Data stack. Since all of the data platform has been deployed in a microservice architecture, our step-by-step and end-to-end experimentation was done over the API endpoints provided by the platform. To showcase the potential of the deployed stack in terms of scalability, we performed the experimentation using three usage scenarios for each step of the process by gradually increasing the requests made toward the platform. To have more accurate results in the case of the incident prediction, 20 different scenarios are tested due to the small data size. First, we identified and abstractly described the provided datasets of Table 1. Then we moved on with evaluating them against the Vs of Big Data, and we identified the data flow each dataset will follow in the stack, denoting the steps of this flow. Finally, using a Python script that simulated bursts of this data flow throughout the data platform stack, we monitored and reported our chosen metrics for this benchmark in real time. We employed Metricbeat, i.e., the proposed tool for monitoring the scalability of Big Data platform, over our Elastic Stack to monitor and report the chosen metrics. The methodology and the results of the platform's evaluation are presented in detail in a previous study of our team [15].

According to the results of the performance evaluation, in terms of the dataset ingestion step, the data platform had a good performance with respect to the completion time as well as the CPU, network, and memory usage. It is a step that can be easily made with a high degree of concurrency without seriously affecting the rest of the stack. The incident prediction demonstrated a very good performance with respect to the completion time as well as the CPU, network, and memory usage. Performance increased by lowering the volumes of data handled. Increasing data needs more time, more CPU, and memory to be trained. Similar behavior is observed in risk assessment step.

11 Discussion

Building a Big Data platform that can collect and process all the available global food safety data requires an architecture that is scalable and can accommodate the dynamic requirements in functionalities, data, and speed of analysis. As presented in this chapter the starting point of the design is a robust internal data schema that can scale to many different data types and can support data standards, i.e., standard properties and semantic vocabularies. Putting significant effort to define the internal data model will give important advantages in the long run.

As presented in Sect. 10, an important aspect that needs to be carefully tested is the scalability of a Big Data platform both in terms of volume and velocity of data. To that direction we have conducted several experiments for realistic scenarios with real and synthetic datasets [15]. The parts of the Big Data platform that are more computational intensive are the data processing and enrichment processes. Increased data volume and velocity can be supported by expanding the infrastructure with more computational and memory resources but still keeping the platform very efficient.

It is critical to have an open architecture that can easily support new data sources and data types but also new processes. Using a data platform that is based on microservices enables such an open and scalable architecture. We have validated this in the case of the food safety sector through many new requirements for new data types and data sources in order to provide descriptive and predictive analytics that will help the experts to take data-informed decisions.

In terms of future work, the Big Data platform will be expanded by mechanisms that will assign global identifiers to entities such as companies, batch numbers of the products, hazards, ingredients, and raw materials. Relying on such global identifiers will enhance the traceability information that will be managed and processed by the platform.

Furthermore, as the scope of the Big Data platform expands from managing only open data to also managing and processing private data, it is important to look at the data security aspects that will enable the setup of public-private data trusts [17]. We are already working in the context of the Food Safety Market H2020 project (https://foodsafetymarket.eu/) on the deployment of authentication and authorization methods as well of data anonymization techniques. We start by deploying secure data exchange services for very critical processes like food certification. Combining AI, data, and security technologies, we aim at creating a data space for the food safety sector.

Finally, the development of marketplace services that will allow data sharing, data discovery, and data monetization is a next step that will open up new possibilities in the food safety sector [18].

12 Conclusions

This chapter presented a Big Data platform that efficiently collects, processes, links, and mines global food safety data to enable the generation of insights and predictions that can help food safety experts to make critical decisions and make our food safer. Using data standards and deploying mature and state-of-the-art Big Data technologies, we managed to develop a data platform that is open and scalable. Best practices on the deployment of data platforms such as identification of records, orchestrating pipelines, automating the aggregation workflow, securing exchange, and monitoring of a Big Data platform were shared.

Acknowledgment This work is funded with the support by the European Commission and more specifically project TheFSM "The Food Safety Market: an SME-powered industrial data platform to boost the competitiveness of European food certification" (grant no. 871703) (https://foodsafetymarket.eu/), which is funded by the schema "Innovation Actions (IA)" under the work program topic "H2020-EU.2.1.1.—INDUSTRIAL LEADERSHIP—Leadership in enabling and industrial technologies—Information and Communication Technologies (ICT)." This publication reflects the views only of the authors, and the Commission cannot be held responsible for any use, which may be made of the information contained therein.

References

1. Sklare, S., Kontogiannis, T., & Stoitsis, G. (2020). The case of chocolate: Lessons for food authenticity and big data. In *Building the future of food safety technology: Blockchain and beyond*.
2. Cavin, C., Cottenet, G., Cooper, K. M., & Zbinden, P. (2018). Meat vulnerabilities to economic food adulteration require new analytical solutions. *Chimia, 72*(10), 697–703. https://doi.org/10.2533/chimia.2018.697
3. Spink, J., Elliott, C., Dean, M., & Speier-Pero, C. (2019). Food fraud data collection needs survey. *NPJ Science of Food, 3*, 8. https://doi.org/10.1038/s41538-019-0036-x
4. RASFF – Rapid Alert System for Food and Feed. (1979). https://ec.europa.eu/food/safety/rasff-en
5. HorizonScan. (2015). *FERA*. https://horizon-scan.fera.co.uk/
6. gComply. (2015). *Decernis*. https://decernis.com/solutions/gcomply/
7. DigiComply. (2017). *SGS*. https://www.digicomply.com/
8. Zillner, S., Curry, E., Metzger, A., Auer, S., & Seidl, R. (2017). *European big data value strategic research & innovation agenda*. Big Data Value Association.
9. BDVA, CLAIRE, ELLIS, EurAI and euRobotics, *Joint Strategic Research Innovation and Deployment Agenda (SRIDA) for the AI, Data and Robotics Partnership*, September 2020.
10. Jin, C., Bouzembrak, Y., Zhou, J., Liang, Q., van den Bulk, L. M., Gavai, A., Liu, N., van den Heuvel, L. J., Hoenderdaal, W., & Marvin, H. J. P. (2020). Big Data in food safety - A review. *Current Opinion in Food Science, 36*, 24–32. ISSN 2214-7993
11. Indrasiri, K., & Siriwardena, P. (2018). *Microservices for the enterprise - Apress, Berkeley*. Springer.
12. EFSA. (2015). *The food classification and description system FoodEx2* (revision 2). EFSA supporting publication EN-804. 90 p.
13. FAO. (1994). *Commodity categories*. http://www.fao.org/fao-who-codexalimentarius/codex-texts/dbs/pestres/commodities/en/

14. GS1. (2016). *EPC Information Services (EPCIS) Standard*. https://www.gs1.org/sites/default/files/docs/epc/EPCIS-Standard-1.2-r-2016-09-29.pdf
15. Polychronou, I., Stoitsis, G., Papakonstantinou, M., & Manouselis, N. (2022). Stress-testing big data platform to extract smart and interoperable food safety analytics. *International Journal of Metadata Semantics and Ontologies*.
16. Taylor, S. J., & Letham, B. (2018). Forecasting at scale. *The American Statistician, 72*(1), 37–45.
17. U.S Food and Drug Administration. (2020). *New era of smarter food safety, FDA's Blueprint for the Future*.
18. Otto, B., & Jarke, M. (2019). Designing a multi-sided data platform: Findings from the International Data Spaces case. *Electron Markets, 29*, 561–580. https://doi.org/10.1007/s12525-019-00362-x
19. Gaona-García, P. A., Stoitsis, G., Sánchez-Alonso, S., & Biniari, K. (2016). An exploratory study of user perception in visual search interfaces based on SKOS. *KO Knowledge Organization, 43*(4), 217–238.
20. K.G. Saur Verlag. (1998). *Functional requirements for bibliographic records: Final report*. KG Saur.
21. Lianou, A., Papakonstantinou, M., Nychas, G.-J. E., & Stoitsis, J. (2021). *Fraud in meat and poultry products, Food Fraud Book*. Academic Press.
22. Polychronou, I., Katsivelis, P., Papakonstantinou, M., Stoitsis, G., & Manouselis, N. (2020) Machine learning algorithms for food intelligence: Towards a method for more accurate predictions. In: Athanasiadis I., Frysinger S., Schimak G., Knibbe W. (eds) Environmental software systems. Data science in action. ISESS 2020. *IFIP Advances in Information and Communication Technology*, vol. 554. Springer. doi:https://doi.org/10.1007/978-3-030-39815-6_16
23. The Food Safety Market: an SME-powered industrial data platform to boost the competitiveness of European food certification. (2020). https://foodsafetymarket.eu/.

Data Space Best Practices for Data Interoperability in FinTechs

Martín Serrano, Edward Curry, Richards Walsh, Gavin Purtill, John Soldatos, Maurizio Ferraris, and Ernesto Troiano

Abstract This chapter focuses on data interoperability best practices related to semantic technologies and data management systems. It introduces a particular view on how relevant data interoperability is achieved and its effects on developing technologies for the financial and insurance sectors. Financial technology (FinTech) and insurance technology (InsuranceTech) are rapidly developing and have created new business models and transformed the financial and insurance services industry in the last few years. The transformation is ongoing, and like many other domains, the vast amount of information available today known as Big Data, the data generated by IoT, and AI applications and also the technologies for data interoperability, which allows data nowadays to be reused, shared, and exchange, will have a strong influence. It is evident the entire financial sector is in a moment of new opportunities with a new vision for substantial growth. This book chapter analyzes the basis of data space design and discusses the best practices for data interoperability by introducing concepts and illustrating the way to understand how to enable the interoperability of information using a methodological approach to formalize and represent financial data by using semantic technologies and information models (knowledge engineering). This chapter provides a state-

M. Serrano (✉)
Insight SFI Research Centre for Data Analytics, University of Galway, Galway, Ireland
e-mail: martin.serrano@nuigalway.ie

E. Curry
Insight SFI Research Centre for Data Analytics, University of Galway, Galway, Ireland
e-mail: edward.curry@nuigalway.ie

R. Walsh · G. Purtill
Banking and Payments Federation of Ireland – BPFI Ireland, Dublin, Ireland
e-mail: richard.walsh@bpfi.ie; gavin.purtill@bpfi.ie

J. Soldatos
Athens Information Technology – AIT Athens, Athens, Greece

M. Ferraris · E. Troiano
CU Innovation Unit - GFT Italia Srl., Milan, Italy
e-mail: maurizio.ferraris@gft.com; ernesto.troiano@gft.com

of-the-art offer called INFINITECH Way using the discussed best practices and explains how semantics for data interoperability are introduced as part of the FinTechs and InsuranceTech.

Keywords FinTechs · InsuranceTech · Interoperability · Data Spaces · Semantics · Knowledge Graph · Linked Data · Ontologies and services · IoT · AI · Big Data

1 Introduction

Digitization or digital transformation, financial technology (FinTech), and insurance technology (InsuranceTech) are rapidly transforming the financial and insurance services industry [1, 2]. Although it is evident the entire financial sector is in a moment of new opportunities and visible, tangible growth, it is also apparent that this transformation is motivated for the FinTech and InsuranceTech enterprises, which are heavily disrupting the traditional business models, and the volume of relevant investments is a proof: Over $23 billion of venture capital and growth equity has been allocated to FinTech innovations during 2011–2014, while $12.2 billion was deployed in 2014 alone [3]. Moreover, a recent McKinsey & Co study revealed that FinTech start-ups in 2016 exceeded 2.000, from approx. 800 in 2015 [4]. Furthermore, most global banks and investment firms have already planned to increase their FinTech/InsuranceTech investments to yield a 20% average return on their investments. Again, beyond FinTech/InsuranceTech, financial institutions and insurance organizations are heavily investing in their digital transformation as a means of improving the efficiency of their business processes and optimizing their decision making.

Traditionally the financial and insurance services sectors and particularly the banking sector have been quite resistant to technology disruption. This is no longer the case in the current trend of digitizing society and its services and applications. The waves of the digital economy and unified markets demand new paradigms to be designed, implemented, and deployed. The vast majority of services and applications that have been developed for the finance and insurance sectors are data-intensive. This transformation holds for applications in different areas such as retail banking, corporate banking, payments, investment banking, capital markets, insurance services, financial services security, and mail. These applications leverage very large datasets from legacy banking systems (e.g., customer accounts, customer transactions, investment portfolio data), which they combine with other data sources such as financial market data, regulatory datasets, social media data, real-time retail transactions, and more. Disruptive innovation in finance and insurance is already possible today, for example, with the advent of Internet-of-Things (IoT) devices and applications (e.g., Fitbits, smartphones, smart home devices), several FinTech/InsuranceTech applications can take advantage of contextual data associated with finance and insurance services to offer a better quality of service at a more competitive cost (e.g., personalized healthcare insurance based on medical devices and improved car insurance based on connected car sensors). Furthermore, alternative data sources (e.g., social media and online news) provide opportunities for new, more automated, personalized, and accurate services. Moreover, recent advances in data storage and processing technologies (including advances in Artificial Intelligence (AI) and blockchain technologies) provide new opportunities for exploiting the above-listed massive datasets, and they are stimulating more investments in digital finance/insurance services [5].

Financial and insurance organizations can take advantage of Big Data, IoT, and AI technologies to improve the accuracy and cost-effectiveness of their services and the overall value they provide to their corporate and retail customers. Nevertheless, despite early data space deployments, there are still many challenges that have to be overcome before leveraging the full potential of Big Data/IoT/AI in the finance and insurance sectors, which could also act as a catalyst for attracting more investments and for significantly improving the competitiveness of enterprises in these sectors [6].

This book chapter analyzes the basis of data space design and best practices for data interoperability by introducing concepts and illustrating the way to understand how to enable the interoperability of information using a methodological approach to formalize and represent financial data by using semantic technologies and information models (knowledge engineering) [7]. This chapter also focuses on the role that semantic technologies like Linked Data and information interoperability provide for the support of financial and insurance industries in the process of digital transformation.

The organization of this chapter is as follows: Section 2 presents challenges in the data space domain in terms of interoperability in the financial and insurance sectors where information exchange occurs to support Big Data, IoT, and AI-enabled services creation and delivery. Section 3 introduces the best practices for the data exchange approach in developments in several parallel streams. These streams facilitate information interoperability and act as a baseline supporting the information interoperability approach. Section 4 introduces the INFINITECH Way, a design, implementation, and deployment methodology to support FinTech Data Spaces. Section 5 presents the current state of the art and motivations for using semantic technologies in convergence and interoperability. Section 6 describes scalable features about the management of Linked Data and its benefits when used in the financial and insurance sectors. Section 7 presents the summary, and finally, some relevant references used in this chapter are listed.

2 Challenges in Data Space Design

Many of the challenges present in current Data Spaces [8] and information management systems are generated by data sharing and exchange, both considered data interoperability problems. Persistent challenges blocking progress in data space design and deployment are as follows.

2.1 Data Fragmentation and Interoperability Barriers

Nowadays, most of the data collected and possessed by financial organizations reside in a wide array of "siloed" (i.e., fragmented) systems and databases, including operational systems and OLTP (online transaction processing) databases, OLAP (online analytical processing) databases and data warehouses, data lakes (e.g., Hadoop-based systems) with raw data (including alternative data like social media), and others. In this fragmented landscape, heavy analytical queries are usually performed over OLAP systems, which leads financial organizations to transfer data from OLTP, data lakes, and other systems to OLAP systems based on intrusive and expensive extract-transform-load (ETL) processes.

In several cases, ETLs consume 75–80% of the budget allocated to data analytics while being a setup for seamless interoperability across different data systems using up-to-date data. Beyond the lack of integrated OLTP and OLAP processes, financial/insurance organizations have no unified way of accessing and querying vast amounts of structured, unstructured, and semi-structured data (i.e., as part of SQL and NoSQL databases), which increase the effort and cost that are associated with the development of Big Data analytics and AI systems. Moreover, there is a lack of semantic interoperability across diverse datasets that refer to the same data entities with similar (yet different) semantics beyond data fragmentation. This is a setback to sharing datasets across various stakeholders and enabling more connected applications that span multiple systems across the financial supply chain.

2.2 Limitations for Cost-Effective Real-Time Analytics

Most of the existing applications operate over offline collections of large datasets based on ETL (extract-transform-load) operations and fail to fully exploit the potential of real-time analytics, which is a prerequisite for a transition from reactive decisions (e.g., what to do following the detection of fraud) to proactive and predictive ones (e.g., how to avoid an anticipated fraud incident). Also, state-of-the-art near-real-time applications tend to be expensive as they have to persist large amounts of data in memory. Moreover, existing engines for real-time analytics (e.g., state-of-the-art streaming engines with stateless parallelization) have limitations when it comes to executing complex data mining tasks such as AI (deep learning-based) algorithms.

2.3 Regulatory Barriers

Big Data and IoT deployments must respect a complex and volatile regulatory environment. In particular, they must adhere to a range of complex regulations (e.g.,

PSD2 (Second Payment Services Directive), MiFIDII/MiFIDR (Markets in Financial Instruments Directive), 4MLD (fourth EU Money Laundering Directive) for financial/insurance) while at the same time complying with general regulations such as the GDPR (General Data Protection Regulation) and the ePrivacy directive. To this end, several RegTech initiatives aim at establishing regulatory sandboxes (e.g., [9–11]), i.e., specialized environments, that facilitate Big Data/IoT experimentation through ensuring access and processing of data in line with applicable laws and regulations. Nevertheless, the development of regulatory sandboxes is in its infancy and only loosely connected to leading-edge Big Data/IoT/AI technologies.

2.4 Data Availability Barriers

To innovate with IoT and Big Data, financial and insurance organizations (including FinTech/InsuranceTech innovators) need access to experimentation yet realistic datasets (e.g., customer account and payments' datasets) that would allow them to test, validate, and benchmark data analytics algorithms. Unfortunately, such data are hardly available, as their creation requires complex anonymization processes or even tedious processes that can realistically simulate/synthesize them. Hence, innovators have no easy access to data for experimentation and testing of novel ideas [12]. Also, due to the fragmentation of Europe's FinTech/InsuranceTech ecosystems, there are no easy ways to share such resources across financial/insurance organizations and innovators.

2.5 Lack of a Blueprint Architectures for Big Data Applications

Given the existing limitations (e.g., data silos and lack of interoperability), financial organizations are creating ad hoc solutions for their problems at hand. They leverage one or more instances of popular data infrastructures such as data warehouses, data lakes, elastic data stores, and machine learning toolkits in various deployment configurations. However, they have no easy way to create, deploy, and operate such infrastructures through adhering to proven patterns and blueprints that will lower their integration, deployment, and operation efforts and costs.

2.6 No Validated Business Models

Big Data and IoT deployments in finance/insurance have, in several cases, demonstrated their merits on the accuracy, performance, and quality of the resulting services (e.g., increased automation in business processes, improved risk assessment, faster transaction completion for end-users, better user experience). However, there is still a lack of concrete and validated business models that could drive monetization and tangible business benefits for these service improvements. Such business models could foster the rapid development and adoption of Big Data and IoT innovations, including emerging innovations that leverage real-time analytics and AI [13].

3 Best Practices for Data Space Design and Implementation

To address these challenges and leverage the full potential of Big Data (including AI) and IoT in finance/insurance, there is a need for developments in several parallel streams.

3.1 Technical/Technological Developments

At the technical/technological forefront, there is a need for Big Data architectures and toolkits tailored to the needs of data-intensive applications in the finance/insurance sector. These shall include several novel building blocks, including (1) infrastructures for handling arbitrarily large datasets from multiple fragmented sources in a unified and interoperable way; (2) semantic interoperability solutions for the financial/insurance supply chain; (3) novel techniques for real-time analytics and real-time AI; (4) advanced data analytics algorithms (including AI); (5) technologies and techniques for security and regulatory compliance, such as data encryption and anonymization technologies; (6) blueprint architectures for combining the above-listed building blocks with coherent and cost-effective solutions; and (7) open APIs that will facilitate innovators to produce and validate innovative solutions.

3.2 Development of Experimentation Infrastructures (Testbeds)

The development of Big Data, IoT, and AI-based innovations requires significant testing and validation efforts, such as testing for regulatory compliance and optimizing machine learning and deep learning data models. Therefore, there is a need for widely available experimentation infrastructures at the national and EU levels, which shall provide access to resources for application development and experimentation, such as datasets, regulatory sandboxes, libraries of ML (machine learning)/DL (deep learning) algorithms, Open (banking/finance) APIs, and more. Furthermore, such experimentation infrastructures should be available in appropriate testbeds, based on deploying the above-listed technical building blocks in various configurations. The latter should support experimentation and testing of all types of Big Data/AI/IoT applications in the finance and insurance sectors, such as KYC (Know Your Customer) and KYB (Know Your Business), credit risk scoring, asset management recommendations, usage-based insurance applications, personalized portfolio management, automated payment applications, and many more.

3.3 Validation of Novel Business Models

To showcase and evaluate the tangible value of the above-listed technologies and testbeds, there is also a need for validating them in the scope of real-life business cases involving realistic business processes and applications for retail and corporate finance/insurance. The validation shall focus on novel business models, which essentially disrupt existing operations of financial organizations and deliver exceptional business benefits in terms of automation, personalization, cost-effectiveness, and intelligence.

4 The INFINITECH Way to Design/Support FinTech Data Spaces

INFINITECH is the largest joint effort of Europe's leaders in IT and finance/insurance sectors toward providing the technological capabilities, the experimentation facilities (testbeds and sandboxes), and the business models needed to enable European financial organizations, insurance enterprises, and FinTech/InsuranceTech innovators to fully leverage the benefits of Big Data, IoT, and AI technologies. The latter benefits include a shift toward autonomous (i.e., automated and intelligent) processes that are dynamically adaptable and personalized to end-user needs while complying with the sector's regulatory environment. Furthermore, INFINITECH brings together all the stakeholders involving NGOs with their members, financial institutions and insurance companies, research centers, large industry, and SMEs.

4.1 Technological Building Blocks for Big Data, IoT, and AI

INFINITECH looks at the finance and insurance sectors and provides multiple assets, including infrastructures, components and toolkits for seamless data access and querying across multiple fragmented data sources, technologies for cost-effective real-time analytics, advanced analytics algorithms (including AI), technologies for Data Governance and regulatory compliance, technologies for trustful and secure data sharing over blockchain infrastructures, as well as handling of semantic data interoperability across stakeholders of the financial/insurance supply chain. Furthermore, INFINITECH also follows reference architecture (RA) approach for Big Data, IoT, and AI applications in the financial sector [14–17], whose aim is to serve as a blueprint for integrating, deploying, and operating Big Data and IoT infrastructures, including infrastructures that will leverage the above-listed building blocks. Furthermore, the reference architecture provides the means for integrating and deploying applications that take advantage of leading-edge technologies, including predictive analytics, different instances of AI (e.g., DL, chatbots), and blockchains.

4.2   Tailored Experimentation Infrastructures

INFINITECH provides the necessary mechanisms for creating tailored experi-mentation environments (i.e., testbeds and sandboxes) for different applications (e.g., sandboxes for fraud detection, credit risk assessment, personalized financial assistance) using flexible configurations of the testbed resources. Testbeds and sandboxes are used for different Big Data, IoT, and AI applications in the financial and insurance sectors, enabling innovators to access and share resources for testing, innovation, and experimentation, including previous datasets. The INFINITECH testbeds and sandboxes use the Open API standard for experimentation, and innovation is crucial. This facilitates the adoption and the extension of the designed, deployed, and tested solutions. ML/DL algorithms and regulatory compliance tools play a relevant role in the tailored experimentation testbeds and sandboxes.

INFINITECH uses this concept by deploying testbeds and sandboxes European-wide, thus demonstrating that it is possible to support the FinTech/InsuranceTech partners through experimentation testbeds. INFINITECH includes seven testbeds established at individual banks and one (EU-wide). The testbeds are made available to innovators' communities via the established innovation management structures of the project's partners and through a (virtualized) digital innovation hub (VDIH) set by the project as part of its exploitation strategy.

4.3   Large-Scale Innovative Pilots in Finance and Insurance

The use of a large ecosystem like INFINITECH for texting and validation will lever-age both the technological developments of the project (including the INFINITECH reference architecture) and the testbeds/sandboxes to later deploy and implement as part of commercial solutions the novel and validated use cases. The pilot's target in real-life environments is based on realistic datasets, i.e., either anonymized or synthetic datasets with pragmatic statistical properties. The pilots will span a wide array of areas covering the most prominent processes of the financial and insurance sectors, including KYC and customer-centric analytics, fraud detection and financial crime, credit risk assessment, risk assessment for capital management, personalized portfolio management, risk assessment in investment banking, personalized usage-based insurance, insurance product recommendations, and more. The pilots will demonstrate the added value of the project's technologies and testbeds while at the same time showcasing the project's disruptive impact on Europe's financial and insurance sectors.

4.4 Business Model Development and Validation

In the scope of the innovative pilots and use cases in finance and insurance, notably a novel and replicable business model or a set of them needs to be associated with each of the listed pilots/use cases [18, 19]. A practice to resolve one of the significant issues when developing new technologies based on experimentation is the use of a real exploitation model. These business models will pave the ground for disrupting the financial sector based on advanced Big Data, IoT, and AI infrastructures and applications, thus demonstrating the tangible impact of the project in financial institutions, insurance organizations, and FinTech/InsuranceTech enterprises.

5 Technology Capabilities for Convergence and Interoperability

Financial technology (FinTech) and insurance technology (InsuranceTech) are rapidly developing and have created new business models and transformed the financial and insurance services industry in the last few years. Technological convergence supporting data sharing and exchange between services applications is a barrier that the financial and insurance sectors have recently confronted with the globalization of economies and markets for a long time. This need is becoming more relevant, and today more than ever before, it needs to be addressed. Semantic technologies have played a crucial role as an enabler of many of the applications and services in other domain areas, although not much in the financial domain, and as has already been mentioned in the financial and insurance sectors, it is until just recently that the progress, in terms of implementation, has become more evident requirements; however, the convergence between technological development and interoperability has not entirely run in parallel, mainly due to many complex issues involving non-interoperable aspects where social, economic, and political dimensions are taking place.

5.1 Semantic Interoperability and Analytics

INFINITECH provides a shared semantics solution for the interoperability of diverse finance/insurance datasets. To this end, the project relies on existing ontologies for financial information modeling and representation (such as FIBO, FIGI, and LKIF) [20, 21], which are appropriately extended as required by the INFINITECH-RA and the project's pilots. Moreover, INFINITECH offers a solution for parallel and high-performance analytics over semantic streams, based on the customization of existing solution of semantic linked stream analytics (such as NUIG's Super Stream Collider (SSC) solution [22–25]). The INFINITECH semantic interoperability infrastructure is available in all cases/pilots where semantic reasoning will be required for extra intelligence.

5.2 INFINITECH Building Blocks for Big Data, IoT, and AI

There is always a high demand for integrated systems and technological components that can almost transparently connect and transfer data in the finance and insurance sectors. The integrated environment, including infrastructures, components, and toolkits, shall be designed to support seamless data access and querying across multiple fragmented data sources, technologies for cost-effective real-time analytics, advanced analytics algorithms (including AI), technologies for Data Governance and regulatory compliance, technologies for trustful and secure data sharing over blockchain infrastructures, as well as handling of semantic data interoperability across stakeholders of the financial/insurance supply chain. INFINITECH emerges as an alternative to those technological demands and provides a reference architecture (RA), as shown in Fig. 1. The INFINITECH reference architecture brings together technologies for Big Data, IoT, and AI applications in the financial sector, which will serve as a blueprint for integrating, deploying, and operating Big Data and IoT infrastructures.

INFINITECH provides the means for integrating and deploying applications that take advantage of leading-edge technologies, including predictive analytics, different instances of AI (e.g., DL, chatbots), and blockchains between other technologies. Figure 1 depicts the mentioned innovation-driven functional architecture approach from the INFINITECH ecosystem. It is an overall and FinTech holistic view. It's design and implementation rely on the intelligence plane, a combination of Big Data, IoT, and AI analytics applications. In the INFINITECH reference architecture data analytics plane, the exchange of information facilitates knowledge-driven support and the generation of composing services with operations by enabling interoperable management information.

The INFINITECH approach, in terms of the design approach, uses the design principles introduced in this chapter and looks at implementing the different scenarios and testbeds as described. INFINITECH moves toward converged IP

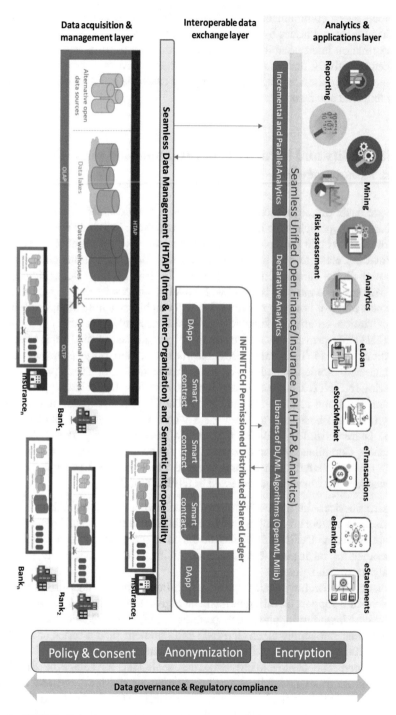

Fig. 1 INFINITECH reference architecture – high-level overview of Big Data/IoT platforms and technological building blocks

and cloud-based communications networks, increasing solutions to a number of significant technical issues by using more standard information exchange, promoting information interoperability, and allowing that the testbeds and sandboxes be managed effectively, and, most importantly, offering new open opportunities for a user knowledge-based service-oriented support can have a fundamental impact on future financial and insurance services.

6   Scalability and Security Considerations for FinTech and InsuranceTech

There are basic characteristics that must be taken into consideration at the time of building new approaches for the FinTech and InsuranceTech industries. Relatively modern ways to build multiple applications are sandboxes and testbeds specialized for providing close to real deployment, and thus implementations can be tested under real digital infrastructure conditions. The Big Data/IoT technologies and sandboxes/testbeds must be coupled with novel business models that will enable a whole new range of novel applications that will emphasize automation, intelligence, personalization, security, stakeholders' collaboration, and regulatory.

INFINTECH provides a 360° coverage of all the issues that hinder financial institutions' and FinTech enterprises' efforts to use and fully leverage IoT and Big Data technologies, including AI. Thus, this section summarizes pragmatically (following INFINITECH experiences) how those efforts, for scalability, are done.

Compliance and at the same time INFINITECH will enable the development, deployment, and business validation of a whole new range of applications that will be characterized by SHARP (Smart, Holistic, Autonomy, Personalized and Regulatory Compliance) characteristics. The following are just short paragraphs describing how SHARP can be implemented and briefly and as implementation reference how they have been addressed in INFINITECH.

- Smart: Services shall take advantage of predictive analytics and AI on Big Data datasets to anticipate changes in financial/insurance contexts and automatically adapt to them. INFINITECH has designed a set of end-to-end, business-to-customer, or business-to-business applications. Those services are based on analytics or ML solutions.
- Holistic: Architectures shall empower a holistic approach to data-driven services, which shall support all different financial applications across all phases of their lifecycle, including applications spanning multiple stakeholders and systems in the financial supply chain. INFINITECH implemented a series of reference architectures in the form of functional components.
- Autonomy: The deployed infrastructures shall take advantage of Big Data, IoT, and AI to significantly reduce error-prone manual processes and decision making through increased automation. The INFINITECH reference architecture presented paves the ground for the fully autonomous processes in the future that

will disrupt the entire finance and insurance sectors. The use of the architecture is a sign of autonomy, and it does contain all the necessary software parts to run and operate services.

- Personalization: Processes where data is processed timely profile customers and subsequently offer them individualized and dynamic services that adapt to their needs. INFINITECH involved KYC/KYB processes to include individuals' characteristics and build profiles based on the available data.
- Regulatory Compliant: Based on the use of particular or general technologies, the financial and insurance enterprises can ensure regulatory compliance by design. INFINITECH will take advantage of data processing to achieve faster, cost-effective, and reliable compliance to regulations.

To ensure scalability and security, permissioned blockchain for data sharing and data trading is required. There are already several use cases in the finance/insurance sector [26] that involve sharing of data across different organizations (e.g., sharing of customer data for customer protection or faster KYC, sharing of businesses' data for improved credit risk assessment, sharing of customer insurance data for faster claims management, and more); these are the ideal scenarios for emerging solutions like DLT (distributed ledger technologies, the baseline for blockchain technologies).

INFINITECH uses a permissioned DLT infrastructure, which provides privacy control, auditability, secure data sharing, and faster operations. Some of the later characteristics are inherent in permissioned DLT's features (i.e., the Hyperledger Fabric by IBM and the Linux Foundation) and can be directly configured in INFINITECH testbeds, sandboxes, and pilots. The core of the DLT infrastructure, i.e., the Fabric, will be enhanced in two directions: (1) integration of tokenization features and relevant cryptography as a means of enabling assets trading (e.g., personal data trading) through the platform and (2) for selected pilots enhancement of the blockchain infrastructure with multi-party computation (MPC) and linear secret sharing (LSS) algorithms from the OPAL (Open Algorithm Project) to enable querying of encrypted data as a means of offering higher data privacy guarantees.

7 Conclusions

This chapter addresses best practices for data space design and implementation identified from the state-of-the-art analysis. These challenges are tested and validated in the context of an H2020 European large-scale ecosystem called INFINITECH. The best practices leverage the full potential of Big Data, IoT, and AI applications in finance/insurance and identify a need for developments in several other areas to support scaling-up applications.

The interoperable data model following the formalization of vocabularies [7, 27] and using the FIBO, FIGI, and LKIF approaches were mentioned. It is the most adaptive interoperability model for the financial and insurance sector. Although the details are out of the scope of this chapter, the references and methods to use them as part of the formal vocabularies to build the FinTech and InsuranceTech lingua franca are already an innovative approach toward data exchange and data sharing capabilities and are introduced as part of the results from the INFINITECH ecosystem approach called INFINITECH Way.

The INFINITECH reference architecture has been introduced, which provides the means for integrating and deploying applications that take advantage of leading-edge technologies, including predictive analytics, different instances of AI (e.g., DL, chatbots), blockchains, and more. In addition, INFINITECH provides technology for semantic interoperability based on shared semantics, along with a permissioned blockchain solution for data sharing across finance/insurance institutions.

This book chapter analyzed the basis of data space design and discussed the best practices for data interoperability by introducing concepts and illustrating the INFINITECH Way to enable interoperability of the information using a reference architecture following the methodological approach and the formalization and representation of financial data by using semantic technologies and information models (knowledge engineering). In addition, the INFINITECH Way introduced discusses best practices and explains how challenges for data interoperability can be overcome using a graph data modeling approach as part of the FinTech and InsuranceTech.

Acknowledgments This work has been partially supported by the H2020 INFINITECH Project (Tailored IoT & BigData Sandboxes and Testbeds for Smart, Autonomous and Personalized Services in the European Finance and Insurance Services Ecosystem) [28], which is co-funded by the European Commission under the H2020 framework program, contract number H2020-ICT-2019-856632-INFINITECH. It is also partially supported by Science Foundation Ireland under grant number SFI/12/RC/2289_2.

References

1. Dietz, M., Khanna, S., Olanrewaju, T., & Rajgopal, K. *Cutting through the noise around financial technology*. McKinsey & Co, February 2016, Accessible online Oct 2021, from https://www.mckinsey.com/industries/financial-services/our-insights/cutting-through-the-noise-around-financial-technology
2. PwC. *Redrawing the lines: FinTech's growing influence on Financial Services*. Global FinTech Report 2017, March 2017. Available at; https://www.pwc.com/jg/en/publications/pwc-global-fintech-report-17.3.17-final.pdf
3. Crofoot, J., Jacobs, P., Sarvady, G., Schultz, A. B., & Willder, J. *McKinsey on payments: Special edition on advanced analytics in banking*, April 2018.
4. McKinsey & Company. *Financial services practice - global payments 2015: A healthy industry confronts disruption*. Available Online Oct 2021 https://www.mckinsey.com/~/media/mckinsey/dotcom/client_service/financial%20services/latest%20thinking/payments/global_payments_2015_a_healthy_industry_confronts_disruption.ashx
5. Brynjolfsson, E., & Mcafee, R. The business of artificial intelligence. *Harvard Business Review (HBR)*, July 17.
6. Zillner, S., Curry, E., Metzger, A., Auer, S., & Seidl, R. (Eds.). (2017). *European big data value strategic research & innovation agenda*. Big Data Value Association. Retrieved from http://www.edwardcurry.org/publications/BDVA_SRIA_v4_Ed1.1.pdf

7. Serrano, J. M., Serrat, J., & Strassner, J. Ontology-based reasoning for supporting context-aware services on autonomic networks. 2007 IEEE/ICC International conference on communications, 24–28 June 2007, Glasgow.
8. Curry, E. (2020). *Real-time linked dataspaces*. Springer International Publishing. https://doi.org/10.1007/978-3-030-29665-0
9. *Industry SandBoxes*. Available on October 2021. http://industrysandbox.org/regulatory-sandboxes/
10. *Financial Conduct Authority, "Regulatory sandbox"*, White Paper, November 2015.
11. Fernandez, R. C., Migliavacca, M., Kalyvianaki, E., & Pietzuch, P. (2014). Making state explicit for imperative big data processing. In *USENIX ATC*.
12. Curry, E., Metzger, A., Zillner, S., Pazzaglia, J.-C., & García Robles, A. (2021). *The elements of big data value*. Springer International Publishing. https://doi.org/10.1007/978-3-030-68176-0
13. Zillner, S., Bisset, D., Milano, M., Curry, E., Hahn, T., Lafrenz, R., et al. (2020). *Strategic research, innovation and deployment agenda - AI, data and robotics partnership. Third Release* (Third). Brussels: BDVA, euRobotics, ELLIS, EurAI and CLAIRE.
14. Pääkkönen, P., & Pakkala, D. (2015). Reference architecture and classification of technologies, products and services for big data systems. *Big Data Res, 2*(4), 166–186. https://doi.org/10.1016/j.bdr.2015.01.001
15. Muan Sang, G., Xu, L., & De Vrieze, P. (2016). *A reference architecture for big data systems* (pp. 370–375). https://doi.org/10.1109/SKIMA.2016.7916249.
16. Muan Sang, G., Xu, L., & De Vrieze, P. (2017). *Simplifying big data analytics systems with a reference architecture* (pp. 242–249). doi: https://doi.org/10.1007/978-3-319-65151-4_23.
17. Schiff, A., & McCaffrey, M.. *Redesigning digital finance for big data*, May 2017.
18. Simitsis, A., Wilkinson, K., Castellanos, M., & Dayal, U. (2012) Optimizing analytic data flows for multiple execution engines. In *Proceedings of the 2012 ACM SIGMOD International Conference on Management of Data*, New York, NY, pp. 829–840.
19. Sun, Z., Strang, K. D., & Yearwood, J. (2014). Analytics service oriented architecture for enterprise information systems (pp. 508–516). doi:https://doi.org/10.1145/2684200.2684358.
20. Object Management Group. *Financial instrument global identifier specification v1.0*, December 2015.
21. Bennett, M. (2013). The financial industry business ontology: Best practice for big data. *Journal of Banking Regulation, 14*(3–4), 255–268.
22. Le-Phuoc, D., Nguyen-Mau, H. Q., Parreira, J. X., & Hauswirth, M. (2012). A middleware framework for scalable management of linked streams. *Journal of Web Semantics, 16*.
23. Neumeyer, L., Robbins, B., Nair, A., & Kesari, A. (2010). S4: Distributed stream computing platform. *IEEE International Conference on Data Mining Workshops, 2010*, 170–177.
24. Nguyen, H., Quoc, M., Serrano, M., Le-phuoc, D., & Hauswirth, M. (2012). Super stream collider linked stream mashups for everyone. In: *Proceedings of the Semantic Web Challenge co-located with ISWC2012*, vol. 1380.
25. Serrano, M., Strassner, J., & ÓFoghlú, M. A formal approach for the inference plane supporting integrated management tasks in the future internet. *1st IFIP/IEEE ManFI International Workshop, In conjunction with*11th IFIP/IEEEIM2009, 1–5 June 2009, at Long Island, NY.
26. Carson, B., Romanelli, G., Walsh, P., & Zhumaev, A., *Blockchain beyond the hype: What is the strategic business value?*, June 2018.
27. Serrano, J. M. (2012). *Applied ontology engineering in cloud services, networks and management systems*. Springer, Hardcover, 222 pages, ISBN-10: 1461422353, ISBN-13: 978-1461422358.
28. H2020 INFINITECH Project. Accessible online October 2021. http://www.infinitech-project.eu

TIKD: A Trusted Integrated Knowledge Dataspace for Sensitive Data Sharing and Collaboration

Julio Hernandez, Lucy McKenna, and Rob Brennan

Abstract This chapter presents the Trusted Integrated Knowledge Dataspace (TIKD)—a trusted data sharing approach, based on Linked Data technologies, that supports compliance with the General Data Privacy Regulation (GDPR) for personal data handling as part of data security infrastructure for sensitive application environments such as healthcare. State-of-the-art shared dataspaces typically do not consider sensitive data and privacy-aware log records as part of their solutions, defining only how to access data. TIKD complements existing dataspace security approaches through trusted data sharing that includes personal data handling, data privileges, pseudonymization of user activity logging, and privacy-aware data interlinking services. TIKD was implemented on the Access Risk Knowledge (ARK) Platform, a socio-technical risk governance system, and deployed as part of the ARK-Virus Project which aims to govern the risk management of personal protection equipment (PPE) across a group of collaborating healthcare institutions. The ARK Platform was evaluated, both before and after implementing the TIKD, using both the ISO 27001 Gap Analysis Tool (GAT), which determines information security standard compliance, and the ISO 27701 standard for privacy information. The results of the security and privacy evaluations indicated that compliance with ISO 27001 increased from 50% to 85% and compliance with ISO 27701 increased from 64% to 90%. This shows that implementing TIKD provides a trusted data

This research has received funding from the ADAPT Centre for Digital Content Technology, funded under the Science Foundation Ireland Research Centres Programme (Grant 13/RC/2106_P2) and under the SFI Covid Rapid Response Grant Agreement No. 20/COV/8463, and co-funded by the European Regional Development Fund and the European Union's Horizon 2020 Research and Innovation Programme under the Marie Sklodowska-Curie grant agreement No. 713567. For the purpose of Open Access, the author has applied a CC BY public copyright license to any Author Accepted Manuscript version arising from this submission.

J. Hernandez (✉) · L. McKenna · R. Brennan
ADAPT Centre, School of Computing, Dublin City University, Dublin, Ireland
e-mail: julio.hernandez@adaptcentre.ie; lucy.mckenna@adaptcentre.ie; rob.brennan@dcu.ie

security dataspace with significantly improved compliance with ISO 27001 and ISO 27701 standards to share data in a collaborative environment.

Keywords Dataspace · Knowledge Graph · Trusted data · Personal data handling

1 Introduction

This chapter relates to the technical priority of data management from the European Big Data Value Strategic Research and Innovation Agenda [23]. It addresses the horizontal concern of data management from the BDV Technical Reference Model and the vertical concerns of cybersecurity. Additionally, this chapter relates to the data for AI enablers of the AI, Data, and Robotics Strategic Research, Innovation, & Deployment Agenda [22].

Sharing sensitive data, between healthcare organizations, for example, can facilitate significant societal, environmental, and economic gains such as medical diagnoses and biomedical research breakthroughs. However, as this data is sensitive, organizations understand the importance (and increasing compliance requirements) of securely sharing, storing, managing, and accessing such data. Here, sensitive data is specified to include personal data or personally identifiable information (PII), GDPR special category personal data,[1] and business confidential or restricted data that does not normally leave an organization. Most recent works in sensitive data sharing have used cryptosystems and blockchain approaches [4, 10, 21]. These approaches were designed to facilitate the sharing of sensitive data, such as sharing patient medical records between healthcare institutions, but need additional infrastructure to support collaborative data sharing environments for the purpose of research or collaborative analysis. This chapter explores the use of a dataspace, a data management framework capable of interrelating heterogeneous data, for the sharing of sensitive data in a collaborative environment. It also illustrates the use of Knowledge Graphs (KGs) in constructing a trusted data sharing environment for sensitive data.

In recent years, KGs have become the base of many information systems which require access to structured knowledge [2]. A KG provides semantically structured information which can be interpreted by computers, offering great promise for building more intelligent systems [24]. KGs have been applied in different domains such as recommendation systems, information retrieval, data integration, medicine, education, and cybersecurity, among others [20]. For example, in the medical domain, KGs have been used to construct, integrate, and map healthcare information [24]. A dataspace integrates data from different sources and heterogeneous formats, offering services without requiring upfront semantic integration [6]. It follows a "pay-as-you-go" approach to data integration where the priority is to quickly set up the fundamental aspects of the dataspace functionality, such as

---

[1] GDPR Art.9-1.

dataset registration and search, and then improve upon the semantic cohesion of the dataspace over time [6, 8]. The dataspace services offered over the aggregated data do not lose their surrounding context, i.e., the data is still managed by its owner, thus preserving autonomy [5].

A dataspace requires security aspects, such as access control and data usage control [2, 17, 18], to avoid data access by unauthorized users. In this sense, access control is a fundamental service in any dataspace where personal data is shared [2, 3, 13, 15, 17, 18]. According to Curry et al. [2], a trusted data sharing dataspace should consider both personal data handling and data security in a clear legal framework. However, there is currently a lack of solutions for dataspaces that consider both the privacy and security aspects of data sharing and collaboration (see Sect. 3). This work explores the following research question: to what extent will the development of a multi-user and multi-organization dataspace, based on Linked Data technologies, personal data handling, data privileges, and data interlinking, contribute to building a trusted sharing dataspace for a collaborative environment? In response, this work proposes the Trusted Integrated Knowledge Dataspace (TIKD)—an approach to the problem of secure data sharing in collaborative dataspaces.

The TIKD is a multi-user and multi-organization Linked Data approach to trustworthy data sharing between an organization's users. The security access to data follows a context-based access control (CBAC) model, which considers the user and data context to authorize or deny data access. The CBAC implementation is based on the Social Semantic SPARQL Security for Access Control Ontology [19] (S4AC) which defines a set of security policies through SPARQL ASK queries. TIKD defines a privacy protecting user log, based on the PROV ontology, to create user history records. User logs are securely stored following a pseudonymized process based on the Secure Hash Algorithm 3 (SHA-3). The TIKD also provides personal data handling, based on the data privacy vocabulary[2] (DPV), to comply with the General Data Protection Regulation (GDPR). It implements an interlinking process to integrate external data to the KG based on the Comprehensive Knowledge Archive Network[3] (CKAN) data management tool. The contributions of this research are:

1. A trusted dataspace, based on Knowledge Graph integration and information security management, for collaborative environments such as healthcare
2. An information security management system to securely handle organizational data sharing, personal data, user history logs, and privacy-aware data interlinking by means of a context-based access control that includes data privileges and applies a pseudonymization process for user logs

This work extends TIKD from the former work [7] by updating the access control model, improving the personal data handling process, describing the data

---

[2] https://dpvcg.github.io/dpv/.

[3] https://ckan.org/.

classification mechanism, and incorporating a new evaluation process based on the privacy information ISO 27701 standard.

The structure of the remainder of this chapter is as follows: the Use Case section defines the requirements of the ARK-Virus Project. The Related Work section presents the state of the art in dataspace data sharing approaches. The Description of the TIKD section details the services of the dataspace. The Evaluation section presents the results from the ISO 27001 Gap Analysis Tool (GAT) and the ISO 27701 control requirements. Finally, the Conclusion section presents a summary of this research and its future directions.

2   Use Case—Sensitive Data Sharing and Collaboration for Healthcare in the ARK-Virus Project

The ARK-Virus Project.[4] extends the ARK Platform to provide a collaborative space for use in the healthcare domain—specifically for the risk governance of personal protective equipment (PPE) use for infection prevention and control (IPC) across diverse healthcare and public service organizations [12]. The consortium consists of the ARK academic team (ADAPT Centre, Dublin City University, and the Centre for Innovative Human Systems, Trinity College Dublin) and a community of practice which includes safety staff in St. James's Hospital Dublin, Beacon Renal, and Dublin Fire Brigade. Staff across all three organizations are involved in trialing the ARK Platform application which is hosted in Trinity College Dublin. This creates many overlapping stakeholders that must be appropriately supported when handling sensitive information.

The ARK Platform uses Semantic Web technologies to model, integrate, and classify PPE risk data, from both qualitative and quantitative sources, into a unified Knowledge Graph. Figure 1 illustrates the ARK Platform's data model supporting the collaborative space for PPE. This model is expressed using the ARK Cube ontology[5] and the ARK Platform Vocabulary[6] [9, 12]. The Cube ontology is used in the overall architecture of the ARK Platform to support data analysis through the Cube methodology—an established methodology for analyzing socio-technical systems and for managing associated risks [1, 11]. The ARK Platform Vocabulary allows for the modeling of platform users, access controls, user permissions, and data classifications.

Through the ARK-Virus Project a set of security requirements for the ARK Platform were defined (see Table 1). These requirements included data interlinking, data accessibility (privacy-aware evidence distillation), and secure evidence publication (as linked open data), as priority security aspects. The ARK Platform implements

---

[4] https://openark.adaptcentre.ie/.

[5] Available at https://openark.adaptcentre.ie/Ontologies/ARKCube.

[6] Available at https://openark.adaptcentre.ie/Ontologies/ARKPlatform.

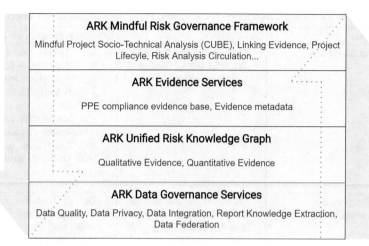

Fig. 1 The ARK Platform data model

the TIKD to cope with these requirements (see Table 1) and to provide secure management of personal data, pseudonymized data (for compliance with the GDPR, explained later in this chapter), and security logs (for history records).

3   Related Work

This section compares available data sharing approaches with the ARK-Virus requirements (see Table 1) in order to establish their suitability. The approaches analyzed can be divided into two main techniques: dataspace-based and blockchain-based, where blockchain is an Internet database technology characterized by decentralization, transparency, and data integrity [14].

Dataspace approaches to data sharing services are primarily associated with the Internet of Things (IoT) [2, 15, 17, 18], where data integration from heterogeneous devices and access control are the main objective. On the other hand, blockchain approaches [4, 10, 21] integrate cryptography techniques as part of the data management system in order to share data between agents (users or institutions). Table 2 provides a comparison of the state of the art and TIKD in relation to the requirements of the ARK-Virus Project.

Data sharing approaches based on blockchain methods [3, 4, 10, 21] use a unified scheme. In most cases records must be plain text, avoiding the integration of data in different formats, and usage policies, which restrict the kind of action that an agent can perform over data, are not defined. Even when the main concern of these approaches is to keep a record's information secure, they do not propose any agent information tracking for activity records. TIKD implements an authorization access control based on security policies that consider context information, security roles,

Table 1 ARK-Virus Project requirements, description, and solution proposed with TIKD

#	Requirement	Description	Solution proposed with TIKD
1	Data encryption	– Data stored on the ARK Platform will be encrypted and stored securely	– The TIKD model will be deployed in an encrypted system based on a Linux Unified Key Setup (LUKS). – The information security management considers the ISO 27001 Gap Analysis Tool[a] (GAT) as base reference
2	Privacy-aware evidence distillation	– Users will be able to manually classify the information they enter into the platform as public, internal, confidential, or restricted. – Data will be securely stored in an evidence base of risk and only accessible to users with the appropriate level of clearance	– The TIKD model defines a usage control to set grants over data, determining who can collaborate (write access) in the project and who can access it (read-only access). – The TIKD defines a pseudonymized log process to securely keep a user history records
3	Data interlinking	– An interlinking function will allow users of the ARK-Virus Platform to interlink risk management data with evidence stored within the platform and, in future iterations, related data held in external authoritative databases	The TIKD model integrates an interlinking service to provide anonymous risk evidence integration. The evidence is integrated by means of data catalogues (DCAT), describing the evidence metadata
4	Evidence publication as linked open data	– Data stored in the ARK-Virus Platform database that has been classified as "Public" will be made available as LOD via an open data portal	– The usage control service defines how data could be accessed based on their classification level. The data classified as public could be searched and collected for publication purposes

[a]https://www.itgovernance.eu/en-ie/shop/product/iso-27001-gap-analysis-tool

Table 2 Comparison of data sharing in dataspace and trusted data sharing approaches

Title	Year	Data encryption (req. #1)	Privacy-aware evidence dist. (req. #2)	Data interlinking (Req. #3)	Evidence publication as LOD (req. #4)
A risk-based framework for biomedical data sharing [3]	2017	No	Security levels	No	No
MedBlock: efficient and secure medical data sharing via blockchain [4]	2018	Private and public keys	Blockchain	No	No
A method and application for constructing an authentic data space [17]	2019	No	Business rules and security levels	No	No
An architecture for providing data usage and access control in data sharing ecosystems [13]	2019	No	Policies	No	No
A real-time linked dataspace for the Internet of Things: enabling "pay-as-you-go" data management in smart environments [2]	2019	No	Roles	No	No
International Data Spaces [15]	2019	No	Roles	No	No
A blockchain-based medical data sharing and protection scheme [10]	2019	Private and public keys	Blockchain	No	No
An IoT data sharing privacy-preserving scheme [18]	2020	No	Policies	No	No
Medical data sharing scheme based on attribute cryptosystem and blockchain technology [21]	2020	Private and public keys	Blockchain	No	No
A Trusted Integrated Knowledge Dataspace (TIKD)	2021	Linux Unified Key Setup, ISO 27001, and ISO 27701	Roles and security levels	Data catalogs (DCAT)	Access control and data classification

and data classification (explained in the next section) in order to share data between users in the same or different organizations.

Typical state-of-the-art dataspaces implement security features such as access control authentication methods [13, 17], defined access roles [2, 15], user attributes [18]), and usage control [13, 15] in order to provide data sharing services. In addition to security aspects, dataspace approaches with sharing services cope with data fusion [17], usage control between multiple organizations[13], real-time data sharing [2], and privacy protection [18]. However, these approaches do not provide mechanisms for personal data handling in compliance with GDPR, privacy-aware log records, or privacy-protected interlinking with external resources. TIKD is based on a set of Linked Data vocabularies that support these aspects, e.g., the Data Protection Vocabulary (DPV) to cope with personal data handling, the Data Catalog Vocabulary[7] (DCAT) to cope with interlinking external resources, and PROV[8] to cope with user logs.

4   Description of the TIKD

The Trusted Integrated Knowledge Dataspace (TIKD) was designed in accordance with the ARK-Virus Project security requirements (see Sect. 2). The TIKD services (Fig. 2) define data permissions (Knowledge Graph integration, subgraph sharing, and data interlinking), user access grants (security control), and external resource integration (data interlinking) to provide a trusted environment for collaborative working.

TIKD is a multi-user and multi-organization dataspace with the capability of securely sharing information between an organization's users. The security control module asserts that only granted users, from the same organization, can access KGs, shared information, and interlinked data. This module follows a context-based approach considering security roles and data classifications (explained later in this section), i.e., access to the organization's data is determined by the user's context and the target data classification. The next subsections explain each of these services.

4.1   Knowledge Graph Integration

The Knowledge Graph integration service (Fig. 2, Knowledge Graph integration) is a central component of the TIKD. This service defines a dataspace where i) multiple users can work on a KG within an organization, ii) multiple organizations can create

---

[7] https://www.w3.org/TR/vocab-dcat-2/.

[8] https://www.w3.org/TR/prov-o/.

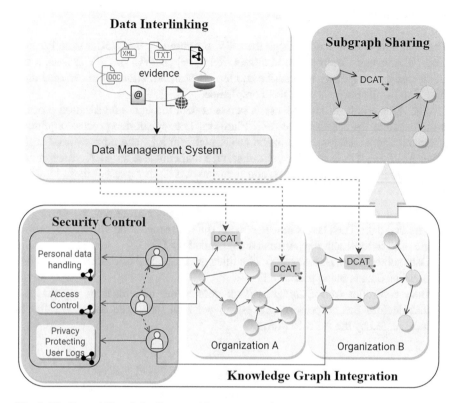

Fig. 2 The Trusted Knowledge Integrated Dataspace services

KGs, iii) linking to datasets by means of DCAT, instead of graphs/data, is supported, iv) fine-grained record linkage via DCAT records is supported, and v) evidence and KG integration/linking are supported.

4.2 Security Control

The security control service (Fig. 2, security control) is the main service of the TIKD. This service makes use of Linked Data vocabularies to handle personal data, access control context specification, and privacy protecting user logs. The following subsections explain in detail each one of these services.

4.2.1   Personal Data Handling

Personal data is described through the DPV, proposed by the W3C's Data Privacy Vocabularies and Controls Community Group [16] (DPVCG). DPV defines a set of classes and properties to describe and represent information about personal data handling for the purpose of GDPR compliance.

The ARK Platform collects user's personal data through a registration process which enables the access to the ARK Platform. The registration process requires a username, email address, organization role, platform role, and a password. On the other hand, the TIKD security control service authenticates an ARK user through their username, or email address, and their password. To represent these kinds of personal data, the following DPV classes (Fig. 3) were used:

- Personal data category (*dpv:PersonalDataCategory*): identifies a category of personal data. The classes *dpv:Password*, *dpv:Username*, and *dpv:EmailAddress* are used to represent the personal data handled by TIKD.
- Data subject (*dpv:DataSubject*): identifies the individual (the ARK user) whose personal data is being processed.
- Data controller (*dpv:DataController*): defines the individual or organization that decides the purpose of processing personal data. The data controller is represented by the ARK Platform.

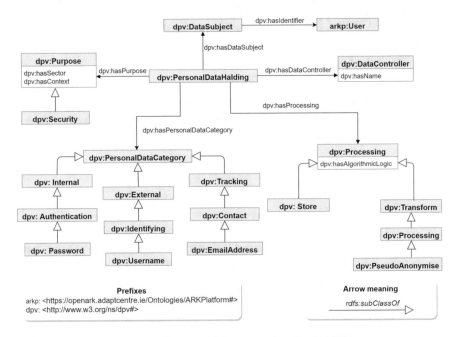

Fig. 3 DPV classes used to describe personal data annotations for the TIKD

Table 3 Data classification access level alongside availability release and unauthorized access impact

Data classification	Availability	Unauthorized access impact
Public	Open access	Low
Internal	Organization members	Low
Confidential	Selected members	Medium
Restricted	Selected members	High

- Purpose (*dpv:Purpose*): defines the purpose of processing personal data. The security class (*dpv:Security*) is used to define the purpose.
- Processing (*dpv:Processing*): describes the processing performed on personal data. In this sense, the ARK Platform performs the action of storing (*dpv:Store*) the ARK user's personal data and TIKD performs the action of pseudonymizing[9] (*dpv:PseudoAnonymise*) the data to perform log actions.

4.2.2 Data Classification

The ARK Platform uses different data classification levels to define the visibility, accessibility, and consequences of unauthorized access to an access control entity[10] (ACE). An ACE defines a KG representing an ARK Project or an ARK Risk Register.[11] Table 3 describes each data classification access level. Considering the data classification levels, a public ACE can be accessed by the general public and mishandling of the data would not impact the organization. Conversely, the impact of unauthorized access or mishandling of a restricted ACE would seriously impact the organization, staff, and related partners. The integration of data classification to the TIKD provides certainty about who can access which data based on the constraints of the data itself.

An ACE can be associated with one or more data entities. A data entity[12] represents an individual unit (data) or aggregate of related data (group of data), each of which can have its own data classification. The data classification of data entities follows a hierarchical structure whereby the ACE represents the root node and the data entities represent a child or sub-node. In line with this hierarchy, sub-nodes cannot have a less restrictive access level than the root/parent node, i.e., if the ACE data classification is defined as internal, then its data entities cannot be classified as public.

---

[9] The action of replacing personal identifiable information with artificial identifiers.

[10] https://openark.adaptcentre.ie/Ontologies/ARKPlatform/index-en.html#AccessControlEntity.

[11] These terms are explained later in this section.

[12] https://openark.adaptcentre.ie/Ontologies/ARKProjects/index-en.html#DataEntity.

4.2.3   Access Control

The access controls (AC) were designed to meet the privacy-aware evidence distillation requirement (Table 1) of providing access to users with the appropriate level of clearance. The AC follows a context-based approach, alongside data classifications, to allow or deny access to an ACE.

Considering the security role, the AC mediates every request to the ARK Platform, determining whether the request should be approved or denied. TIKD defines a context-based access control (CBAC) model, based on context and role specification, where data owners can authorize and control data access. In a CBAC model, policies associate one or more subjects with sets of access rights, pertaining to users, resources, and the environment, in order to grant or deny access to resources. In this sense, the set of policies consider the current user's context information to approve or deny access to ACEs. The AC takes into account the following authorization access elements (Fig. 4):

- ARK user: an ARK user has associated an organization role, a platform status, and a security role. The security role is assigned after creating or relating an ARK user with an ACE.
- ARK Platform status: defines the user's status in the ARK Platform, e.g., active, pending, update pending, and update approved.
- Organization role: each organization has the facility to define their own organization and security role hierarchy independently. The ARK Platform contains some predefined security roles (admin, owner, collaborator, and read-only) and platform roles (frontline staff, clinical specialist, and safety manager, among others). However, these roles can be extended according to the organization's requirements.

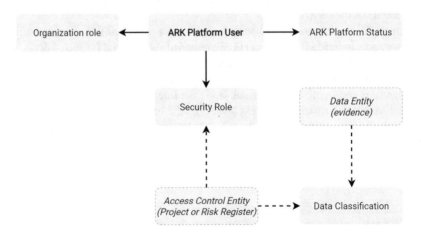

Fig. 4 Access authorization elements

- Security role: an ARK user is associated with an ACE through their security role. In this sense, an ARK user could take one of the following predefined security roles: admin, owner, collaborator, or read-only, where owner and admin are the highest level roles.
- Data classification: defines the data visibility of ACEs and data entities considering the rules from Table 3.
- Data entity (evidence): refers to interlinked data. A user can interlink data from external sources to enrich an ACE. In the ARK Platform context, this interlinked data is considered "evidence." The evidence is under the owning organization's jurisdiction, i.e., only users from the same organization have access. Additionally, the evidence can take any of the data classification access level, i.e., an evidence could be defined as public, internal, confidential, or restricted.

The TIKD AC (Fig. 5) is based on the Social Semantic SPARQL Security for Access Control Ontology (S4AC). The S4AC is a fine-grained access control over Resource Description Framework (RDF) data. The access control model provides the users with means to define policies to restrict the access to specific RDF data

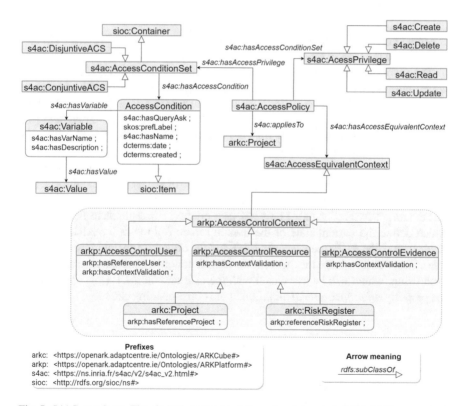

Fig. 5 S4AC ontology. The dashed rectangle defines the integrated ARK Platform context information

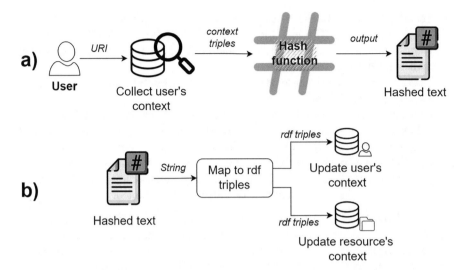

Fig. 6 Hash string generation process

at named graphs or triple level. It reuses concepts from SIOC,[13] SKOS,[14] WAC,[15] SPIN,[16] and the Dublin Core.[17]

The main element of the S4AC model is the access policy (Fig. 5). An access policy defines the constraints that must be satisfied to access a given named graph or a specific triple. If the access policy is satisfied, the user is allowed to access the data, but if not, access is denied. TIKD access policies consider ARK user context (the ARK Platform status, the security role, organization role) and the data classification of the target resource (an ACE or a data entity).

The TIKD AC integrates the *arkp:AccessControlContext* class to the S4AC to define the ARK Platform context information. The ARK user's context information is represented as a hash string to validate the relationship between the ARK user and the target ACE (Fig. 6a). The ARK user context corresponds to the attributes which define the current state of the user in relationship with the ARK Platform (their status), the ACE (their security role), and the organization (their organization's role). These attributes are the input for the hash function to generate a corresponding hash string, which will be associated with the user and the ACE (Fig. 6b), through the property *arkp:hasContextValidation* in the corresponding class.

---

[13] http://rdfs.org/sioc/spec.

[14] http://www.w3.org/TR/skos-reference.

[15] http://www.w3.org/wiki/WebAccessControl.

[16] http://spinrdf.org.

[17] http://dublincore.org/documents/dcmi-terms.

Table 4 Data classification

Data classification	Security role
Public	– Admin – Owner – Collaborator – Read-only
Internal	– Admin – Owner – Collaborator – Read-only
Confidential	– Admin – Owner – Selected users
Restricted	– Admin – Owner – Selected users

4.2.4 Policy Specification

The TIKD AC defines two kinds of policies: global and local. The global policy and context policy compare the ARK user's context hash string against the hash string from the target ACE (ARK Project or ARK Risk Register). If both are the same, access to the ACE is granted; otherwise, it is denied. The local policy considers the data classification of ACEs and data entities to grant or deny access to an ARK user. Table 4 describes the data classification and the security role required to access the data. Local polices check if an ARK user's security role has the correct permissions to access the requested data.

A TIKD AC policy is defined by the tuple $P = < ACS, AP, R, AEC >$, where ACS stands for the set of access conditions, AP for the access privilege (create, delete, read, update), R for the resource to be protected, and AEC for the access evaluation context. An access condition is defined through a SPARQL ASK query, representing a condition to evaluate a policy or policies. The AEC is represented by the hash string value produced from the ARK user context.

The policy specification process selects the corresponding global and local policies. After an ARK user sends a request to access an ACE (Fig. 7a), the global policy is selected (Fig. 7b, c). The local policies include the ACE and their data entity data classification configuration (Fig. 7d), which defines data authorization access; according to this configuration, the corresponding ASK queries are selected.

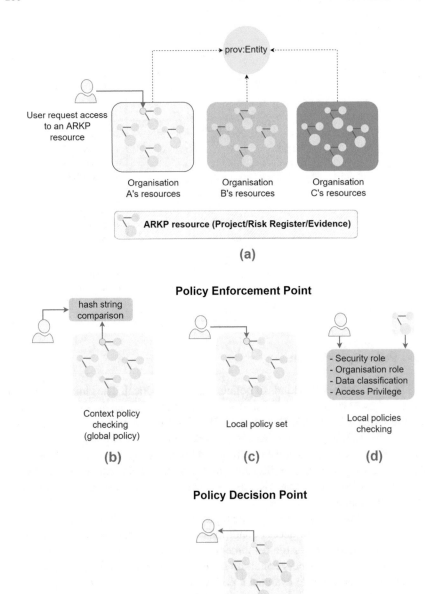

Fig. 7 Policy enforcement and decision process

4.2.5 Policy Enforcement

The policy enforcement process executes the corresponding ASK queries and returns the decision to grant or deny access to the ACE (Fig. 7e). The global policies are executed first, and if the ASK query returns a true value, then the local policies are executed. In the ARK Platform, the user context could change at any moment by several factors, e.g., update to organization role, organization change, update to security role, update to platform status, etc. The global policy validates the ARK user context with the target ACE. A correct validation means that the user is granted access to the ACE. On the other hand, the local policy defines a fine-grained data access for data entities allowed to be accessed by the user.

4.2.6 Privacy Protecting User Logs

Finally, the privacy protecting user logs record the actions performed by users during their sessions on the ARK Platform for historical record purposes. User information is pseudonymized in the log data, using the SHA-3 algorithm, by combining the username, email, and registration date parameters.

The user logs record user activities on the platform and the results retrieved by the system (failure, success, warning, etc.) during a session, e.g., if the user tries to modify the KG but their role is read-only, the privacy protecting user log process will record this activity as well as the failure response from the system. The PROV ontology[18] is used to implement the privacy protecting user logs following an agent-centered perspective i.e., focusing on the people or organizations involved in the data generation or manipulation process.

4.3 Data Interlinking

TIKD supports the integration of KGs and also provides special support for the integration of potentially sensitive external resources (a data interlinking requirement of the ARK-Virus Project), by means of an interlinking service (Fig. 2 data interlinking).

The data interlinking service allows users to add data from an external source as evidence to a risk management project. Evidence is used as supporting data for the KG, providing findings or adding valuable information to enrich the content of the KG. The multi-user and multi-organizational nature of the ARK Platform requires an access restriction to evidence. In this sense, the access control service restricts access to evidence only to users from the same organization.

---

[18] https://www.w3.org/TR/2013/NOTE-prov-primer-20130430/.

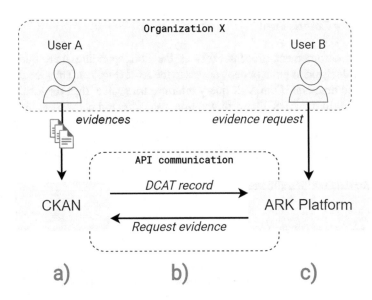

Fig. 8 Data interlinking process

The TIKD data interlinking process was implemented through CKAN, a data management system which enables organizations and individuals to create and publish datasets, and associated metadata, through a web interface. CKAN is an open-source community project, thus providing a rich number of extensions/plugins.

The data interlinking process (Fig. 8) consists of three main steps: (i) dataset creation, (ii) API communication, and (iii) evidence integration. In step one, a user creates a dataset, containing evidence resources, using CKAN (Fig. 8a). In step two, the API communication (Fig. 8b) handles the evidence requests, i.e., the ARK Platform requests evidence metadata via the CKAN API which returns the requested information as a DCAT record. In step three, (Fig. 8c), users request access to evidence metadata through the ARK Platform, which validates the user's grants based on the access control, in order to interlink the evidence to the project KG.

Datasets created using CKAN can be classified as public or private—public datasets are visible to everyone and private datasets are visible only to users of the owning organization. Private datasets align with the internal classification of the ARK data classification model.

As the ARK-Virus requirements define the visibility of data through a more complex structure than CKAN, the default data classification of CKAN will be altered to align with the ARK data classifications. This will be achieved through CKAN extensions that allow for dataset access to be more restricted than the current private/internal visibility level.

4.4 Data Sharing

TIKD provides the functionality to share data between users from the same organization, considering the ARK-Virus security requirements. Data sharing is performed by means of the data interlinking service and data classifications.

The sharing mechanism allows users from the same organization to share evidence through CKAN. The data classification of the shared evidence remains under the control of the owner or the admin user, i.e., the data classification of shared evidence is not transferable between projects.

The data interlinking service and the sharing mechanism allow organizations to reuse data between projects. Evidence data is shared under a secured scenario where the access control and the data classification determine the visibility of the evidence.

4.5 Subgraph Sharing

The ARK-Virus Project defines a collaborative environment where users can share data from ACEs using a privacy-aware sharing mechanism whereby confidential or sensitive data cannot be shared outside an organization. This sharing functionality helps to reuse information to enrich related ACEs. In this sense, the subgraph sharing service (Fig. 2, subgraph sharing) helps to extend or complement information from one ACE to another.

The subgraph sharing process (Fig. 9) considers the access control policies, from the security control service, to determine which data is accessible to an organization's users and which data is not, e.g., ACE data defined as public (P-labeled nodes) could be reused by any member of the same organization, whereas restricted data (R-labeled node) cannot be shared with any other member of the organization, i.e., the data defined as restricted is enabled only for the owner of the data, the organization admin, and other explicitly specified users. The accessibility is defined by the data classification (Table 4) of the ACE and its data entities. If the user's request is allowed, the corresponding subgraph is returned.

The sharing methods defined by TIKD enable collaboration between members from the same organization. The subgraph sharing enables the reuse of data between ACEs. These sharing functionalities are handled by the access control policies which determine whether the requester (user) is able to access evidence or subgraph information.

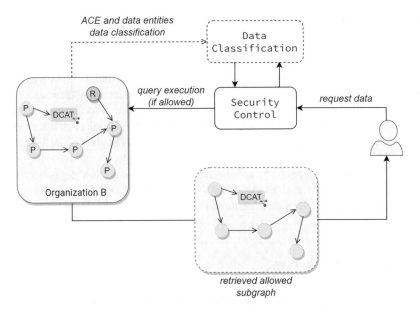

Fig. 9 Subgraph sharing process. P-labeled nodes represent public data, while R-labeled nodes represent restricted nodes

5 Security and Privacy Evaluations of the ARK Platform

This section presents a security evaluation of the ARK Platform considering the requirements of the ISO 27001 (ISO/IEC 27001) standard and the privacy control requirements of the ISO 27701 (ISO/IEC 27701). The ISO 27001[19] is a specification for information security management systems (ISMS) to increase the reliability and security of systems and information by means of a set of requirements.

The second standard considered for the evaluation of TIKD is the ISO 27701.[20] The ISO 27701 is the international standard for personally identifiable information (PII). This standard defines a privacy information management system (PIMS) based on the structure of the ISO 27001. The standard integrates the general requirements of GDPR, the Information Security Management System (ISMS) of ISO 27001, and the ISO 27002 which defines the best security practices.

The requirements of the ISO 27701 include 114 security controls of Annex A of ISO/IEC 27001 and the guide of ISO/IEC 27002 about how to implement these security controls. The ISO 27701 defines specific security controls that are directly related to PII, which are grouped into two categories: PII processors (Annex A) and PII controllers (Annex B).

---

[19] https://www.iso.org/isoiec-27001-information-security.html.

[20] https://www.iso.org/standard/71670.html.

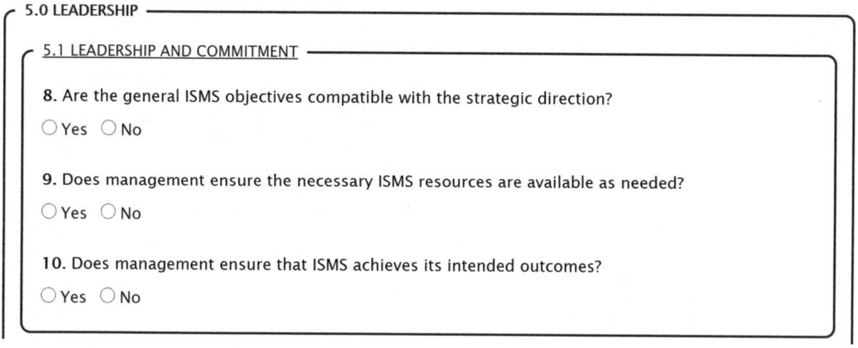

```
┌─ 5.0 LEADERSHIP ──────────────────────────────────────────────

│  ┌─ 5.1 LEADERSHIP AND COMMITMENT ─────────────────────────

│  │  8. Are the general ISMS objectives compatible with the strategic direction?

│  │  ○ Yes  ○ No

│  │  9. Does management ensure the necessary ISMS resources are available as needed?

│  │  ○ Yes  ○ No

│  │  10. Does management ensure that ISMS achieves its intended outcomes?

│  │  ○ Yes  ○ No
```

Fig. 10 Excerpt of the ISO 27001 GAT

5.1 Security Evaluation

The security evaluation of the ARK Platform[21] was conducted using the ISO 27001 GAT. The ISO 27001 GAT can be used to identify gaps in ISO 27001 compliance.

The ISO 27001 GAT consists of 41 questions divided into 7 clauses. Each clause is divided into sub-clauses, containing one or more requirements (questions). For example, the "Leadership" clause is divided into three sub-clauses: the first sub-clause is *leadership and commitment* which contains three requirements. The first requirement is: "are the general ISMS objectives compatible with the strategic direction?"; a positive answer means that the ISMS supports the achievement of the business objectives. (Figure 10 illustrates this example.)

The ISO 27001 GAT was conducted on the ARK Platform both before and after implementing TIKD. Before implementing TIKD, the ARK Platform only used access control, based on authentication process, to provide access to the platform. The results of both evaluations can be seen in Table 5 where *#Req.* defines the number of requirements for each sub-clause, *Impl* defines the number of implemented requirements, and *%Impl.* defines the percentage of implemented requirements.

It can be seen that compliance with the ISO 27001 standard increased, from 54% to 85%, after implementing the TIKD on the ARK Platform. There was a notable increase in the "Operation" and "Performance evaluation" clauses after the TIKD was employed. However, there are still some requirements that are yet to be addressed in order to achieve an increased level of compliance with the ISO 27001 standard. Table 6 outlines these unaddressed requirements as well as the action needed to implement them.

---

[21] The evaluation was performed by three computer scientists with strong backgrounds in Linked Data and security systems. The first evaluation was performed in February 2021 and the second was performed in April 2021.

Table 5 ARK Platform security evaluation, before and after implementing the TIKD, based on the ISO 27001 GAT

Clause	Sub-clause	#Req.	Before TIKD		After TIKD	
			Impl.	%Impl.	Impl.	%Impl.
Context of the organization	Understanding the organization and its context	3	2	66.67%	2	66.67%
	Understanding the needs and expectations of interested parties	2	2	100%	2	100%
	Determining the scope of the information security management system	1	0	0%	1	100%
	Information security management system	1	0	0%	1	100%
Leadership	Leadership and commitment	3	3	100%	3	100%
	Policy	2	0	0%	2	100%
	Organizational roles, responsibilities, and authorities	1	1	100%	1	100%
Planning	Actions to address risks and opportunities	3	1	33.33%	3	100%
	Information security objectives and planning to achieve them	2	1	50%	2	100%
Support	Resources	1	1	100%	1	100%
	Competence	1	1	100%	1	100%
	Awareness	1	1	100%	1	100%
	Communication	1	1	100%	1	100%
	Documented information	3	2	66.67%	3	100%
Operation	Operational planning and control	3	3	100%	3	100%
	Information security risk assessment	1	0	0%	1	100%
	Information risk treatment	2	0	0%	1	50%
Performance evaluation	Monitoring, measurement, analysis, and evaluation	2	0	0%	2	100%
	Internal audit	2	0	0%	1	50%
	Management review	2	2	100%	2	100%
Improvement	Nonconformity and corrective action	3	0	0%	0	0%
	Continual improvement	1	1	100%	1	100%
Total and average (%)		41	22	53.66%	35	85.37%

Table 6 Unaddressed clauses and the action needed to comply with the ISO 27001 requirement

Clause	Requirement	Action to perform
Context of the organization	– Did the organization determine how internal and external issues could influence the ISMS ability to achieve its intended outcomes?	– Organization stakeholders will define how internal and external issues can affect the security model
Operation	– Is there a documented list with all controls deemed as necessary, with proper justification and implementation status?	– A set of actions, to address risks, will be established. – This will be periodically reviewed, tested, and revised where practicable
Performance evaluation	– Are internal audits performed according to an audit program, results reported through an internal audit report, and relevant corrective actions raised?	– An internal audit will be performed
Improvement	– Does the organization react to every nonconformity? – Does the organization consider eliminating the cause of the nonconformity and, where appropriate, take corrective action? – Are all non-conformities recorded, together with corrective actions?	– Nonconformity data will be collected from the stakeholders. – A document with the procedure(s) to address non-conformities, including the identification of causes and actions to prevent recurrence, will be prepared. – The results will be recorded for future reference and corresponding documentation will be prepared

5.2 Privacy Information Evaluation

The privacy information evaluation of the ARK Platform[22] was conducted considering the clauses defined in the ISO/IEC 27701:2019[23] Annex A and B, concerned with the personal data handling. Annex A, PIMS-specific reference control objectives and controls, defines the control requirements for PII controllers. Annex B, PIMS-specific reference control objectives and controls, defines the control requirements for PII processors.

The ISO 27701 evaluation followed the same configuration as the ISO 27001 GAT evaluation (conducted before and after TIKD). For this evaluation, before the implementation of TIKD, the ARK Platform had documented personal data

---

[22] The evaluation was performed by the same three computer scientists from the first evaluation. The first evaluation was performed in June 2021 and the second was performed in August 2021.

[23] https://www.iso.org/obp/ui/#iso:std:iso-iec:27701:ed-1:v1:en.

Table 7 ARK Platform privacy information evaluation, before and after implementing the TIKD, based on the ISO 27701 Annex A and B

Annex	Clause	#Ctrl.	Before TIKD		After TIKD	
			Impl.	%Impl.	Impl.	%Impl.
A.7.2	Conditions for collection and processing	8	7	87.5%	7	87.5%
A 7.3	Obligations to PII principals	10	4	40%	10	100%
A 7.4	Privacy by design and privacy by default	9	9	100%	9	100%
A 7.5	PII sharing, transfer, and disclosure	4	2	50%	3	75%
B 8.2	Conditions for collection and processing	6	6	100%	6	100%
B 8.3	Obligations to PII principals	1	1	100%	1	100%
B 8.4	Privacy by design and privacy by default	3	0	0%	3	100%
B 8.5	PII sharing, transfer, and disclosure	8	2	25%	4	50%
Total and average (%)		49	32	64.4%	44	90.63%

handling; however, some elements were not fully implemented. After implementing TIKD on the ARK Platform, all personal data handling elements were included. Table 7 shows the evaluation results, where the first and second columns represent the Annex and the target clause. The third column defines the number of control requirement for the corresponding clause. The before TIKD group of columns defines the number and percentage of the implemented control requirements for the corresponding Annex clause. The same applies for the after TIKD group of columns.

According to the evaluation results, Annex A results (A 7.2–7.5) show a compliance improvement after implementing TIKD, mainly in A 7.3 and A 7.5. In the case of A 7.3, obligations to PII principals, the ARK Platform before TIKD was less accurate than the ARK Platform after TIKD implementation as some control requirements related to implementation aspects were only covered by the latter. In A 7.5, PII sharing, transfer, and disclosure, the ARK Platform before TIKD complied with the documented control requirements; meanwhile, the ARK Platform after TIKD complied with both the documented and implementation control requirements. In this clause, both versions did not comply with the control requirement of "Countries and international organizations to which PII can be transferred are identified and documented" as sharing information with international organizations is beyond the scope of the ARK Platform.

Similar to Annex A, the Annex B results (B 8.2–8.5) show a compliance improvement after implementing TIKD. In B 8.5, PII sharing, transfer, and disclosure control requirements, the low percentage in the ARK Platform after TIKD is due to the fact that the ARK-Virus Project does not define subcontractors for processing personal data. Additionally, the control requirements of B 8.5 are related to countries and international organizations—this is beyond scope of the ARK-Virus Project. In B 8.4, privacy by design and privacy by default, the ARK Platform after TIKD satisfies the control requirements; however, the before TIKD version did not comply with any of the control requirements as they are all related to implementation aspects which were not covered by this version.

6   Conclusions

In this chapter the Trusted Integrated Knowledge Dataspace (TIKD) was presented as an approach to securely share data in collaborative environments by considering personal data handling, data privileges, access control context specification, and a privacy-aware data interlinking.

TIKD was implemented in the ARK Platform, considering the security require-ments of the ARK-Virus Project, to explore the extent to which an integrated sharing dataspace, based on Linked Data technologies, personal data handling, data privileges, and interlinking data, contributes to building a trusted sharing dataspace in a collaborative environment. In comparison with state-of-the-art works TIKD integrates solutions for security aspects in compliance with the ISO 27001 security information standard and GDPR-compliant personal data handling in compliance with the ISO 27701 privacy information standard as part of the data security infrastructure.

The TIKD evaluation considers the requirements of the security standard ISO 27001 and the control requirements of the privacy information standard ISO 27701. The security evaluation of the ARK Platform was conducted using the ISO 27001 Gap Analysis Tool (GAT). The evaluation compared two versions of the ARK Platform, a version before TIKD implementation and a version after TIKD implementation. According to the results, the implementation of the TIKD achieved an 85% ISO 27001 compliance score, improving the security aspects of the ARK Platform as compared to the version before TIKD implementation (54% ISO 27001 compliance score).

The privacy information evaluation was conducted considering the control requirements defined by the ISO/IEC 27701:2019 standard and following the same configuration as the security evaluation. According to the results, the ARK Platform after implementing TIKD achieved a 91% ISO 27701 compliance score, improving the privacy information aspects defined by the standard when compared to the version before TIKD implementation (64% ISO 27701 compliance score).

Future work will focus on addressing the remaining ISO 27001 standard requirements. Additionally, the TIKD will be evaluated by the project stakeholders and their feedback will be used to distill further requirements.

References

1. Corrigan, S., Kay, A., O'Byrne, K., Slattery, D., Sheehan, S., McDonald, N., Smyth, D., Mealy, K., & Cromie, S. (2018). A socio-technical exploration for reducing & mitigating the risk of retained foreign objects. *International Journal of Environmental Research and Public Health, 15*(4). https://doi.org/10.3390/ijerph15040714
2. Curry, E., Derguech, W., Hasan, S., Kouroupetroglou, C., & ul Hassan, U. (2019). A real-time linked dataspace for the Internet of Things: Enabling "pay-as-you-go" data management in smart environments. *Future Generation Computer Systems, 90*, 405–422. https://doi.org/10.1016/j.future.2018.07.019

3. Dankar, F. K., & Badji, R. (2017). A risk-based framework for biomedical data sharing. *Journal of Biomedical Informatics, 66*, 231–240. https://doi.org/10.1016/j.jbi.2017.01.012

4. Fan, K., Wang, S., Ren, Y., Li, H., & Yang, Y. (2018). Medblock: Efficient and secure medical data sharing via blockchain. *Journal of Medical Systems, 42*(8), 1–11. https://doi.org/10.1007/s10916-018-0993-7

5. Franklin, M., Halevy, A., & Maier, D. (2005). From databases to dataspaces: A new abstraction for information management. *SIGMOD Record, 34*(4), 27–33. https://doi.org/10.1145/1107499.1107502

6. Franklin, M., Halevy, A., & Maier, D. (2008). A first tutorial on dataspaces. *Proceedings of the VLDB Endowment, 1*(2), 1516–1517. https://doi.org/10.14778/1454159.1454217

7. Hernandez, J., McKenna, L., & Brennan, R. (2021). Tikd: A trusted integrated knowledge dataspace for sensitive healthcare data sharing. In *2021 IEEE 45th Annual Computers, Software, and Applications Conference (COMPSAC)* (pp. 1855–1860). IEEE.

8. Jeffery, S. R., Franklin, M. J., & Halevy, A. Y. (2008). Pay-as-you-go user feedback for dataspace systems. In *Proceedings of the 2008 ACM SIGMOD International Conference on Management of Data, SIGMOD '08* (pp. 847–860). New York, NY: Association for Computing Machinery. https://doi.org/10.1145/1376616.1376701

9. Junior, A. C., Basereh, M., Abgaz, Y.M., Liang, J., Duda, N., McDonald, N., & Brennan, R. (2020). The ARK platform: Enabling risk management through semantic web technologies. In: J. Hastings & F. Loebe (Eds.) *Proceedings of the 11th International Conference on Biomedical Ontologies (ICBO)*, Italy, September 17, 2020, *CEUR Workshop Proceedings* (Vol. 2807, pp. 1–10). CEUR-WS.org (2020). http://ceur-ws.org/Vol-2807/paperM.pdf

10. Liu, X., Wang, Z., Jin, C., Li, F., & Li, G. (2019). A blockchain-based medical data sharing and protection scheme. *IEEE Access, 7*, 118943–118953. https://doi.org/10.1109/ACCESS.2019.2937685

11. McDonald, N. (2015). The evaluation of change. *Cognition, Technology and Work, 17*(2), 193–206. https://doi.org/10.1007/s10111-014-0296-9

12. McKenna, L., Liang, J., Duda, N., McDonald, N., & Brennan, R. (2021). Ark-virus: An ark platform extension for mindful risk governance of personal protective equipment use in healthcare. In *Companion Proceedings of the Web Conference 2021 (WWW '21 Companion)*, April 19–23, 2021, Ljubljana, Slovenia. New York, NY: ACM (2021). https://doi.org/10.1145/3442442.3458609

13. Munoz-Arcentales, A., López-Pernas, S., Pozo, A., Alonso, Á., Salvachúa, J., & Huecas, G. (2019). An architecture for providing data usage and access control in data sharing ecosystems. *Procedia Computer Science, 160*, 590–597 (2019). https://doi.org/10.1016/j.procs.2019.11.042

14. Nakamoto, S. (2009). Bitcoin: A peer-to-peer electronic cash system. http://www.bitcoin.org/bitcoin.pdf

15. Otto, B., Hompel, M. T., & Wrobel, S. (2019). *International data spaces* (pp. 109–128). Springer, Berlin (2019). https://doi.org/10.1007/978-3-662-58134-6_8

16. Pandit, H. J., Polleres, A., Bos, B., Brennan, R., Bruegger, B. P., Ekaputra, F. J., Fernández, J. D., Hamed, R. G., Kiesling, E., Lizar, M., Schlehahn, E., Steyskal, S., & Wenning, R. (2019). Creating a vocabulary for data privacy—the first-year report of data privacy vocabularies and controls community group (DPVCG). In *On the Move to Meaningful Internet Systems*, 2019, Rhodes, Greece, October 21–25, 2019, Proceedings, *Lecture Notes in Computer Science* (Vol. 11877, pp. 714–730). Springer (2019). https://doi.org/10.1007/978-3-030-33246-4_44

17. Sun, W., Huang, Z., Wang, Z., Yuan, Z., & Dai, W. (2019). A method and application for constructing a authentic data space. In *2019 IEEE International Conference on Internet of Things and Intelligence System, IoTaIS 2019*, Bali, Indonesia, November 5–7, 2019 (pp. 218–224). IEEE. https://doi.org/10.1109/IoTaIS47347.2019.8980430

18. Sun, Y., Yin, L., Sun, Z., Tian, Z., & Du, X. (2020). An IoT data sharing privacy preserving scheme. In *39th IEEE Conference on Computer Communications, INFOCOM Workshops 2020*, Toronto, ON, Canada, July 6–9, 2020 (pp. 984–990). IEEE. https://doi.org/10.1109/INFOCOMWKSHPS50562.2020.9162939

19. Villata, S., Delaforge, N., Gandon, F., & Gyrard, A. (2011). An access control model for linked data. In: R. Meersman, T. Dillon, & P. Herrero (Eds.), *On the Move to Meaningful Internet Systems: OTM 2011 Workshops* (pp. 454–463). Berlin: Springer.
20. Xu, Z., Sheng, Y.-P., He, L. R., & Wang, Y. F. (2016). Review on knowledge graph techniques. *Journal of University of Electronic Science and Technology of China, 45*(dzkjdxxb-45-4-589), 589. https://doi.org/10.3969/j.issn.1001-0548.2016.04.012
21. Yang, X., Li, T., Pei, X., Wen, L., & Wang, C. (2020). Medical data sharing scheme based on attribute cryptosystem and blockchain technology. *IEEE Access, 8,* 45468–45476. https://doi.org/10.1109/ACCESS.2020.2976894
22. Zillner, S., Bisset, D., Milano, M., Curry, E., García Robles, A., Hahn, T., Irgens, M., Lafrenz, R., Liepert, B., O'Sullivan, B., & Smeulders, A.E. (2020). Strategic research, innovation and deployment agenda—AI, data and robotics partnership. Third release. BDVA, euRobotics, ELLIS, EurAI and CLAIRE (Vol. 3)
23. Zillner, S., Curry, E., Metzger, A., Auer, S., & Seidl, R. E. (2017). European big data value strategic research & innovation agenda.
24. Zou, X. (2020). A survey on application of knowledge graph. *Journal of Physics: Conference Series, 1487,* 012016. https://doi.org/10.1088/1742-6596/1487/1/012016

Toward an Energy Data Platform Design: Challenges and Perspectives from the SYNERGY Big Data Platform and AI Analytics Marketplace

Fenareti Lampathaki, Evmorfia Biliri, Tasos Tsitsanis, Kostas Tsatsakis, Dimitris Miltiadou, and Konstantinos Perakis

Abstract Today, the need for "end-to-end" coordination between the electricity sector stakeholders, not only in business terms but also in securely exchanging real-time data, is becoming a necessity to increase electricity networks' stability and resilience while satisfying individual operational optimization objectives and business case targets of all stakeholders. To this end, the SYNERGY energy data platform builds on state-of-the-art data management, sharing, and analytics technologies, driven by the actual needs of the electricity data value chain. This paper will describe the layered SYNERGY Reference Architecture that consists of a Cloud Infrastructure, On-Premise Environments, and Energy Apps and discuss the main challenges and solutions adopted for (a) the design of custom pipelines for batch and streaming data collection and for data manipulation and analytics (based on baseline or pre-trained machine learning and deep learning algorithms) and (b) their scheduled, on-event, or real-time execution on the cloud, on-premise and in gateways, toward an energy data space. Particular focus will be laid on the design of the SYNERGY AI analytics marketplace that allows for trustful sharing of data assets (i.e., datasets, pipelines, trained AI models, analytics results) which belong to different stakeholders, through a multi-party smart contract mechanism powered by blockchain technologies.

Keywords Energy Data Spaces · Energy data platform · Architecture · Data pipelines · Data lifecycle management · Data sharing · Data analytics

F. Lampathaki (✉) · E. Biliri · T. Tsitsanis · K. Tsatsakis
Suite5 Data Intelligence Solutions, Limassol, Cyprus
e-mail: fenareti@suite5.eu

D. Miltiadou · K. Perakis
UBITECH, Thessalias 8 & Etolias, Chalandri, Greece

E. Curry et al. (eds.), *Data Spaces*, https://doi.org/10.1007/978-3-030-98636-0_14

1 Introduction

The European electricity sector is undergoing a major fundamental change with the increasing digitalization and roll-out of smart meters. This advent of the electricity sector modernization comes together with the fact that the power system is becoming more thoroughly monitored and controlled from "end to end" and through the whole value chain of stakeholders involved in the electricity system operation. This is a huge shift away from traditional monitoring and control approaches that have been applied exclusively over the transmission and distribution networks, since the smart electricity grid era is pushing sensing, control, and data collection at the edge of electricity networks, which needs to be further re-defined due to the wide penetration of distributed energy resources (DERs), such as renewable energy sources (RES), smart home devices and appliances (IoT-enabled), distributed storage, smart meters, and electric vehicles (EVs).

Distributed smart grid resources are associated with the generation of vast amounts of data, spanning SCADA systems information (generation, transmission, and distribution), smart metering and sub-metering information (demand), IoT device information, distributed generation and storage data, electric vehicle, and electricity market information, altogether characterized by continuously increasing growth rate, multi-diverse spatio-temporal resolutions, and huge volume. Such large datasets provide significant opportunities for better "end-to-end" monitoring, control, and operation of electric grids by enabling better understanding and offering further insights on all aspects affecting (directly or indirectly) the operation of the networks (and DERs, as new individual connected components of smart electricity grids) toward optimizing their performance (individually and network-wide), through advanced big energy data analytics [1, 2]. However, while the industry may now recognize the potential of Big Data, it struggles to translate that into action. A recent study from CapGemini [1] found that only 20% of smart grid stakeholders have already implemented Big Data analytics. There is a significant group (41%) with no Big Data analytics initiatives which compares quite unfavorably with take-up levels in other sectors.

The analytics opportunity for electricity sector stakeholders is there, and benefits are significant; however, recent studies have pointed out that electricity sector actors are reluctant to make the move due to high upfront costs and sheer complexity of data [1]. Taking data management and analytics away from their hands (in a trustful manner, thus reducing complexity and changing their mindsets) and offering to them easily digestible intelligence extracted from the advanced processing and analysis of highly diverse, variable, and volatile data streams (through ready-to-use trained algorithms that can be easily utilized in different contexts and business cases) could be the first step forward, toward enabling the realization of data-driven optimization functions that can pave a ROI-positive path to effectively solving operational and business challenges and highlighting the value of the big distributed data generated at the wealth of end points of the power system.

The value of similar approaches and implementation references has been already showcased in relevant reference implementations, mainly in the USA, where the introduction of advanced (and near-real-time) data analytics in the electricity sector proved to facilitate the early detection of anomalies, trends, possible security breaches, and other costly business interruptions and enable the avoidance of undesired costs, along with the creation of new profit opportunities [3]. The cost, investment, and resulting value of Big Data analytics and data-driven insights have different impacts on the grid's various major stakeholders. Each stakeholder has different roles and responsibilities in managing, storing, processing, protecting, owning, and using data. For instance, the value of Data Spaces and analytics for electricity grid operators lies on the fact that they can further optimize the operational stability and resilience of their network through improved demand and generation forecasting, advanced predictive maintenance, and management of their owned assets (lines, transformers, sub-station equipment, etc.), improve power quality and continuity of supply by avoiding interruptions due to equipment failures, optimize scheduling of maintenance activities, and enhance physical security of critical distribution network infrastructure.

In this context, this paper introduces the SYNERGY Reference Architecture that aims to allow electricity value chain stakeholders to simultaneously enhance their data reach and improve their internal intelligence on electricity-related optimization functions while getting involved in novel sharing/trading models of data sources and intelligence, in order to gain better insights and shift individual decision-making at a collective intelligence level. The SYNERGY Reference Architecture is based on state-of-the-art approaches from a technology perspective (in terms of data management, data analytics, data sharing, and data security techniques and technologies), as well as from a market perspective (considering the different data platforms that are introduced in Sect. 2). The different workflows that are enabled though the SYNERGY Reference Architecture are discussed highlighting the core challenges that have been jointly identified by representatives of the electricity data value chain and technology experts.

This chapter relates to the technical priorities of the European Big Data Value Strategic Research and Innovation Agenda [4], addressing the horizontal concerns "Data Management," "Data Protection," "Data Processing Architectures," "Data Analytics," and "Data Visualization" and the vertical concern "Industrial Data Sharing Platforms" of the BDV Technical Reference Model. In addition, the chapter relates to the "Knowledge and Learning" and "Reasoning and Decision Making" enablers of the AI, Data and Robotics Strategic Research, Innovation and Deployment Agenda [5].

2   Data Platforms

The unprecedented supply of data and the technological advancements in terms of storage and processing solutions, e.g., offered through on-demand computing power as a service through the cloud, are among the forces fueling the emergence of data as a new tradable good online. Data marketplaces and Data Spaces are the infrastructures through which this new market is realized. The market's growth however cannot be attributed solely to the technical innovations, notwithstanding their enabling role, but should be examined under the prism of demand and supply. The abundance of data created every day and the way data analysis can transform them to insights for more informed decision-making create incentives for businesses to develop a data sharing mentality and investigate data monetization approaches. Technical, motivational, economic, political, legal, and ethical challenges in fostering a data sharing mentality in an industry environment are numerous, yet realizing the prospective benefits from disrupting the current data siloed situation is an important first step toward seeking ways to overcome the aforementioned barriers.

A more concrete definition of a marketplace would be that of a "multi-sided platform, where a digital intermediary connects data providers, data purchasers, and other complementary technology providers" [6]. In practice, functionalities of data marketplaces extend beyond the implementation of the data trading action.

2.1 Generic-Purpose Data Hubs and Marketplaces

A number of leading data marketplaces have emerged over the last years, demonstrating significant diversity in the provided offerings, stemming from the target domain and scope and the underlying technologies. The data marketplace concept is inherently interdisciplinary, in the sense that it brings together technological, legal, and business knowledge in order to successfully capture and satisfy the underlying demand and supply data needs. In many cases, the marketplace services constitute an application of an underlying technology, built to support the data trading functionalities, but also independently exploited.

Indicative examples of data marketplaces are briefly presented below in order to give a comprehensive overview of the current status and future perspectives of these platforms and outline ways in which they could create an innovation environment for new digital business models:

- Datapace (https://datapace.io/) is a marketplace for IoT sensor data with technical and policy-based data verification and access to a worldwide network of sensors. It supports micropayments using a custom token (namely, the TAS which is native to the platform and has no use externally to it) and offers smart contracts based on a permissioned enterprise blockchain. The Datapace storage encrypts and anonymizes the access to the submitted data streams.
- The DX Network (https://dx.network/) is one of the largest blockchain-based business data marketplaces. It is API-based, therefore can be easily integrated into any data-enabled services, and focuses on real-time data streams, allowing asset trading at data point granularity which is based on its custom query language that leverages Semantic Web technologies.

- Dawex (https://www.dawex.com/en/) is a leading data exchange technology company and the operator of one of the largest global data marketplaces. Its global marketplace provides customizable data access control mechanisms, supports various data formats, and provides visualizations to evaluate data quality and contents. Representative data samples are created through custom algorithms to support this process. Data are hosted encrypted, and the platform has certification from independent data protection authorities to ensure regulatory compliance. Dawex also enables organizations to create their own data exchange platforms using its technology. Apart from the core data trading services, the platform offers machine learning algorithms to match data supply and demand, allowing for proactive suggestions to members.
- IOTA (https://data.iota.org/#/) is an open, feeless, and scalable distributed ledger, designed to support frictionless data and value transfer. IOTA's network, called Tangle, immutably records exchanges and ensures that the information is trustworthy and cannot be tampered with or destroyed and was designed to address blockchain inefficiencies in terms of transaction times and scalability. It is a secure data communication protocol and zero-fee microtransaction system for the IoT/M2M.
- Qlik DataMarket (https://www.qlik.com/us/products/qlik-data-market) offers an extensive collection of up-to-date and ready-to-use data from external sources accessible directly from within the company's data analytics platform Qlik Sense. It provides current and historical weather and demographic data, currency exchange rates, as well as business, economic, and societal data, addressing data augmentation needs in the contextualization and analysis of business data leveraging external sources. Integration is in this context effortless, and validation, profiling, and quality measures are provided to evaluate the data available in the market.
- Streamr (https://streamr.network/marketplace) offers a marketplace for real-time data, leveraging blockchain and Ethereum-based smart contracts for security-critical operations like data transfers. It provides tools and libraries to (a) create, process, visualize, and sell real-time data and (b) acquire and ingest real-time data to enable business intelligence. The marketplace is an application of the Streamr network, a massively scalable peer-to-peer network for transporting machine data in real time with the PubSub pattern. It also offers crowdsourcing functionalities to incentivize gathering of previously unavailable data.
- MADANA (https://www.madana.io/vision.html) aims to create a self-governing and community-driven market for data analysis through a platform that connects data providers, data analysis providers (called plugin providers in the platform's terminology), and consumers/buyers for data analysis results. Beyond a marketplace, MADANA aspires to be a platform for data analysis which provides secured computation, data monetization, and the outsourcing of analytics on demand. Purchases are based on smart contracts and the platform's custom cryptocurrency called MADANA PAX. Upon collection, data are encrypted and kept in a distributed storage. Access is not foreseen to be provided to raw data, so only analysis results can be purchased.

European projects have been also active in this promising field. The ICARUS (www.icarus2020.aero) [7] marketplace offers brokerage functionalities specialized in aviation data assets conforming to a common data and metadata model and provides smart contracts based on Ethereum. Safe-DEED (https://safe-deed.eu/) explores how technology, e.g., in the fields of cryptography and data science, can foster a data sharing mentality, incentivizing businesses and innovating business models.

Distributed ledger technology (DLT) applications are also extremely popular, showcasing numerous additional data marketplaces, e.g., (a) Wibson (https://wibson.org/), a decentralized blockchain-based marketplace allowing members of the public to profit from securely and anonymously selling their personal data, and (b) Datum (https://datum.org/), which enables decentralized storage of structured data on a smart contract blockchain and data brokerage using a smart token.

Depending on the type of data, both in terms of content and formats, the prospective buyers and sellers, the target industries, the employed technologies, etc., a long list of platforms offering data marketplace services, either exclusively or as a side product of their core/other businesses, can be compiled. When traded commodities extend beyond data to other data-based assets, e.g., processed data and extracted insights, the number of platforms that can be considered as relevant can easily explode. Identifying and examining all data marketplaces is not possible and would largely be out of scope for the current work. However, different approaches used in literature to study and group data marketplaces have been extensively studied in [8–11].

For many-to-many data marketplaces, additional attributes could be selected for a more fine-grained analysis, e.g., the choice between a centralized and a decentralized design. This architecture decision entails implementation implications and affects the overall marketplace operation in various ways. Indicatively, [11] highlight that in the centralized setting, the market intermediary trades off quality (provenance control) for lower transaction costs. In a decentralized setting, e.g., one implemented through a distributed ledger technology, transaction costs are higher and bottlenecks may emerge, yet there is increased provenance control and transparency.

An important attribute of data marketplaces is the contract drafting and enforcement process, which is typically one of the services provided by such platforms and is an integral part of asset trading. Stringent enforcement of contract terms in this scope is challenging, and several factors, including technical limitations and legal implications, need to be examined. Data protection and security mechanisms, as well as data privacy and confidentiality, should be ensured to foster trust among the platform members and to comply with applicable regulations, e.g., the General Data Protection Regulation (GDPR). Technical advancements can also help in this direction, e.g., multi-party computation (MPC), a cryptographic technique that enables joint data analyses by multiple parties while retaining data secrecy, is explored as a way to increase industry's willingness to participate in data marketplaces. Auditability should also be possible in industry data trading agreements, yet anonymity in transactions may also be required. Furthermore, licensing, ownership, and IPR of data and data products are contentious issues requiring careful definition and adjudication, which may not be possible to capture within blanket agreements [11]. License compatibility, in the case of combination and derivation, e.g., data assets as a result of data integration from multiple sources and/or data analysis processes, is also challenging. On a final note, for a marketplace to establish a vibrant business ecosystem that will render it sustainable, data interoperability achieved through agreed data and metadata models and common semantics is required. Especially in the case of data marketplaces connecting numerous suppliers and consumers, data discoverability, timely and secure exchange, effortless ingestion, and (re-)usability across diverse data sources, all facilitated by an appropriate level of automation, will allow the marketplace to scale and foster opportunities on monetizing data. Such considerations were taken into consideration in the scope of the SYNERGY positioning.

2.2   Energy Data Hubs and Marketplaces

In the energy domain, there is no mature state of the art about the role and potential data marketplaces. The recently evolving energy data hubs, though, provide some insights from research initiatives about the potential of the energy data marketplaces in the future.

The term energy data hub, or energy data space, is defined as an on-demand, back-end repository of historical and current energy data. The objective is to streamline energy data flows across the sector and enable consumers, authorized agents on consumer's behalf, and other users to access energy data. While there is an increasing interest about the penetration of energy data hubs, following the increased installation of smart equipment and the deregulation of the market in the energy value chain, the number of the existing implementations is rather narrow. The data hubs are mainly focusing on specific business stakeholders and business processes in the energy value chain [12], and thus a business-driven taxonomy of the different energy hubs is considered as follows:

- *Retail data/smart meter hubs* are defined as the data hubs at EU country level which are responsible for the management of smart metering data. Retail data hubs are introduced to address two primary issues: (a) secure equal access to data from smart metering and (b) increase efficiency in the communication between market parties, especially between network operators and retails for billing and switching purposes. There are many region-level implementations around the world considering the smart meter's deployment with the most prominent examples being:

  - The Central Market System (CMS) aka ATRIAS started in 2018, as the centralized data hub to facilitate the data exchange between market parties in Belgium. The CMS focuses on the data exchange between the DSOs and retail businesses and thus connects the databases of the network operators (who collect the data from the smart meters) with the relevant and eligible market parties. Other parties, like the transmission system operators and third-party service providers, may access the data as well.

– In Norway [13], the ElHub (Electricity Hub) facilitates the data exchange between market parties in Norway. ElHub is operated by the national TSO with the smart metering data to be collected via the DSOs and stored in the ElHub together with consumer data from the retailers. The customers are in full control of their data, which they can access via an online tool and thereby manage third-party access to their datasets.

- *Smart market data hubs* are defined as the data hubs at EU country level responsible for the management of energy market data. The major electricity market operators in Europe are handling energy market data hubs to share data with the different business stakeholders. Special reference can be made to the following market operators:

  – Nord Pool (https://www.nordpoolgroup.com/services/power-market-data-services/) which runs the leading power market in Europe and offers day-ahead and intraday markets to its customers. The day-ahead market is the main arena for trading power, and the intraday market supplements the day-ahead market and helps secure balance between supply and demand. Access on real-time market data is available online, though fine-grained data services (access on data per country, product, means of access, etc.) are offered by the company. More specifically, customized power data services may be provided to external interest parties, setting that way a market-based framework for data exchange.

  – EPEX SPOT energy market data hub (https://www.epexspot.com/en) which offers a wide range of datasets covering the different market areas, available through different modalities: from running subscriptions for files that are updated daily to access to one-shot historical data.

- *Smart grid data hubs*: This is a step beyond the currently deployed smart meter data hubs. Through their evolving role around Europe, the network operators aim to act as data hub providers beyond smart meter data, while their data hubs will be used to provide services for the network operators (e.g., data exchange between the DSO and the TSO) as well as for new market entrants with new business models (e.g., related to behind-the-meter services). Therefore, the role of network operators as grid-related data managers is expanding. Under this category, there are some very promising initiatives, which are further presented below:

  – At country/regional level, there are state network operators responsible to publish their data required for the normal operation of the grid. Toward this direction, the European Network of Transmission System Operators for Electricity (ENTSOE) is operating a Transparency Platform (https://transparency.entsoe.eu/) where the data from the national TSOs are published in order to facilitate the normal operation of the transmission grid in Europe.

- At the regional level, the distribution network operators have started making their data public to help other stakeholders and market parties with, e.g., better decision-making, create new services, and promote synergies between different sectors. As not all DSO data are suitable to be made publicly available due to potential breaches of security or violations of privacy regulations, it is important for DSOs to have a common understanding. For that reason, E.DSO made recently available a policy brief to illustrate the possibilities of open data from each member state, in terms of meaningful use cases [14]. Key highlights of open data repositories from DSOs (EDP in Portugal, ENEDIS in France) are to be considered for the future expansion of open data repositories in the EU.
- Moving beyond the national level is the PCI project of Data Bridge (now defined as an Alliance of Grid Operators, https://www.databridge.energy/) with the goal to ensure the interoperability of exchanging different types of data between a variety of stakeholders (like system operators, market operators, flexibility providers, suppliers, ESCOs, end customers). Types of data may include smart meter data (both low-voltage and high-voltage meter data), sub-meter data, operational data, market data required for functioning flexible energy market, reliable system operation, etc.

From the aforementioned analysis, it is evident that the main focus of the energy actors in the data management landscape is about establishing functional energy data hubs that will be able to provide useful information to selected stakeholders of the energy value chain in a unified way. The concept of enabling the different energy stakeholders to match and trade their energy data assets and requirements in a marketplace environment does not exist yet at large scale. There are some early implementations of generic data marketplaces that enable management of data from the energy sector, which include (in addition to Dawex that has been already analyzed in Sect. 2.1 and includes an energy-specific solution with focus on smart home and renewable source data):

- Snowflake data marketplace (https://www.snowflake.com/datasets/yes-energy) is a data hub that enables data providers to leverage and monetize their data. In this platform, Yes Energy, the industry leader in North American power market data and analytic tools, acts as a data provider in the platform by collecting, managing, and continuously delivering real-time and historical power market data series including market data, transmission and generation outages, real-time generation and flow data, and load and weather forecasts.
- The re.alto marketplace (https://realto.io/) represents the first mature attempt to provide a European API marketplace for the digital exchange of energy data and services. Established in 2019, re.alto data marketplace enables companies to capture, organize, and share data and services easily, quickly, and securely. So far, the datasets available in the platform span between energy market data, asset generation data, weather data, energy metering, and smart home data. In addition, AI applications such as generation and demand forecasts, price forecasts, etc. are made available through the marketplace.

- The ElectriCChain (http://www.electricchain.org) is defined as an Open Solar data marketplace with an initial focus on verifying and publishing energy generation data from the ten million solar energy generators globally on an open blockchain. The ElectriCChain project supports the development of open standards and tools to enable generation asset owners to publish solar electricity generation data and scientists, researchers, and consumers to have access on the data and insights they need.

On the other hand, in the field of IoT solutions (as the wider term that covers the smart assets deployed in the electricity network spanning from network devices, smart meters, home automation solutions, DER data loggers, etc.), there is an ongoing discussion about the importance of the data and the way to put IoT data to work and cash, offering the information to third parties through data marketplaces. There are many small-scale/proof-of-concept initiatives of IoT data marketplaces to collect sensor data which data providers source from smart home appliances and installations in people's homes and smart cities, while companies looking to understand consumer behavior can leverage such machine data directly from the marketplaces in real time. The most prominent solutions include Datapace (https://datapace.io/) that offers blockchain-powered secure transactions and automated smart contracts to sell and buy data streams from any source, physical assets, autonomous cars, drones, and the IOTA marketplace that has been also mentioned in Sect. 2.1.

From the aforementioned analysis, it is evident that the concept of regulated and standardized energy data marketplaces is new for a domain that is still undergoing its digital transformation. There is an ongoing work to design and develop standards-based data hubs to ensure interoperability of exchanging different types of data between a variety of energy stakeholders, but still the value of such data that can be made available via data platforms and marketplaces remains largely unexplored.

3   SYNERGY Reference Architecture

In an effort to leverage such unique data-driven opportunities that the electricity data value chain presents, our work is focused on the development of an all-around data platform that builds on state-of-the-art technologies, is driven by the actual needs of the different stakeholders, and turns over a new leaf in the way data sharing and data analytics are applied. Taking into consideration the different use cases and requirements of the different energy stakeholders as well as the state of play described in Sect. 2, the reference architecture of the overall SYNERGY platform has been conceptually divided into three main layers as depicted in Fig. 1:

- The **SYNERGY Cloud Infrastructure** that consists of (a) the *Core Big Data Management Platform*, essentially including the Energy Big Data Platform and the AI Analytics Marketplace which are instrumental for all functionalities that SYNERGY supports at all layers, and (b) the *Secure Experimentation*

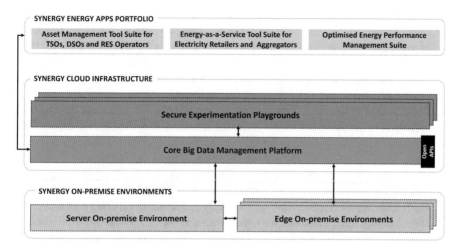

Fig. 1 SYNERGY three-layered high-level architecture

Playgrounds (SEP) which are realized in the form of dedicated virtual machines that are spawned per organization to ensure that each electricity data value chain stakeholder is able to execute Big Data analytics in isolated and secure environments in the SYNERGY Cloud Infrastructure.

- The **SYNERGY On-Premise Environments (OPE)** which are executed in the energy stakeholders' premises for increased security and trust and can be distinguished in the server environment and the edge environments that are installed in gateways. The On-Premise Environments are not self-standing, but always communicate with the SYNERGY Cloud Infrastructure to deliver their intended functionality.

- The **SYNERGY Energy Apps Portfolio** that embraces the set of applications addressed to the needs of (a) DSOs (distribution system operators), TSOs (transmission system operators), and RES (renewable energy sources) operators in respect to grid-level analytics for optimized network and asset management services, (b) electricity retailers and aggregators for portfolio-level analytics toward energy-as-a-service (EaaS) solutions, and (c) facility managers and ESCOs (energy service companies) toward building/district-level analytics from the perspective of optimized energy performance management.

In order to deliver the intended functionalities toward the different electricity data value chain stakeholders who at any moment may assume the role of data asset providers and/or data asset consumers, the high-level architecture consists of the following data-driven services bundles that have well-defined interfaces to ensure their seamless integration and operation within the SYNERGY integrated platform:

- **Data Collection Services Bundle** which enables the configuration of the data check-in process by the data provider at "design" time in the Core Big Data Management Platform and its proper execution in the SYNERGY Cloud Infrastructure and/or the On-Premise Environments. Different data ingestion, mapping, and transformation and cleaning services are invoked to appropriately handle batch, near-real-time, and streaming data collection.
- **Data Security Services Bundle** that is responsible for safeguarding the data assets in the overall SYNERGY platform (i.e., Core Big Data Management Platform and On-Premise Environments for end-to-end security) through different ways, e.g., by anonymizing the sensitive data (from an individual or business perspective), by selectively encrypting the data, and by applying access policies over the data assets that allow a data provider to control who can even view them.
- **Data Sharing Services Bundle**, essentially providing the SYNERGY Core Big Data Management Platform with the functionalities expected from a data and AI analytics marketplace in terms of sharing and trading data assets (embracing datasets, pre-trained AI models, analytics results) in a secure and trustful manner, powered by the immutability and non-repudiation aspects that are available in distributed ledger technologies.
- **Data Matchmaking Services Bundle** that delivers exploration and search functionalities (in the SYNERGY Core Big Data Management Platform) over data assets that the data consumers are eligible to view and potentially acquire while providing recommendations for additional data assets of interest or for electricity data value chain stakeholders who could potentially have/create the requested data asset.
- **Data Analytics Services Bundle** which lies at the core of the design of data analytics pipelines including the data manipulation configuration, the basic and baseline (pre-trained) machine learning and deep learning algorithms configuration, and the visualization/results configuration, in the SYNERGY Core Big Data Management Platform, while allowing for the execution of the defined pipelines in the Secure Experimentation Playgrounds and the On-Premise Environments.
- **Data Storage Services Bundle** that offers different persistence modalities (ranging from storage of the data assets, their metadata, their indexing, the algorithms and pipelines, the contracts' ledger, etc.) depending on the scope and the type of the data in the SYNERGY Cloud Infrastructure (in the Core Cloud Platform and the Secure Experimentation Playgrounds) and the On-Premise Environments.
- **Data Governance Services Bundle** that provides different features to support the proper coordination and end-to-end management of the data across all layers of the SYNERGY platform (cloud, on-premise).
- **Platform Management Services Bundle** which is responsible for resources management, the security and authentication aspects, the notifications management, the platform analytics, and the Open APIs that the SYNERGY platform provides.

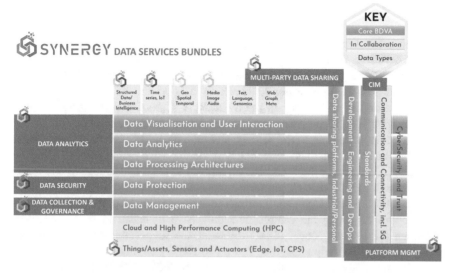

Fig. 2 SYNERGY Data Services Bundles in relation to the BDVA Reference Model

As depicted in Fig. 2, the overall SYNERGY Big Data Platform and AI Marketplace, along with its different Data Services Bundles, is well aligned to the BDVA Reference Model defined in the European Big Data Value Strategic Research and Innovation Agenda [15]. On the one hand, topics around Data Management are appropriately addressed through the SYNERGY Data Collection and Data Governance Service Bundles. Data Protection is considered from an all-around perspective in the SYNERGY Data Security Service Bundle. Data Processing Architectures, Data Analytics, and Data Visualization and User Interaction aspects have a similar context and orientation as in the SYNERGY Data Analysis Services Bundle. On the other hand, the Data Sharing Platforms are indeed tackled through the SYNERGY Data Sharing Services Bundle that is innovative in introducing the concept of multi-party sharing. Development, Engineering, and DevOps aspects are well embedded in the SYNERGY Platform Management Services Bundle. Finally, the Standards dimension is addressed within the SYNERGY Common Information Model that builds upon different energy data standards, ontologies, and vocabularies.

It needs to be noted that the SYNERGY architecture was designed taking into consideration the SGAM philosophy and design patterns [16, 17] even though in a more loosely coupled manner.

3.1 SYNERGY Cloud Infrastructure Layer

As depicted in Fig. 3, the *SYNERGY Core Big Data Management Platform* (or SYNERGY Core Cloud Platform in abbreviation) is the entry point for any user (as representative of an electricity data value chain stakeholder) in the overall SYNERGY platform. In order to check in data to the SYNERGY platform, the **Data Handling Manager** in the SYNERGY Core Cloud Platform provides the user interfaces to properly configure and manage the data check-in jobs at "design" time, according to the settings and preferences of each data provider for uploading batch data as files; collecting data via third-party applications' APIs, via open data APIs, or via the SYNERGY platform's APIs; and ingesting streaming data (through the SYNERGY platform's mechanisms or through the stakeholders' PubSub mechanisms). Upon configuring the data ingestion step, the data providers need to properly map the sample data they have uploaded to the SYNERGY Common Information Model (CIM) following the suggestions and guidelines of the **Matching Prediction Engine**. The SYNERGY Common Information Model is built on different standards, such as IEC 61968/61970/62325, IEC 61850, OpenADR2.0b, USEF, and SAREF, and aims to provide a proper representation of the knowledge of the electricity data value chain, defining in detail the concepts to

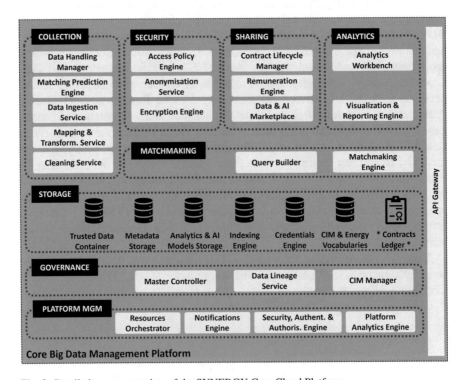

Fig. 3 Detailed component view of the SYNERGY Core Cloud Platform

which the datasets that are expected to be uploaded in the SYNERGY marketplace will refer and taking into consideration the standards' modelling approaches.

Optionally, the data providers are able to also configure the cleaning rules, the anonymization rules, and the encryption rules that need to be applied over the data. The **Access Policy Engine** provides the opportunity to define access policies based on different attributes in order to fully control which stakeholders can potentially view the specific data asset's details in the SYNERGY platform.

The data check-in job execution is triggered by the **Master Controller** according to the schedule set by the data providers and in the execution location they have set (i.e., Cloud Platform or On-Premise Environment). The Master Controller communicates with the **Resources Orchestrator** to ensure the necessary compute and memory resources (esp. in the SYNERGY Cloud Infrastructure) and orchestrates the appropriate list of services among the **Data Ingestion Service**, the **Mapping and Transformation Service**, the **Cleaning Service**, the **Anonymization Service**, and the **Encryption Engine** that are invoked in a sequential manner while forwarding them the data check-in job's configuration. The data are stored in **Trusted Data Containers** in the Data Storage Services Bundle, and a set of metadata (in alignment with the SYNERGY metadata schema built on DCMI and DCAT-AP) are either extracted automatically during the previous steps (e.g., in the case of temporal coverage, temporal granularity, spatial coverage, and spatial granularity metadata that can be extracted from the data, as well as the data schema mapped in the SYNERGY CIM) or manually defined by the data providers in the **Data and AI Marketplace** (such as title, description, tags, and license-related metadata) and persisted in the **Metadata Storage**.

The Data and AI Marketplace is essentially the one-stop shop for energy-related data assets from the electricity data value chain stakeholders as it enables secure and trusted data asset sharing and trading among them. It allows them to efficiently search for data assets of interest through the **Query Builder** and provides them with the help of the **Matchmaking Engine** with recommendations for data assets or data assets' providers (that may potentially have/create the specific data asset). The Data and AI Marketplace allows data consumers to navigate to the available data assets, preview their offerings, and proceed with their acquisition through smart data asset contracts that are created, negotiated, and signed among the involved parties in the **Contract Lifecycle Manager** and stored in each step in the **Contracts Ledger**. The contract terms of use, the cost and VAT, the contract effective date, the contract duration, the data asset provider, and the data asset consumer are among the contract's details that are stored in hash in the blockchain. In order for a signed contract to be considered as active, the respective payment needs to be settled with the help of the **Remuneration Engine**.

In order for electricity data value chain stakeholders to leverage the potential of data analytics over data that they own or have acquired, the **Analytics Workbench** gives them the opportunity to design data analysis pipelines according to their needs and requirements. Such pipelines may consist of (a) different data manipulation functions, (b) pre-trained machine learning or deep learning algorithms that have been created for the needs of the energy domain, or (c) simple algorithms that are offered in an out-of-the box manner wrapping the Spark MLlib algorithms, the sk-learn algorithms, and the TensorFlow (over Keras) algorithms. The execution settings are defined by the data asset consumers that define when and how the data analysis pipeline should be executed and how the output will be stored. In this context, the **Visualization and Reporting Engine** allows the data asset consumers to select, customize, and save appropriate visualizations to gain insights into the analytics results, but also to create simple reports to potentially combine results.

The **API Gateway** allows the authorized SYNERGY energy applications and any application to retrieve from the SYNERGY platform's Open APIs the exact raw data or analytics results they need according to filters they are able to set. The overall platform's security, organization's and user's registration, and authorization decisions are dependent on the **Security, Authentication and Authorization Engine**.

The SYNERGY cloud platform is complemented by the **Data Lineage Service** to provide provenance-related views over the data assets; the **Notifications Engine** to send notifications about the ongoing processes that are related to a user or organization; the **Platform Analytics Engine** that provides insights into the added value of the data assets in the SYNERGY platform, but also on the overall platform's services progress; and the **CIM Manager** that is behind the evolution and propagation of changes of the Common Information Model across the involved services in the whole SYNERGY platform.

The execution of a data analysis job in the SYNERGY Cloud Platform is performed in *Secure Experimentation Playgrounds* which are essentially sandboxed environments that become available per organization. The data that belong to an organization or have been acquired by an organization (based on a legitimate data asset contract) are transferred through the **Data Ingestion Service** based on the instructions provided by the Master Controller, are decrypted upon getting access to the decryption key in the **Encryption Engine** (with the help of the Master Controller and the Security, Authentication and Authorization Engine), and are stored in **Trusted Data Containers**. Any data analysis pipeline that needs to be executed is triggered according to the organization's preferences by the Master Controller that invokes the **Data Manipulation Service** and the **Analytics Execution Service**. The **Secure Results Export Service** is responsible to prepare the results for use by the respective organization in different ways (e.g., as a file, exposing them via an API, sharing them in the Data and AI Marketplace). Finally, the **Data Lineage Service** provides an overview of the relations and provenance of the data assets stored in the Secure Experimentation Playground (as depicted in Fig. 4).

3.2 SYNERGY On-Premise Environments Layer

The *SYNERGY Server On-Premise Environment* is responsible for (a) preparing the data assets, which an organization owns, "locally" to ensure end-to-end security (especially when encryption is required in the data check-in job configuration)

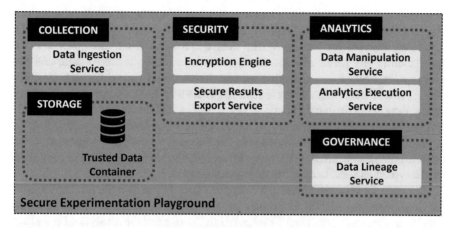

Fig. 4 Detailed component view of the SYNERGY Secure Experimentation Playground

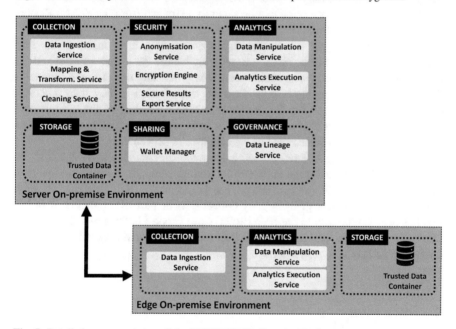

Fig. 5 Detailed component view of the SYNERGY On-Premise Environments

prior to uploading them in the SYNERGY Core Cloud Platform; (b) preparing and storing the own data assets "locally" in case they are not allowed to even leave a stakeholder's premises; and (c) running analytics "locally" over data that are also stored "locally."

As depicted in Fig. 5, according to the instructions received by the Master Controller in the SYNERGY Core Cloud Platform, a data check-in job is executed in the Server On-Premise Environment as follows: the **Data Ingestion Service** is responsible for collecting the necessary data, the **Mapping and Transformation Service** for processing the data (to ensure their alignment with the CIM), the **Cleaning Service** for increasing the data quality, the **Anonymization Service** for handling any personally identifying or commercially sensitive data, and the **Encryption Engine** for encrypting the data. Then, the data are either stored locally in the Trusted Data Container or transferred to the SYNERGY Core Cloud Platform where they are permanently stored. It needs to be noted that in case an active smart data asset contract's terms allow it, the data assets that have been acquired by an organization can be also downloaded in the Server On-Premise Environment to be used to complement an analysis, again through the Data Ingestion Service, and shall be decrypted with the help of the Encryption Engine.

In order to execute a data analysis job "locally" in the Server On-Premise Environment, the Master Controller of the SYNERGY Core Cloud Platform appropriately invokes the **Data Manipulation Service** and the **Analytics Execution Service** to run all necessary steps of the designed pipeline. The results are stored in the Trusted Data Container and can be securely extracted from the Secure Results Export Service of On-Premise Environment Server Edition.

The **Wallet Manager** allows the organizations that have installed the On-Premise Environment Server Edition to securely handle the ledger account and the cryptocurrency funds of their organization. It is practically used to send payments for smart asset data contracts that allow an organization to buy data, but also to receive reimbursement for data assets that have been sold by the organization (especially in the context of a multi-party smart asset contract). The Data Lineage Service again allows a better view of the data asset's provenance.

The *Edge On-Premise Environment* has limited functionalities in respect to the Server On-Premise Environment due to the limited compute, memory, and storage capacity it can leverage in any gateway. It has a light version of (a) the Data Ingestion Service to ensure that a gateway may collect data as part of a data check-in job that has been configured in the SYNERGY Core Cloud Platform and (b) the Data Manipulation Service and the Analysis Execution Service that may run limited data analysis pipelines with restrictions.

4   Discussion

During the design of the SYNERGY Reference Architecture and iterative discussions performed in different technical meetings, the need to bring different stakeholders on the same page with regard to certain core end-to-end functionalities of the SYNERGY platform emerged. To this end, the basic workflows that the SYNERGY Cloud Infrastructure and On-Premise Environments will support from the user-oriented perspective of data asset providers and data asset consumers were designed and extensively discussed in dedicated workshops, focusing on the main challenges that are expected to be encountered:

- The **data check-in workflow (I)** allowing data asset providers to make available their data in the SYNERGY Energy Big Data Platform and the AI Analytics Marketplace.

  - *Challenge I.1*: Complexity of a fully flexible data check-in job configuration vs user friendliness. There is an explicit need for guidance and for setting certain quality thresholds in order to properly configure all steps since the settings for data mapping, cleaning, and anonymization cannot be fully and automatically extracted, but instead always have to rely on the expertise of the data provider who is uploading the data.
  - *Challenge I.2*: Performance vs security trade-off. When executing demanding pre-processing steps like Mapping, Cleaning, Anonymization, and especially Encryption over a dataset, certain restrictions need to apply (to avoid ending up with inconsistent data in a datastore), while real-time access to the processed data cannot be guaranteed. Increased security requires data replication and decryption in the different secure spaces of the data consumers, which cannot be instantly completed either.
  - *Challenge I.3*: Data profiling completeness vs status quo. In order to facilitate search, full profiles of different datasets need to be provided which requires significant attention by a data provider. Data licenses profiling in particular appears as a pain-point in an industry who is not used in sharing their own data. Although fine-grained access data access policies are considered as instrumental in ensuring the business interests of the demo partners toward their competitors, their configuration needs to be straightforward explaining the exact implications.

- The **data search and sharing workflow (II)** allowing data asset consumers to find data of interest in the SYNERGY Energy Big Data Platform and the AI Analytics Marketplace and acquire them in a trustful and reliable manner based on smart data asset contracts.

  - *Challenge II.1*: Search performance over Big Data vs the metadata of encrypted data. Search functionalities need to be always adapted to different cases of how and where the data are stored and indexed.
  - *Challenge II.2*: Multi-party contracts as a necessity vs a necessary "evil." In order to properly handle the chain of licenses and IPR that are associated with analytics results that can be traded in the marketplace, the SYNERGY platform needs to act as a "man in the middle" that creates bilateral contracts with the data asset consumer and each involved data asset providers under a broad multi-party contract. Facilitating the data asset consumer in this case comes at the cost of complexity on the platform side. In order to properly handle multi-party contracts, payments over a cryptocurrency (supported by SYNERGY) are also enforced which may lower the entry barrier for the potential stakeholders, but also potentially decrease their trust.
  - *Challenge III.3*: Limitations on data access and retrieval. Retrieval of appropriate data assets is not contingent only on the existence of an active data

asset contract, but also on the actual location of the data (cloud vs on-premise environment of the data provider) and the terms that dictate the data transfer. Although cloud presence of unencrypted data ensures that they can be retrieved via user-defined retrieval queries, encrypted data and on-premise data can be potentially (if there is a provision for offline/off-platform storage) only retrieved as full files through the SYNERGY platform APIs.

- The **data analytics workflow (III)** allowing data asset providers and consumers to run analytics over their own and the acquired data assets in the SYNERGY Energy Big Data Platform and the AI Analytics Marketplace and gain previously unattainable insights.

  - *Challenge III.1: Pipeline configuration for a business user vs a data scientist.* When trying to design a solution that allows the design of analytics pipelines, different perspectives need to be considered: the perspective of a business user who needs to easily create pipelines and gain insights over data and the perspective of data scientists that expect more advanced functionalities for feature engineering, model training, and evaluation.
  - *Challenge III.2*: Customizable pipelines for basic vs pre-trained energy data analytics algorithms across different execution frameworks (ranging from Spark and Python/sk-learn to TensorFlow over Keras). Since the input data to run an analytics pipeline are available as uploaded by their stakeholders, they need to be easily manipulated through an interactive user experience in order to be fit as input to an ML/DL model.
  - *Challenge III.3*: Data and model versioning affect the execution of any analytics pipeline. The expected impact on performance in "real-time" data and analytics when the data are originally stored in an encrypted form or only on premise (with limited resources) cannot be disregarded.
  - *Challenge III.4*: Running analytics with data that are never allowed to leave their provider's premises (according to the applicable data asset contract terms) render secure multi-party computations as a necessity (despite their inherent limitations in terms of analysis richness).

It needs to be noted that the aforementioned challenges represent an extract of the challenges identified during interactive workshops in which the technical partners were requested to discuss the technical challenges they expect to be associated with each feature/requirement and comment on their technical feasibility, according to their experience. In parallel, the different end users (across five demo countries) were requested to evaluate (a) the actual importance/added value for own organization (by rating the importance and added value of the specific feature for their business operations) and (b) the perceived importance/added value for the electricity data value chain (by rating the importance and added value that they perceive the specific feature brings to their stakeholder category and the overall electricity data value chain). For the assessment, a scale between 1 (little or no added value/importance/impact) and 5 (extremely high added value/importance/impact) was put into use, as indicatively depicted in Fig. 6.

Step	Functionality / Feature	Actual Importance / Added Value for own Organization	Perceived Importance / Added Value for the electricity data value chain
I-1. Data Ingestion Configuration	Upload of file(s) (csv, json, xml, other) that have been already extracted by the energy stakeholder's back-end systems	4.31	4.15
	Data ingestion via energy stakeholders' APIs	4.00	4.23
	Data ingestion via Open Data APIs (e.g. ENTSO E Transaparency Platform)	3.62	3.92
	Data ingestion via the SYNERGY Platform APIs	4.38	4.00
	Streaming Data Ingestion via the SYNERGY Core Cloud Platform (SYNERGY PubSub mechanism)	3.25	3.33
	Streaming Data Ingestion via stakeholders' own (PubSub) mechanisms	2.75	3.58
I-2. Data Mapping & Harmonization Configuration	Automatic mapping predictions from the data sample to the SYNERGY Common Information Model		
	Easy navigation and search for missing concept mappings in the SYNERGY Common Information Model		
	Manual updates and corrections of mappings from the data sample to the SYNERGY Common Information Model	4.31	4.31
	Manual definition of transformation rules (for measurement units, datetime format and timezones, etc)		
	Manual definition of advanced transformation rules (e.g. for merging data columns to comply with the SYNERGY Common Information Model)		
	Semantic enrichment of the data concepts with machine-processable information (e.g. information from codelists, vocabularies)		
	Manual suggestions for missing concepts in the SYNERGY Common Information Model		
I-3. Data Cleaning Configuration	(Semi-)Automatic quality check of the data and assessment of quality level	4.31	4.31
	Recommendation for cleaning rules to improve data quality		
	Definition of cleaning rules (for handling outliers, for handling missing data values, etc.)		
I-4. Data Anonymization Configuration	Semi-automatic check for identifying the "privacy-risky" columns of a dataset	4.31	4.31
	Definition of anonymization rules for "identifying", "quasi-identifying" and "sensitive" columns/ variables		
I-5. Data Encryption Configuration	Encrypt the whole dataset	4.08	4.23
	Encrypt manually-selected "columns" of the dataset	3.67	3.92
	No encryption is needed for the dataset	2.31	2.15
I-6. Data Storage Configuration	Upload in the SYNERGY Core Platform	3.92	4.00
	Instantly replicate in my organization's Secure Experimentation Playground in the SYNERGY Core Platform	3.62	3.77
	Store in my organization's On-Premise Environment	4.23	4.23
I-7. Metadata Enrichment	Detailed data profiling of the data to be uploaded according to the SYNERGY metadata schema	3.33	3.33
	Quick check-in for confidential data assets (without filling in all metadata)		
I-8. Data License Definition	Detailed license profiling of the data to be uploaded (e.g. IPR, terms & conditions, pricing)	3.77	3.92
	Step-by-step guidance on how to define the appropriate license of the data to be uploaded	3.62	3.62
	Creation of a "license" contract between the data provider and the SYNERGY platform	4.08	4.08
I-9. Data Access Policies Definition	Definition of access policies regarding who is permitted to access the data (e.g. Organizations from Greece can access the data)	4.23	4.38
	Definition of access policies regarding who is denied access to the data (e.g. No DSO can access the data)	4.31	4.46
I-10. Data Collection Execution	Data uploading in SYNERGY Core Platform through the On-Premise Environment, according to the configuration provided	3.83	3.83
	Data uploading in SYNERGY Core Platform directly, according to the configuration provided		
	Data uploading in the On-Premise Environment only, according to the configuration provided		
	Automatic and on-demand data transfer to the organization's Secure Experimentation Playground		
	Scheduled execution of API data retrieval (e.g. every minute/ hour/ day, at defined schedules)		
	Real-time execution of streaming data retrieval		

Fig. 6 SYNERGY platform features assessment by the demo partners for own and industry business value for " data check-in" for a data provider

5 Conclusions

This paper focused on the motivation and state of play behind energy data platforms, marketplaces, and essentially Data Spaces in order to present the SYNERGY Reference Architecture that aims to facilitate electricity data value chain stakeholders to (a) attach value to their own data assets; (b) gain new insights over their data assets; (c) share and trade their own data assets in a trustful, legitimate manner; and (d) enjoy the benefits of the reuse of their own data assets. The different layers of the architecture as well as the different components across the SYNERGY Data Services Bundles have been elaborated, while the core technical challenges have been introduced.

The next steps of our work include the finalization of the beta release of the SYNERGY integrated platform (which is currently on its alpha, mockup version) and its demonstration and use by different electricity data value chain stakeholders.

Acknowledgments This work was carried out within the SYNERGY project. The SYNERGY project has received funding from the European Union's Horizon 2020 research and innovation program under grant agreement no. 872734.

References

1. CapGemini. (2015). *Big Data BlackOut: Are utilities powering up their data analytics?*.
2. Bhattarai, B. P., et al. (2019). Big data analytics in smart grids: State-of-the-art, challenges, opportunities and future directions. *IET Research Journals*, 1–15.
3. Daneshmand, M., et al. (2017). Big challenges for big data in the smart grid era. *ECN Magazine*.
4. Zillner, S., Curry, E., Metzger, A., Auer, S., & Seidl, R. (2017). *European big data value strategic research & innovation agenda*. Big Data Value Association.
5. Zillner, S., Bisset, D., Milano, M., Curry, E., García Robles, A., Hahn, T., Irgens, M., Lafrenz, R., Liepert, B., O'Sullivan, B., & Smeulders, A., (eds) (2020). *Strategic research, innovation and deployment agenda - AI, data and robotics partnership. Third Release.* September 2020, Brussels. BDVA, eu-Robotics, ELLIS, EurAI and CLAIRE.
6. HBR. (2006). *Strategies for two-sided markets.* https://hbr.org/2006/10/strategies-for-two-sided-markets
7. Biliri, E., Pertselakis, M., Phinikettos, M., Zacharias, M., Lampathaki, F., & Alexandrou, D. (2019). Designing a trusted data brokerage framework in the aviation domain. *PRO-VE, 2019*, 234–241.
8. Schomm, F., Stahl, F., & Vossen, G. (2013). Marketplaces for data: An initial survey. *ACM SIGMOD Record, 42*(1), 15–26.
9. Stahl, F., Schomm, F., Vomfell, L., & Vossen, G. (2015). *Marketplaces for digital data: Quo vadis? (No. 24).* Working Papers, ERCIS-European Research Center for Information Systems.
10. Stahl, F., Schomm, F., Vossen, G., & Vomfell, L. (2016). A classification framework for data marketplaces. *Vietnam Journal of Computer Science, 3*(3), 137–143.
11. Koutroumpis, P., Leiponen, A., & Thomas, L. D. (2017). *The (unfulfilled) potential of data marketplaces (No. 53).* ETLA Working Papers.

12. EURELECTRIC. (2016). *The power sector goes digital - Next generation data management for energy Consumers.* https://www.eurelectric.org/media/2029/joint_retail_dso_data_report_final_11may_as-2016-030-0258-01-e.pdf
13. NordREG. (2018). *Implementation of data hubs in the Nordic countries.* http://www.nordicenergyregulators.org/wp-content/uploads/2018/08/NordREG-Status-report-on-datahubs-June-2018.pdf
14. European Distribution System Operators. (2018) *E.DSO Policy Brief on Open Data.* https://www.edsoforsmartgrids.eu/wp-content/uploads/EDSO-Open-Data-Policy-Brief_1812_final-1.pdf
15. BDVA. (2017). *European big data value - Strategic research and innovation agenda, Version 4.0.* Retrieved from https://bdva.eu/sites/default/files/BDVA_SRIA_v4_Ed1.1.pdf. Accessed December 30, 2020.
16. CEN-CENELEC-ETSI Smart Grid Coordination Group. (2012). *Smart grid reference architecture.* Retrieved from https://ec.europa.eu/energy/sites/ener/files/documents/xpert_group1_reference_architecture.pdf. Accessed December 30, 2020.
17. EC. (2011). *Smart Grid Mandate Standardization Mandate to European Standardisation Organisations (ESOs) to support European Smart Grid deployment (M/490 EN).* Retrieved from https://ec.europa.eu/energy/sites/ener/files/documents/2011_03_01_mandate_m490_en.pdf. Accessed December 30, 2020.
18. Bröring, A. (2019). *Future data marketplaces.* https://www.bdva.eu/node/1220.

Part III
Future Directions

Privacy-Preserving Techniques for Trustworthy Data Sharing: Opportunities and Challenges for Future Research

Lidia Dutkiewicz, Yuliya Miadzvetskaya, Hosea Ofe, Alan Barnett, Lukas Helminger, Stefanie Lindstaedt, and Andreas Trügler

Abstract One of the foundations of data sharing in the European Union (EU) is trust, especially in view of the advancing digitalization and recent developments with respect to European Data Spaces. In this chapter, we argue that privacy-preserving techniques, such as multi-party computation and fully homomorphic encryption, can play a positive role in enhancing trust in data sharing transactions. We therefore focus on an interdisciplinary perspective on how privacy-preserving techniques can facilitate trustworthy data sharing. We start with introducing the legal landscape of data sharing in the EU. Then, we discuss the different functions of third-party intermediaries, namely, data marketplaces. Before giving a legal perspective on privacy-preserving techniques for enhancing trust in data sharing, we briefly touch upon the Data Governance Act (DGA) proposal with relation to trust and its intersection with the General Data Protection Regulation (GDPR). We continue with an overview on the technical aspects of privacy-preserving methods in the later part, where we focus on methods based on cryptography (such as homomorphic encryption, multi-party computation, private set intersection) and link

L. Dutkiewicz · Y. Miadzvetskaya
KU Leuven Centre for IT & IP Law – imec, Leuven, Belgium

H. Ofe
TU Delft, Faculty of Technology, Policy and Management, Delft, The Netherlands

A. Barnett
OCTO Research Office, Dell Technologies, Ovens, County Cork, Ireland

L. Helminger · S. Lindstaedt
Graz University of Technology, Graz, Austria

Know-Center GmbH, Graz, Austria

A. Trügler (✉)
Know-Center GmbH, Graz, Austria

Graz University of Technology, Graz, Austria
e-mail: atruegler@know-center.at

them to smart contracts. We discuss the main principles behind these methods and highlight the open challenges with respect to privacy, performance bottlenecks, and a more widespread application of privacy-preserving analytics. Finally, we suggest directions for future research by highlighting that the mutual understanding of legal frameworks and technical capabilities will form an essential building block of sustainable and secure data sharing in the future

Keywords Data law · Data sharing · Trust · Data Governance Act · Privacy-enhancing techniques · Homomorphic encryption · Multi-party computation · Cryptography · Private set intersection · Federated learning · GDPR · Data marketplace · Data Governance · Smart contracts · Secure enclave

1 Introduction

One of the backbones of data sharing intermediaries and European Data Spaces is privacy, especially in view of the advancing digitalization and global economic and socioeconomic developments. New research breakthroughs and the possibilities of privacy-preserving technologies have to comply with data protection laws to enable a secure and sustainable data economy.

In this chapter, we therefore focus on an interdisciplinary perspective on how privacy-preserving techniques can facilitate trustworthy data sharing. We start with introducing the legal landscape of data sharing in the European Union and give an overview on the technical aspects of privacy-preserving methods in the later part. We discuss the main principles behind these methods and highlight the open challenges with respect to privacy and suggestions for future research for data platforms.

The chapter relates to the technical priorities of data processing architecture of the European Big Data Value Strategic Research and Innovation Agenda [1]. It addresses the horizontal concern of data protection of the BDV Technical Reference Model, and it addresses the vertical concerns of Marketplaces, Industrial Data Platforms, and Personal Data Platforms.

The chapter relates to the Knowledge and Learning, Reasoning, and Decision-Making enablers of the AI, Data and Robotics Strategic Research, Innovation and Deployment Agenda [2].

1.1 Data Sharing Now: A Legal Patchwork

Advances in ICT have had and continue to have fundamental impacts on society. A vital aspect of this trend is the vast amount of data collected and used as data-related technologies impact the socioeconomic life of companies and individuals. Data is often referred to as a new oil, new resource, new infrastructure, and the fifth freedom of the EU internal market. This trend toward treating data as an economic asset just like goods, capital, and services is known as a "commodification of data."

An estimated amount of 33 zettabytes of data was generated worldwide in 2018, and according to the European Data Strategy, this amount of data is expected to rise to 175 zettabytes in 2025. The EU's data economy value is estimated to reach 550 billion euros by 2025 [3]. The free movement of personal and non-personal data is therefore of strategic importance for fostering the EU data-driven economy. However, one of the main difficulties for this economic opportunity to materialize resides in the fact that data transactions are regulated in the EU by a legal patchwork. The intersections between those legal instruments are often a subject of controversies.

First of all, there is the General Data Protection Regulation (GDPR)[1] that applies since 25 May 2018 and constitutes the cornerstone of the EU personal data-related framework. The GDPR touches upon a few data protection-related questions particularly relevant to data market ecosystems such as this of TRUSTS.[2] These include, e.g., the determination of controllership and the ensuing allocation of data protection responsibilities and the legal basis for processing personal data [4].

Second, the Regulation on the free flow of non-personal data[3] is another building block of the EU data-related legal patchwork. According to its Article 1, the Regulation ensures the free flow of data other than personal data within the Union. The Regulation aims at removing obstacles to the free movement of non-personal data across the EU, notably data localization requirements, unless they are justified on grounds of public security (Article 4 of the Regulation) and vendor lock-in practices in the private sector.

At the same time, it remains unclear how to delineate what qualifies as personal data and what remains outside the scope of the personal data protection regime. In accordance with Article 4 of the GDPR, the notion of personal data is rather broad and encompasses "any information relating to an identified or identifiable natural person." It is not excluded that technological developments will make it possible to turn anonymized data into personal data and vice versa.[4] Thus, it is always safer to treat any data as personal.

Another difficulty concerns a mixed data set composed of both personal and non-personal data. The Regulation on the free flow of non-personal data applies only to the non-personal data part of the data set. Where data sets are inextricably linked, the GDPR shall prevail in accordance with Article 2(2) of the Regulation. The Commission also published informative guidance on the interaction between

---

[1] Regulation (EU) 2016/679 of the European Parliament and of the Council of 27 April 2016 on the protection of natural persons with regard to the processing of personal data and on the free movement of such data, and repealing Directive 95/46/EC (General Data Protection Regulation), OJ 2016 L 119/1.

[2] Trusted Secure Data Sharing Space, Horizon 2020, https://www.trusts-data.eu/

[3] Regulation (EU) 2018/1807 of the European Parliament and of the Council of 14 November 2018 on a framework for the free flow of non-personal data in the European Union, OJ 2018 L 303/59.

[4] GDPR, rec. 9.

the Regulation on the free flow of non-personal data and the GDPR where it clarified which rules to follow when processing mixed data sets and explained the concept of data sets "being inextricably linked" [5].

The Open Data Directive[5] in force since 2019 is another building block of the EU data-related framework. Its main aim is to allow free re-use of data that are held by national public sector bodies. This is meant to foster the emergence of new businesses that offer digital products and services. The Directive aims at increased re-use of data held by public sector bodies and certain public undertakings. However, the Open Data Directive does not apply to documents for which third parties hold intellectual property rights or that constitute commercial secrets. The Open Data Directive does not prevail over the GDPR in accordance with its Art. 1(4) and only applies to data that is not personal.

Moreover, there is a vast amount of EU legislation indirectly applicable to data sharing consisting of general and horizontal legislation (e.g., Database Directive, Copyright DSM Directive, Trade Secrets Directive, Software Directive, Regulation of B2B unfair commercial practices) and sector-specific rules (e.g., the PSD2 and the AML). For absence of a horizontal legal framework regulating B2B data sharing, the EU has been active in elaborating soft law guidelines for businesses [6].

Up to this date, the legal patchwork for data transactions does not sufficiently address the commodification of data and leaves some uncertainties when it comes to applicable rules.

However, recently, the EU has shifted its focus to other ways of regulating data transactions, notably data sharing, data re-use, and making the data available. In the European Data Strategy, the European Commission emphasized that the development of data marketplaces is a key policy instrument to revitalize the full potential of the value of data generated across member states [4]. The broad aim of the strategy is to "create a genuine single market for data, where personal and non-personal data, including confidential and sensitive data, are secure and where businesses and the public sector have easy access to huge amounts of high-quality data to create and innovate" [4].

1.2 Data Marketplaces

In spite of the economic potential data is suggested to have, data sharing between companies has not taken off at sufficient scale. This is, among others, due to a "lack of trust between economic operators that the data will be used in line with contractual agreements, imbalances in negotiating power, the fear of misappropriation of the data by third parties, and a lack of legal clarity on who can do what with the data" [4].

---

[5] Directive (EU) 2019/1024 of the European Parliament and of the Council of 20 June 2019 on open data and the re-use of public sector information, OJ 2019 L 172/56.

To address these challenges, the trusted third-party intermediaries (e.g., data marketplaces) come into play. Data marketplaces are defined as platforms that provide services for buying and selling of data products [7]. They bring data suppliers and data users together to exchange data in a secure online platform. Based on the matching function they perform, data marketplaces can range from one to one, one to many, many to one, and many to many [6]. For example, one-to-one data marketplaces enable bilateral exchanges between two parties, while many-to-many are multi-lateral marketplaces [6].

Data marketplaces can also be characterized based on the functions they perform. As indicated by the European Commission, a data marketplace is a specific type of intermediary which may have the following functions [6]:

Match-Making Between Potential Data Supplier and Data Buyer

In that scenario, the platform matches the supply and demand between the potential suppliers and potential buyers and facilitates data sharing between the parties. From an economic perspective, it lowers transaction costs through combining different data sources [9].

The Actual Transfer of the Data and Trust Creation

For businesses, data trading is quite sensitive since they become vulnerable to competitors or adverse effects. Platforms may therefore rely on the usage of privacy-preserving technologies, perform screening of data sharing partners, supervise and protocol the individual transactions, as well as enforce usage constraints.

Provider of the Technical Infrastructure

Data marketplaces may be defined as an "architecture allowing programmability and reuse of content and data, typically through API, and organizing modularity between a stable core and variable components" [10].

Data intermediaries can also provide additional services and functionalities such as model contract clauses or (pseudo)anonymization services (if personal or confidential data are exchanged), privacy-preserving data analytics, etc.

The variety of data marketplaces and the functions they can perform raise the question of how to regulate the activities of data sharing intermediaries.

1.3   Data Governance Act ("DGA")

In November 2020, the European Commission put forward a proposal for a regulation on European Data Governance[6] (Data Governance Act, "DGA") that provides for the rules aimed at facilitating the re-use of publicly held data, regulating the activities of data sharing intermediaries, fostering data altruism, and preventing international access to EU-based data by foreign governments and

---

[6] Proposal for a Regulation of the European Parliament and of the Council on European Data Governance (Data Governance Act) COM/2020/767 final, Brussels, 25.11.2020.

entities. According to the Impact Assessment of the European Commission, the overall objective of the DGA proposal is to set the conditions for the development of common European Data Spaces and strengthen trust in data sharing and in data intermediaries.

With the DGA proposal, in order to increase trust in such data sharing services, the EC aims to create an EU-wide regulatory framework, which would set out highly harmonized requirements related to the trustworthy provision of data sharing services. According to the proposal, a key element to bring trust and more control for data holder and data users in data sharing services is the neutrality of intermediaries—data sharing service providers.[7] The Regulation proposes a number of measures to increase trust in data sharing, including the structural separation between the data sharing service and any other services provided and a notification regime for data sharing providers.

Moreover, the intersection between the GDPR and DGA raises a number of questions. First of all, data processing principles, enshrined in the GDPR, such as purpose limitation and data minimization, are difficultly compatible with the objective of stimulating data sharing in the EU. Secondly, the sharing of personal data by data subjects requires trust in data controllers and data users to prevent any cases of misuse of personal data for different purposes than those communicated at the moment of data collection or sharing.

Finally, the DGA provides for techniques enabling privacy-friendly analyses where personal data are involved, such as anonymization, pseudonymization, differential privacy, generalization, or suppression and randomization. The application of these privacy-enhancing technologies and compliance with the GDPR are meant to ensure the safe re-use of personal data and commercially confidential business data for research, innovation, and statistical purposes.[8] Against this background, this chapter argues that privacy-preserving techniques, such as multi-party computation and fully homomorphic encryption, can play a positive role as enablers of trust in data sharing in compliance with fundamental rights to privacy and data protection. In the next section, we will provide a legal perspective on different privacy-preserving techniques and their impact on leveraging trust for data transactions.

2 Legal Perspective on Privacy-Preserving Techniques for Enhancing Trust in Data Sharing

2.1 What Is Trust?

Trust is a fundamental aspect of social interactions. It is generally understood as a relationship in which an agent (the trustor) decides to depend on another agent's (the

---

[7] DGA, rec. 26.

[8] DGA, rec. 6.

trustee) foreseeable behavior in order to fulfil his expectations [11]. Trust is a much-discussed concept in ethics of digital technologies. In recent years, the concept of trust in digital contexts—known as e-trust—has come to the fore [12]. According to Taddeo, "e-trust occurs in environments where direct and physical contacts do not take place, where moral and social pressures can be differently perceived, and where interactions are mediated by digital devices." However, it is beyond the scope of this chapter to further elaborate on this concept. Our objective is to explore the relations between trust and data markets and how trust could be put into effect in the data markets.

2.2   The Role of Trust in Data Markets

A study on data sharing between companies in Europe identified key characteristics of a thriving data-driven economy. They include, among others, the availability of data sets from actors across the economy and the necessary infrastructure, knowledge, and skills within companies that would make possible to engage in data sharing and re-use. Other features included the existence of trust between independent economic operators, appropriate cybersecurity measures, and the development of common standards for technologies and data interoperability [13].

Trust between data suppliers and data users is one of the success factors for data sharing between companies (ibid., 83). There are different visions to successfully build trust, such as high security levels, enabling communication between data suppliers and users, and providing clarity with respect to what will be ultimately done with users' data (ibid.). Other ways include "empowering data suppliers and giving them full control over their datasets" and providing "comprehensive licensing agreements outlining data usage conditions and restrictions" (ibid.). Finally, informing data users about the origin of the data and lawfulness of data sharing activities have also been identified as key in building trust (ibid.).

In the context of data marketplace, enhancing trust requires a trusted third-party intermediary who brings data suppliers and data users together to exchange data in a secure online platform. TRUSTS goal is to create such a secure and trustworthy European data market. Against this background, how can one ensure that a data marketplace fulfils its role of the "trustworthy" intermediary?

2.3   Privacy-Preserving Techniques as a Means to Bring More Trust in Data Sharing

Privacy-preserving techniques play a crucial role for bringing trust to data markets and ensuring that personal data remains under the control of data subjects and is further shared with no harm on fundamental rights and freedoms of individuals.

Traditionally, the applicable legal regime will depend on the nature of data (personal/non-personal) at stake. In order to assess whether data on which privacy-preserving or re-identification techniques have been performed are considered as personal data or as anonymous information, the following criteria shall be used. First, the personal or non-personal character of the data depends on the identifiability of the individual (the data subject). The identifiable natural person is an individual who can be identified, directly or indirectly, in particular by reference to an identifier, inter alia a name, an identification number, location data, or an online identifier.[9]

Second, identifiability also depends on the capacity of actors to reverse an anonymization process with a decryption key or direct identifiers.[10] The identifiability is a dynamic concept. While it may not be possible to identify someone today with all the available means, it may happen at a later stage due to a technological progress. To determine whether an individual is identifiable, Recital 26 of the GDPR underlines that account should be taken of all the means reasonably likely to be used to identify the natural person directly or indirectly. This includes all objective factors, such as the costs of and the amount of time required for identification, taking into consideration the available technology at the time of the processing.[11]

Furthermore, according to the CJEU, the abovementioned concept of "means reasonably likely to be used" does not imply that all the information enabling the identification of the data subject is in the hands of one person, i.e., the data controller.[12] Where additional data are required to identify the individual, what matters is the means reasonably likely to be used in order to access and combine such additional data (ibid.). As an illustration, dynamic IP addresses constitute personal data for online media service providers that can legally obtain required additional information held by internet service providers to identify an individual behind a dynamic IP address at a specific moment of time (ibid. para 47–48).

On the one hand, there is an absolute approach supporting that data on which privacy-preserving techniques have been applied will almost always remain personal as long as it is possible to reverse the process and identify the individual. Furthermore, it is also claimed that no technique is "perfect" and endurable against future technological developments [14]. On the other hand, a relative, risk-based approach builds on the criterion of "means that are reasonably likely to be used" in order to identify an individual.[13] Following the latter, privacy-preserving techniques provide for different degrees of re-identification taking into account contextual

---

[9] GDPR, Art. 4 (1).

[10] Article 29 Working Party, 'Opinion 4/2007 on the concept of personal data' (WP 136, 20 June 2007) p. 19–20.

[11] GDPR, Rec. 26.

[12] CJEU 19 October 2016 C582/14 Patrick Breyer v Bundesrepublik Deutschland ECLI:EU:C:2016:779 ('Breyer case') para 43–45.

[13] Article 29 Working Party, 'Statement on the role of a risk-based approach in data protection legal framework' (WP 218, 30 May 2014).

elements, such as the technical process, the safeguards restricting access to the data, and the overall realistic risk of re-identification. In other words, if excessive effort, in technical, organizational, and financial terms, is required for reversing privacy-enhancing techniques, the re-identification of the natural person may not be considered as likely.

Anonymization, for instance, is considered to provide for different levels of re-identification. If we apply the absolute approach, only data that have been irreversibly anonymized and whose original raw data set has been deleted may be considered as data that are no longer personal.[14]

When it comes to encryption, the GDPR does not define "encrypted data" or "encryption" but refers to encryption in several provisions as a risk mitigation measure. Encryption is listed as one of the "appropriate safeguards" of Article 6(4)(e) GDPR and is mentioned as an appropriate technical and organizational measure to ensure the security of processing.[15]

Since the GDPR does not define "encrypted data," it has to be examined whether encrypted data are anonymous or pseudonymous data. As it has been mentioned above, the answer to this question depends on whether an absolute or a relative approach regarding the identifiability of a data subject is applied. When personal data are encrypted, the data will always remain personal to the holders or to the authorized users of the decryption key. However, encrypted data may even be considered as personal if there are means reasonably likely to be used by others for decrypting them [15]. If encryption prevents an unauthorized party from having access to data, then the data in question no longer refer to an identified or identifiable person [14]. Consequently, it has to be examined which level of encryption is sufficient for the encrypted personal data to be considered as anonymous. Such an evaluation of the encryption method should take account of objective factors. These include the level of security of encrypted data and decryption prevention, such as the strength of the encryption algorithm used, the length of the encryption key, and the security of the key management [15].

Importantly, we have to distinguish between encrypted transfer of data (e.g., via end-to-end encryption) and encrypted storing of data (e.g., in a cloud) [14]. Processing of stored encrypted data is possible by using fully homomorphic encryption (FHE) or secure multi-party computation (MPC). In such a scenario, for the processing of the data, no decryption and thus no knowledge of the private key is needed. Moreover, the result of the processing is encrypted and can only be decrypted by the user and not by the cloud provider. The cloud provider will never see the data in plaintext. Thus, when processing personal data with the use of FHE, the GDPR is not applicable to the cloud provider which consequently does not process personal data (ibid.).

---

[14] Article 29 Working Party, 'Opinion 05/2014 on Anonymisation Techniques' (WP 216 10 April 214).

[15] GDPR, Article 32 (1)(a).

Therefore, encrypted personal data will be anonymous data, when it would require an excessively high effort or cost or it would cause serious disadvantages to reverse the process and re-identify the individual. It has to be considered whether there are reasonably likely means which could give a third party a potential possibility of obtaining the key. For instance, MPC allows data to be shared in a secret form (i.e., encrypted), while at the same time meaningful computations are performed on these data. Once the data have been divided into the shares, it is stored on different servers. At no point in this process, parties involved in data sharing and computing on the data—other than the data controller—can have access to the data [16].

Spindler et al. rightly argue that when applying an *absolute* approach on the identifiability of data subjects, these data shares would have to be considered as personal data. It is theoretically possible that all fragments of the data are gathered and put together; however, in practice, this is highly unlikely (ibid.). This unreasonable chance of collusion may lead to ruling out the applicability of the GDPR.

In addition to these concepts, the GDPR has introduced the notion and definition of "pseudonymization." More specifically, pseudonymization refers to the processing of personal data in such a manner that the personal data can no longer be attributed to a specific data subject without the use of additional information. Furthermore, such additional information shall be kept separately and shall be subject to technical and organizational measures preventing the identifiability of a natural person.[16] Pseudonymization is commonly perceived as a data security measure that reduces linkability by replacing any identifying characteristic or attribute by another identifier, a pseudonym.[17] According to the GDPR, pseudonymized data are personal data.[18] Thus, data could be considered pseudonymized, and hence personal, insofar as the technical process they have undergone is reversible.

Nevertheless, it remains questionable whether reversibly anonymized, encrypted, and split data will be considered as personal, pseudonymized data or whether they will be referred to as anonymous toward the parties that cannot access the additional information, reverse the technical process, and identify the individual [14].

In the next section, we will provide a detailed technical description of these privacy-preserving techniques.

---

[16] GDPR, Art. 4 (5).

[17] Article 29 Working Party, 'Opinion 05/2014 on Anonymisation Techniques' (WP 216 10 April 2014).

[18] GDPR, Rec. 26.

3   Methods for Privacy-Preserving Analytics

Throughout the centuries, cryptographic ciphers have been designed to protect stored data or, with the emergence of modern information transmission, also to protect data in transmission. These scenarios usually follow an all-or-nothing principle where, e.g., two parties can access full information and outsiders nothing or where only the data owner has full information and nobody else. In reality, trust relationships are often a lot more complicated and diverse of course as we have seen in the previous sections, especially when it comes to outsourcing computations or accessing pre-trained machine learning models. Some of the very successful cryptosystems like RSA, for example, also have a special and usually unwanted property that allows to do limited calculations on the encrypted ciphertexts while preserving structure (called homomorphic property) to the unencrypted data. This means adding two ciphertexts yields the encrypted version of the plaintext sum, for example. These partial homomorphic properties led to a quest for new cryptosystems which turn the unwanted side effect into an advantage and allow unlimited manipulations and calculations on encrypted data. This opened up a new era of cryptography that allows to evaluate functions on encrypted, unknown data and to anchor cryptographic privacy-preserving methods in modern data analytics. The applications of such privacy-preserving techniques are widespread and range from evaluations of medical data [17, 18], over data mining [19] to applications in finance [20]. In this section, we give an overview of two main cryptographic protocols and primitives, FHE [21] and MPC [22], and discuss their links to data platforms and data sharing spaces. Additionally, we also introduce private set intersection (PSI) as a special MPC case.

3.1   Homomorphic Encryption

The introduction of "A fully homomorphic encryption system" by Craig Gentry [21] is regarded as one of the biggest advances in modern cryptography. Since then, many variations and improvements of (fully) homomorphic encryption have been developed. The main principle behind FHE is to start from a Somewhat Homomorphic Encryption (SHE) scheme that allows a limited number of operations. Gentry then introduced a technique called *bootstrapping* to refresh the ciphertexts to allow additional operations. Repeating the process opened the door for unlimited operations resulting in the change from somewhat to fully homomorphic encryption.

The starting point of all cryptographic protocols are mathematical problems that are very hard to solve (at least given appropriate constraints regarding time or computational power). The modern versions of FHE are based on such hard problems called Learning with Errors or an optimized variant thereof, which are formulated on mathematical lattices [22]. The security in these schemes comes from the introduction of random noise into the ciphertexts, which is removed again during the decryption process. The main bottleneck of such approaches is that this noise starts to grow for each computed operation, e.g., adding two ciphertexts results roughly in doubling the original noise. Once a certain threshold has been reached, the resulting ciphertext cannot be decrypted anymore because the randomness prevails over the actual encrypted information. Before this point is reached, the bootstrapping process comes into play and allows to start over with a fresh noise budget by re-encrypting the original ciphertext into a new ciphertext with lower noise. This leads to a high-performance overhead for bootstrapping, and in several libraries, this functionality is therefore not even implemented at the moment. Instead, SHE is much more efficient and already sufficient for typical encrypted evaluations. Very complex evaluations cannot be realized with SHE because the number of calculations is limited.

In general, one of the main advantages of homomorphic encryption is the ability to outsource computation without giving up any privacy. Sensitive data can be homomorphically evaluated on a data platform or cloud, and only the data owners can decrypt computed results. Suppose you want to benefit from the evaluation of a machine learning model from a service provider, but you don't want to share your data with anyone outside your company. Setting up an FHE framework will allow you to do this without having to trust the service provider since they are not able to access the actual content of your data. An example of such a platform for medical data has been developed by researchers and engineers from the École Polytechnique Fédérale de Lausanne and the Lausanne University Hospital, for example [24]. They also use multi-party computation which we discuss in the next section. Another example of the advantages of FHE is the connection of human mobility to infectious diseases, where typically sensitive and private data have to be jointly evaluated to link these two fields. An efficient FHE implementation of a protocol where two parties can securely compute a Covid heatmap without revealing sensitive data was recently published [25, 26].

3.2   Secure Multi-Party Computation

Secure multi-party computation is a subfield of cryptography that enables privacy-preserving computations between multiple participants. It first appeared in computer science literature around 1980. In recent years, secure multi-party computation has become practical due to extensive ongoing research and exponential growth in computing power. Every traditional computation involving two or more participants can be made privacy-preserving through secure multi-party computation. However, this transformation's computational overhead varies depending on the underlying computation and sometimes can be prohibitive. To illustrate the privacy and confidentiality guarantees offered by secure multi-party computation, we consider the case of anti-money laundering. As with most anti-fraud activities, anti-money laundering benefits from collaboration. However, financial institutions are reluctant to share data because of competition and data protection regulations.

Fig. 1 Secure multi-party computation

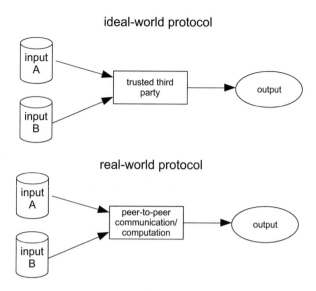

A secure multi-party anti-money laundering computation would flag suspicious transactions without revealing any other information. To understand what this means, imagine an ideal world where there exists a hypothetical trusted third party. In this ideal world, every institution sends its data to the trusted third party which performs the anti-money laundering computation and reports back to the institutions about any detected suspicious behavior. Because the trusted third party cannot be corrupted, nothing except the output of the computation gets shared between institutions.

Secure multi-party computation provides similar confidentiality and privacy in the real world, where one cannot fully trust third parties. Therefore, what can be achieved in the ideal world can also be done by applying secure multi-party computation, as illustrated in Fig. 1.

3.2.1 Private Set Intersection

Private set intersection is a special-purpose secure multi-party computation. It allows two participants to compute the intersection of their data sets. Thereby, neither participant learns information from the protocol execution, except for the data entries in the intersection. For instance, private set intersection enables two companies to find out common customers privately—information that can subsequently be used for a joint advertising campaign. Note that, in Fig. 2, the output of the protocol is John, but company A would not know about company B's customers Marlene and Elsie. Private set intersection is the most mature secure multi-party protocol, and computational overhead is small. Therefore, when parties

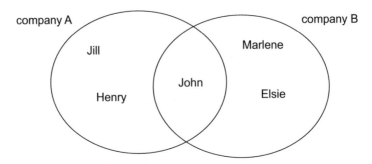

Fig. 2 Basic principle of private set intersection

engage in a private set intersection protocol, they do not have to expect significant performance issues.

4 Privacy-Preserving Technologies for Smart Contracts

Smart contracts are another example of where privacy-preserving techniques can be applied. They enact digital transactions that in a manner are similar to a physical transaction underpinned by a legal contract. Smart contract applications in a blockchain environment function within the context of the blockchain. Blockchains were not originally designed for preserving privacy; their original purpose was to verify integrity and legitimacy via transaction chains rooted in cryptographic hashes. In a public blockchain, data is available to all participants in unencrypted form – a problematic design for privacy preservation; off-chain smart contracts with hashes stored on-chain for verification purposes are a notable solution to this design problem [27].

Some blockchain variants can mitigate privacy concerns. Private and consortium blockchains utilize one or many managing authorities, and only approved authority members can access the blockchain data, but these infrastructures are typically much smaller than their public counterparts. The large, decentralized nature of public blockchains typically offers stronger security and integrity while foregoing the privacy and confidentiality controls of private and consortium blockchain variants [28].

4.1 Encrypted On-Chain Data with Homomorphic Encryption

This approach stores personal data on-chain in encrypted form. Applications cannot typically process encrypted data, and all participants on the blockchain will have visibility of any decryption operation, revealing both data and cryptographic keys. Homomorphic encryption, described earlier, enables operations on encrypted data, preserving the privacy of on-chain smart contracts. However, the mentioned performance bottlenecks of FHE are currently a limiting factor for enterprise-level blockchain scenarios, and more research is needed in this regard.

4.2   Smart Contracts Based on Multi-party Computation

MPC splits personal data into specific subsets, ensuring that each subset is meaningless individually. The data owner sends each subset to a separate actor for processing. Processing only one data subset renders each processing actor unable to infer any further understanding of the source data, but the data owner can recombine the computational results from each actor into a complete output. MPCs are theoretically highly collusion resistant as every actor must collude to infer the source data's meaning. Personal smart contract data could, as such, be safely computed using MPC.

4.3   Secure Enclaves

Secure enclaves, or SEs, conceal program state and segregate enclaved code from external access. SEs are provided by trusted execution environments (TEEs)—secure CPU sections supported on several modern CPUs. Coupling SEs and asymmetric-key cryptography enables encryption of smart contracts using an SEs' public key, with the private key held in the SE; thus, the smart contract ciphertext can only be decrypted within that SE.

A chief issue with SEs is certain companies dominating the TEE hardware space, which creates a reliance on a less diverse set of chip architectures; this increases the possible impact of any security flaw found in one such widely adopted architecture—further compounded by past practical attacks, such as "Meltdown" and "Spectre," targeting such architectures. Another argument against TEEs purports that commercial TEE implementations are not necessarily publicly visible and, in these cases, can't be as rigorously analyzed as, say, public specifications from the Trusted Compute Group on which such implementations are based [29].

5   Conclusion: Opportunities and Future Challenges

The notion of enhancing trust in data sharing is present in various European Commission's documents, including the European strategy for data and the proposal for the Data Governance Act. The Commission intends to continue its work on the setting up of common rules for EU-wide common interoperable Data Spaces which would address issues of *trust*. First, clear and trustworthy rules for data sharing and

Data Governance are needed. However, it remains to be seen whether the DGA and other Commission's initiatives will fulfil its promise to "increase trust in data sharing services."

Second, data transaction involving personal data would benefit from further explanation in the text of the DGA on how privacy-preserving techniques could increase the level of trust and control of data holders over their personal data in their personal Data Spaces.

Regarding the technical aspects of privacy-preserving methods, future research should address the current performance bottlenecks to allow efficient and secure computations also for complex scenarios. This will enable also a more widespread application of privacy-preserving analytics for data sharing spaces and beyond. With the possible rise of quantum computers, there is also a growing need for long-term secure systems; methods like FHE that rely on lattice-based problems are already regarded as quantum-secure.

In general, the mutual understanding of legal frameworks, the benefits of data sharing spaces, and the corresponding technical capabilities will form an essential building block of a sustainable and secure European economy in the future.

Acknowledgments This work was supported by EU's Horizon 2020 project TRUSTS under grant agreement n°871481 and by the "DDAI" COMET Module within the COMET (Competence Centers for Excellent Technologies) Programme, funded by the Austrian Federal Ministry for Transport, Innovation and Technology (bmvit), the Austrian Federal Ministry for Digital and Economic Affairs (bmdw), the Austrian Research Promotion Agency (FFG), the province of Styria (SFG), and partners from industry and academia. The COMET Programme is managed by FFG.

References

1. Zillner, S., Curry, E., Metzger, A., Auer, S., & Seidl, R. (Eds.). (2017). *European big data value strategic research & innovation agenda*. Big Data Value Association.
2. Zillner, S., Bisset, D., Milano, M., Curry, E., García Robles, A., Hahn, T., Irgens, M., Lafrenz, R., Liepert, B., O'Sullivan, B. and Smeulders, A. (Eds.) (2020, September). *Strategic research, innovation and deployment agenda—AI, data and robotics partnership. Third release*. BDVA, euRobotics, ELLIS, EurAI and CLAIRE.
3. Glennon, M., et al. (2020). *The European data market monitoring tool*. EU Publications.
4. European Commission. (2020). *A European strategy for data COM/2020/66 final*. s.l.:s.n.
5. European Commission. (2019). *Communication from the Commission to the European Parliament and the Council. Guidance on the Regulation on a framework for the free flow of non-personal data in the European Union COM/2019/250 final*. s.l.:s.n.
6. European Commission. (2018). *Guidance on sharing private sector data in the European data economy, Accompanying the document "Towards a common European data space"*. s.l.:s.n.
7. Spiekermann, M. (2019). Data marketplaces: Trends and monetisation of data goods. *Intereconomics, 54*(7), 208–216.
8. Koutroumpis, P., Leiponen, A. & Thomas, L. D. (2017). The (unfulfilled) potential of data marketplaces (No. 53). *ETLA Working Papers*.
9. Richter, H., & Slowinski, P. R. (2018). The data sharing economy: On the emergence of new intermediaries. *IIC - International Review of Intellectual Property and Competition Law, 50*(12), 4–29.

10. Plantin, J.-C., Lagoze, C., & Edwards, P. N. (2018). Re-integrating scholarly infrastructure: The ambiguous role of data sharing platforms. *Big Data & Society, 5*(1), 205395171875668.
11. Gambetta, D. (1988). Can we trust trust? In D. Gambetta (Ed.), *Trust: Making and breaking cooperative relations* (pp. 213–237). Blackwell.
12. Taddeo, M. (2009). Defining trust and E-trust. *International Journal of Technology and Human Interaction, 5*(4), 23–35.
13. Arnaut, C. et al. (2018). *Study on data sharing between companies in Europe.* s.l.:s.n.
14. Spindler, G. & Schmechel, P. (2016). *Personal data and encryption in the european general data protection regulation.* s.l.:s.n.
15. Hon, W. K., Millard, C., & Walden, I. (2011). The problem of \textquotesinglePersonal Data\textquotesingle in cloud computing - what information is regulated? The cloud of unknowing, Part 1. *SSRN Electronic Journal.*
16. Roman, D., & Vu, K. (2019). Enabling data markets using smart contracts and multi-party computation. In *Business Information Systems Workshops* (pp. 258–263). Springer International Publishing.
17. Kaissis, G. A., Makowski, M. R., Rückert, D., & Braren, R. F. (2020). Secure, privacy-preserving and federated machine learning in medical imaging. *Nature Machine Intelligence, 2*(6), 305–311.
18. Thapa, C., & Camtepe, S. (2021). Precision health data: Requirements, challenges and existing techniques for data security and privacy. *Computers in Biology and Medicine, 129,* 104130.
19. Lindell, Y., & Pinkas, B. (2000). *Privacy preserving data mining* (pp. 36–54). Springer.
20. Masters, O. et al. (2019). *Towards a homomorphic machine learning big data pipeline for the financial services sector.* s.l.:s.n.
21. Gentry, C. (2009). *Fully homomorphic encryption using ideal lattices* (pp. 169–178). Association for Computing Machinery.
22. Yao, A. C. (1986). *How to generate and exchange secrets* (pp. 162–167). s.l.: s.n.
23. Regev, O. (2009). On lattices, learning with errors, random linear codes, and cryptography. *Journal of the ACM, 56*(9).
24. MedCO. (2020). https://medco.epfl.ch. s.l.:s.n.
25. Bampoulidis, A. et al. (2020). *Privately connecting mobility to infectious diseases via applied cryptography.* s.l.:s.n.
26. Covid-Heatmap. (2021). https://covid-heatmap.iaik.tugraz.at/en/. s.l.:s.n.
27. Neubauer, M. & Goebel, A. (2018). *Blockchain for off-chain smart contracts.* s.l.:s.n.
28. Sharma, T. K. (2020). *Types of blockchains explained.* s.l.:s.n.
29. Anderson, R. (2003). *"Trusted computing" Frequently asked questions.* s.l.:s.n.

Common European Data Spaces: Challenges and Opportunities

Simon Scerri, Tuomo Tuikka, Irene Lopez de Vallejo, and Edward Curry

Abstract Common European data sharing spaces are essential for the implementation of the European digital market. This chapter addresses the challenges and opportunities of Data Spaces identified by the Big Data Value Association community. It brings forward five independent goals, convergence, experimentation, standardization, deployment, and awareness, each targeted toward specific stakeholders in the data sharing ecosystem and presents a timeframe when the goals should take place. Furthermore, we have proposed actions based on BDVA recommendations and mapped them over the five goals.

Keywords Data space · Data ecosystem · Big Data Value · Data innovation

1 Introduction

The digital market is essential for Europe to act concertedly and based on European values, i.e., self-determination, privacy, transparency, security, and fair competition. A legal framework is essential to support the emerging data economy to delineate data protection, fundamental rights, safety, and cybersecurity. One of the EU's key policy deliverables is the harmonization of digital markets. The main tangible document is now the European Strategy for Data, released in early 2020. This will

S. Scerri
metaphacts, Walldorf, Germany

T. Tuikka
VTT Technical Research Centre of Finland, Oulu, Finland
e-mail: tuomo.tuikka@vtt.fi

I. L. de Vallejo
DisCO.coop, Bilbao, Spain

E. Curry (✉)
Insight SFI Research Centre for Data Analytics, University of Galway, Galway, Ireland
e-mail: edward.curry@insight-centre.org

© The Author(s) 2022
E. Curry et al. (eds.), *Data Spaces*, https://doi.org/10.1007/978-3-030-98636-0_16

be followed by the Data Governance Act, aiming to foster data available for use by increasing trust in data intermediaries and strengthening data sharing mechanisms across the EU. Also forthcoming is the Data Act, which is a legislative proposal that aims to create a framework that will encourage business-to-government data sharing.

Data space has been an emerging concept but especially outlined by the European Strategy for Data, which guides European activities toward data economy. The strategy goes so far that it names nine common European Data Spaces: Health, Industrial, Agriculture, Finance, Mobility, Green Deal, Energy, Public Administration, and Skills. Naming Data Spaces gives direction but does not unfold the nature or essence of Data Spaces, ecosystems, which may have their peculiarities and emphasis on layers considered common to all Data Spaces. The BDVA (Big Data Value Association) community of experts has been addressing Data Spaces for many years to understand and reflect on the complexity of the concept. Distinctly, BDVA envisions European data sharing space, which refers to a space that is composed of or connects a multitude of distinct spaces that cut across sectoral, organizational, and geographical boundaries. Data Spaces can also be considered an umbrella term to an ecosystem, benefiting data sharing technologies, a suitable regulative framework, and innovative new business aspects.

This chapter presents the position of the BDVA community [1]. It first provides a short overview of Data Spaces in Sect. 2 and the common European Data Spaces vision in Sect. 3. Section 4 dives into the critical challenges standing in the way of expected value generated by the identified opportunities. These challenges are technical, business, and organizational, legal compliance, and national and regional challenges. Section 5 provides an overview of opportunities, in business, for citizens and science, and government and public bodies. As these challenges and opportunities reflect a large community, they are very informative and manifest their concerns and hopes over the vision. Section 6 presents BDVA's call to action. BDVA has identified five recommended preconditions for successfully developing, implementing, and adopting a European data sharing space: convergence, standardization, deployment, experimentation, and awareness. These recommendations have been translated with the BDVA community into 12 concrete actions, which are placed in a suggested timeframe until the year 2030. Actions can be aligned with the implementation roadmap of Horizon Europe and Digital Europe Programmes. Finally, the chapter concludes in Sect. 7.

2 Data Spaces

Data Spaces is an umbrella term corresponding to any ecosystem of data models, datasets, ontologies, data sharing contracts, and specialized management services (i.e., as often provided by data centers, stores, repositories, individually or within "data lakes"), together with soft competencies around it (i.e., governance, social interactions, business processes). These competencies follow a data engineering approach to optimize data storage and exchange mechanisms, preserving, generating, and sharing new knowledge. In comparison, data platforms refer to architectures and repositories of interoperable hardware/software components, which follow a software engineering approach to enable the creation, transformation, evolution, curation, and exploitation of static and dynamic data [2, 3] in Data Spaces. Although distinct, the evolution of the two concepts goes hand in hand and needs to be jointly considered, as both can be considered the two faces of the same data economy "coin." Their complementary nature means that commercial solutions often do not distinguish between the two concepts. For example, the Siemens MindSphere platform relies on MS Azure data solutions; the Amazon solutions embed the EC2 applications (as the platform) and the S3 storage (space) services. Furthermore, due to the particular requirements for the preservation of individual privacy, a distinction between technology and infrastructures that store and handle personal and other data has emerged. The evolution of industrial data platforms (considered key enablers of overall industrial digitization) and personal data platforms (services that use personal data, subject to privacy preservation, for value creation) has continued to follow different paths [4].

3 Common European Data Spaces Vision

The European strategy for data aims at creating a single market for data that will ensure Europe's global competitiveness and data sovereignty. The strategy aims to ensure:

- Data can flow within the EU and across sectors.
- Availability of high-quality data to create and innovate.
- European rules and values are fully respected.
- Rules for access and use of data are fair, practical, and clear, and precise Data Governance mechanisms are in place.

Common European Data Spaces will ensure that more data becomes available in the economy and society while keeping companies and individuals who generate the data in control [5]. Furthermore, as illustrated in Fig. 1, common European Data Spaces will be central to enabling AI techniques and supporting the marketplace for cloud and edge-based services.

4 Challenges

The BDVA community has identified the most critical challenges (see Fig. 2) that stand in the way of the expected value generated by the identified opportunities [1]. The challenges can be categorized into two main concerns: inter-organizational (lack of suitable data sharing ecosystems) and intra-organizational (issues faced by data producers and consumers, as data sharing participants).

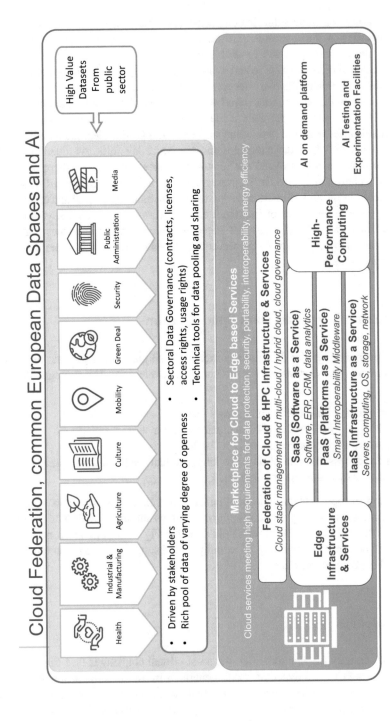

Fig. 1 Overview of cloud federation, common European Data Spaces, and AI [5]

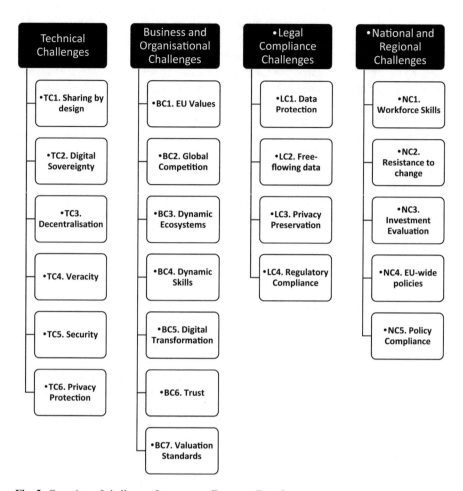

Fig. 2 Overview of challenges for common European Data Spaces

The most pressing inter-organizational concern remains the lack of valuable and trustworthy data sharing ecosystems that inspire immediate large-scale participation. Primary causes include the lack of robust legal and ethical frameworks and governance models and trusted intermediaries that guarantee data quality, reliability, and fair use. This is compounded by the lack of widespread adherence to emerging best practices and standards (e.g., interoperability, provenance, and quality assurance standards), whose maturity pace also continues to fail expectations. From a technical point of view, data sharing solutions need to address European concerns like ethics-by-design for democratic AI, and the rapid shift toward decentralized mixed-mode data sharing and processing architectures also poses significant scalability challenges.

In terms of intra-organizational concerns, the first significant concern is the difficulty determining data value due to a lack of data valuation standards and assessment tools. This problem is further compounded by the highly subjective and party-dependent nature of data value and the lack of data sharing foresight exhibited by most producers. The second concern revolves around the difficulty faced by data producers balancing their data's perceived value (after sharing) against risks exposed (upon its sharing) despite adhering to standard guidelines. Specific examples include the perceived loss of control over data (due to the fluid nature of data ownership, which remains hard if not impossible to define legally), the loss of trade secrets due to unintentional exposure or malicious reverse engineering (in a business landscape that is already very competitive), and the risk of navigating around legal constraint given potential data policies breaches (including GDPR and exposure of private identities).

The subsections below offer different perspectives to these challenges but should be considered neither exhaustive nor in isolation. In addition, the fact that some challenges are more prominent in specific sectors is well known and should also be taken into consideration.

4.1   Technical Challenges

The ambition to realize a cross-border, cross-sectoral sharing data space and enable platforms to process "mixed" proprietary, personal, and open public data introduces new technical challenges and compounds existing ones. The impact of known challenges (e.g., the Vs of Big Data: volume, velocity, variety, veracity) along the data lifecycle needs revisiting following the arising opportunities for data sharing which, in addition to conventional raw data and its transformations along the processing chain, also extend to metadata, models, and processing algorithms. The main challenges are:

- **TC1. Sharing by Design:** Data lifecycle management is not designed around sharing. Most data producers do not yet consider data sharing as a possibility at the data creation stage. Existing data lifecycle management models need to improve how they incorporate all relevant processes, including preparing data for sharing and finding the correct data. The maturity of data services (e.g., cleaning, aggregation) in data sharing ecosystems is as crucial as the availability of the data itself; without them, the data economy will not establish. Furthermore, the above differentiation between multiple kinds of data that can be made available for sharing also raises the complexity of the "variety" challenge, and interoperability solutions need to address this change.
- **TC2. Digital Sovereignty:** Enforcing data usage rights. The realization of a mixed data sharing space will only materialize if data producers are guaranteed to retain their rights as the original owners, enabling them to control who can use their data, for what purpose, and under which terms and conditions. Different ownership models or suitable data rights management frameworks need to be further explored to guarantee digital sovereignty.

- **TC3. Decentralization:** Decentralized data sharing and processing architectures. The need to guarantee that data producers control their data results in setups that forego data porting favoring decentralized data storage architectures. Thus, discussions on data volumes and data velocity (data streams) need to increasingly consider both the scalability of real-time operations over-dispersed data-at-rest in undetermined geographical distributions and the distributed processing of data-in-motion, which requires no intermediate storage. Standard data exchange protocols in decentralized architectures are therefore increasingly required.
- **TC4. Veracity:** Weak verification and provenance support. Data veracity remains crucial for the sustainability of data sharing ecosystems. Data in various stages processing chain will need to carry traceable information about its origins and operations (i.e., metadata about its raw form, algorithms, and operations it was subjected to). Support for advanced provenance is required to raise trust.
- **TC5. Security:** Secure data access and restrictions. To enable a trusted network within which closed (proprietary, personal) data can be unlocked for exchange and sharing, issues like confidentiality and digital rights management must be addressed appropriately. Furthermore, secure access control needs to be guaranteed even within a decentralized peer-to-peer network. Therefore, security solutions and exchange protocols must be standardized across all data sharing space nodes and participants.
- **TC6. Privacy Protection:** Maturity of privacy-preserving technologies for Big Data. Although technical solutions for secure and trustworthy data sharing (e.g., privacy-enhancing and privacy-preserving technologies, including digital identity management) are in continuous development, continued investment is required toward their further maturity to increase their reliability and uptake. More flexible ways of allowing uptake of compliance solutions also need to be explored.

4.2 Business and Organizational Challenges

Challenges in this category have predominantly been observed in industrial applications (non-personal data sharing). The clarity of the GDPR means that Industrial Data Platforms (IDPs) can potentially be extended to handle de-personalized data for added business value. However, legal compliance constitutes a challenge in itself and is covered in the next section. Thus, foreseen business challenges are related to the socioeconomic sustainability of a pan-EU IDP connecting multiple Data Spaces and offering data marketplaces and include the following:

- **BC1. EU Values:** Difficulty establishing EU IDPs in the global market. EU-designed IDPs need to adhere to values such as democracy, open competition, and egalitarian treatment. These characteristics can distinguish them in the global market and eliminate questionable "shortcuts" to the advantage of global competitors. In addition, new business models need to demonstrate clear business value in adhering to EU values and their advantage over existing commercial solutions in this setting.

- **BC2. Global Competition:** Competing in the global market through product-service platforms. The combination of data and service economies represents a significant competitive advantage of the EU in the global marketplace. Thus, value-added data-driven services that could make "Made in EU" products competitive globally must be identified. Furthermore, SMEs (99% of the EU industrial fabric) and the role of PPP mediators like the Digital Innovation Hub (DIH) need to be further clarified, and co-opetition models need further investigation.

- **BC3. Dynamic Ecosystems:** Implementing Data Spaces in dynamic business and data ecosystems. In the industrial domain, shared data ecosystems must guarantee data producers complete control over the access and usage of their data. However, ownership is difficult to define legally (see related legal compliance challenge, LC2). Furthermore, there are no clear guidelines or consensus on implementing data sovereignty in flexible and dynamic business ecosystems (rather than in static hierarchical supply chains). It is also unclear how next-generation peer-to-peer networks can guarantee trust and sovereignty without centralized control.

- **BC4. Dynamic Skills:** Effects of disruptive technology challenges on the job market. There are differing views on the exact impact of new data-driven technology and automation on jobs and employment. Short-term actions include the re-skilling and up-skilling of personnel. However, a complete re-definition of workflows, processes, and human-machine interaction patterns (e.g., "collaborative intelligence" between humans and autonomous systems) is required in the longer term. In addition, the current education system is still not geared toward continuously catering for new and unknown professions.

- **BC5. Digital Transformation:** Challenging organizational impact of the 6Ps digital transformation model. Data-driven transformations are needed at the level of products (and services), processes (and organizations), platforms (and spaces, marketplaces), people (and roles), partnerships (and participatory innovation models), and performance (and data-driven KPIs). Methods and tools to support EU industry with this transformation are required. Although disruptive innovation models benefit start-ups and scale-ups, evolutionary innovation models need to be considered alternatives for the broader industrial ecosystem.

- **BC6. Trust:** Lack of data sharing trust and motivation. Data marketplaces rely on an understanding of the commercial value of data produced by industry at all levels. The lack of confidence in the quality of data available for sharing is in itself a challenge. Without quality standards, widespread, automatic data exchanges will not materialize. Attempts at optimizing data accuracy should also extend to algorithms (e.g., algorithm bias). In addition, preparation costs for sharing data (e.g., cleaning, quality assurance) need to be considered, together with risks (e.g., potential access to trade secrets, intellectual property sharing). In addition, sharing personal data in B2B applications needs to comply with the GDPR strictly. The effective application of anonymization and obfuscation methods can minimize both risks. However, it can generate increasingly synthetic data whose classification can be objective and whose actual value can be extremely low (e.g., critical applications), thus not offering producers an immediate return on investment. Open data models can become a new industry model if the value of open innovation, early involvement of customers, and strategic alliances (even with competitors, as in the case of service ecosystems) are well understood. To set up trusted data networks, ad hoc and on-the-fly B2B data exchange mechanisms and contracts, provided under well-defined data sovereignty principles, must be considered.

- **BC7. Valuation Standards:** Lack of data valuation standards in marketplaces. Data marketplaces introduce new opportunities and business models whose center is valorization or monetization of data assets. New challenges revolve around the pricing of data, e.g., whether this is determined by the producer, by the market demand, or by a broker or third party: whether the value for a specific data asset is universal or depends on the buyer-seller relationship. Guidelines and pricing models need to be established to guide businesses in determining value in participation (refer to last business challenge). New forms of value creation uncovered by new sharing mechanisms need to be explored. In addition, data discovery will need to be better addressed since the value of data assets is materialized upon findability.

4.3 Legal Compliance Challenges

All the different regulations introduced in the last decade in the context of the digital single market make for a complex landscape of policy for data. However, notwithstanding the inherent complex role of data, an increased understanding is needed of how data regulation interplays and connects within data platforms. The following are the most pressing challenges that need to be addressed:

- **LC1. Data Protection:** Tackling inverse privacy and understanding personal data rights. There is a significant gap between the rights introduced by the GDPR (and its 28 national implementations) and the average citizens' and companies' understanding of their implications, what control they can exercise, and how. New business models should not assume that a sufficient portion of private users have the time, expertise, and interest to comprehend these implications fully, but data rights and consent can change. The practice of inversely private data should be discouraged to provide means for individuals to retain control and access to their private data at all times. More guidance is needed from regulators and data platform developers. Developments such as sticky policies and dynamic user consent offer promising avenues (refer to the related technical "Privacy Protection" challenge).
- **LC2. Free-Flowing Data**: Ownership and other legal blockers. Although we speak of the free movement of data as a fifth European freedom, data is far from

flowing freely. Legal questions surrounding data ownership, access, portability, and retention remain pressing topics of attention, even more so in an AI context. Existing legislation (e.g., database rights) are outdated, hampering the use of data in AI and the development of new business models. Furthermore, data ownership is tough to address in a data marketplace setting, as it is difficult to define legally. In the absence of a "GDPR for non-personal data," the principle of data sovereignty can be an answer to confidentiality and security requirements but also poses implementation challenges (see TC2).

- **LC3. Privacy Preservation:** Privacy preservation in an open data landscape. Open data initiatives and public blockchains are driving open innovation in multiple ways. Privacy preservation in this openness is a topic that has to be carefully examined not only in technical terms but also regarding legal compliance at national and European levels.
- **LC4. Regulatory Compliance:** General uncertainty around data policies. Data-driven SMEs and companies that aim to develop data platforms still face questions on how to incorporate and adjust for the effects of the regulatory landscape within the digital single market, e.g., how to be compliant; when, where, and which regulation comes into effect; and how to gather knowledge on implementing the regulation.

4.4 National and Regional Challenges

Industry and academia adopt new and disruptive technology much faster than member states, and the European Commission can adapt their policies and regulations. Amid an emergent data economy facilitated by the convergence of digital technologies, these challenges need to be high in the political agenda:

- **NC1. Workforce Skills:** Public organizations lack digital skills and resources. Digital technology is developing fast, and public organizations have difficulties keeping up with the pace of development (perhaps more so than business; see also business challenge BC4). At the same time, it is difficult to identify what kind of new skills and education public organizations would need. For instance, new digital skills include planning how data is organized and creating value in society. Organizational and individual skill development are also budget issues, which may not be high on the public agenda. The challenge is to use funding wisely and to avoid waste of resources.
- **NC2. Resistance to Change:** Insufficient support for digital transformation in business by public authorities. Digitization will transform processes, and data, along with AI, will build up knowledge of society. Transforming the organization leads to changing personnel's work profiles. Roles will change, leading to employment disruptions and the need for re- and up-skilling. New services are an opportunity, but resources for the transformation are limited. Efficiency and

transparency need data sharing but also investments in order to create new Data Spaces.

- **NC3. Investment Evaluation:** Evaluating public organization efficiency and economic impact in the data era. Public organizations serve society, both citizens and industry alike. The constant requirement of efficiency and impact improvement motivates governments to find out new services based on data. However, decisions on development investments are difficult to make, and quite often, investments are considered risky. Nevertheless, public organizations and their services are an essential part of society and one of the starting points of emerging data ecosystems. From a governmental point of view, the challenge is to evaluate investment in data-centric organizations and ensure that economic results impact the whole society.
- **NC4. EU-Wide Policies:** Lack of common innovation policies. Stepping up from regional innovation policies to EU-level comparisons is challenging. Data provides a means to measure the impact of innovation policies, but regions find it difficult to compare due to varying requirements. For instance, simple dataset timescale variation may give odd results depending on the region.
- **NC5. Policy Compliance:** Translating European-wide policies into tangible measurements. To enable the possibility of real-time, data-driven policy compliance verification, further investments in infrastructure and the certification of data from devices such as IoT appliances and edge nodes are required. Furthermore, when data is needed as evidence for compliance with specific regional and European policies, standard or common approaches recognized and accepted by the respective policies are required to map data, e.g., from IoT device measurements, into compliance levels.

5 Opportunities

As indicated in the previous section, in recent years, considerable interest has been observed by major industrial players, national and European legislative institutions, and other key stakeholders in:

- Alignment and integration of established data sharing technologies and solutions, avoiding reinventing the wheel and supporting scale.
- **Architectures, standards, protocols, and governance models aiming to unlock data silos**, over which (a) fair and secure data exchange and sharing take place, (b) protection of personal data is paramount, and (c) distributed and decentralized solutions enabling new types of data value chains can be explored.
- **Business models that can exploit the value of data assets** (including through the implementation of AI) bilaterally or multilaterally among participating stakeholders that are not limited to industry but include local, national, and European authorities and institutions, research entities, and even private individuals.

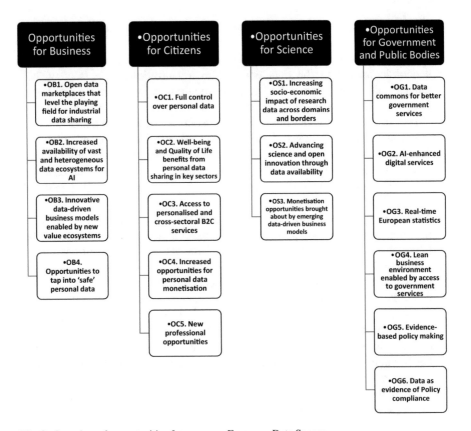

Fig. 3 Overview of opportunities for common European Data Spaces

- **Fostering and accelerating the uptake of data technologies** and the data economy within sectors whose business models are not yet data-driven.
- **Enabling data analytics across a European data sharing ecosystem,** including research centers, industry, government, and multi-national bodies leveraging existing pan-European initiatives and networks (e.g., DIH and i-Spaces).

This section extrapolates current trends to outline opportunities arising over the next decade for common European Data Spaces. As illustrated in Fig. 3, the opportunities are categorized into four primary societal spheres of industry (business), private users (citizens as customers), research and academia (science), and local, national, and European government and public bodies (government). However, the identified opportunities have a broader socioeconomic value, often benefitting multiple spheres, and should therefore not be considered solely within the confines of these categories. Indeed, the possibility to boost the data economy will lead to an improved general economy, thus benefiting society.

5.1 Opportunities for Business

Both SMEs and large industries stand to gain from the following emerging opportunities:

- **OB1. Open data marketplaces that level the playing field for industrial data sharing:** The emergence of large-scale marketplaces whose participation is open to all kinds of data producers and consumers will complement the currently dominant, exclusive data sharing agreements and solutions designed for specific parties. Thus, industrial data can be shared inside and outside of a value network, with guarantees that the producers retain data sovereignty and receive fair compensation. European industrial players of all sizes, who provide both tangible and immaterial services, can tap into data sources that are made available within the rules set by the data producers.
- **OB2. Increased availability of vast and heterogeneous data ecosystems for AI:** Advanced AI applications play a fundamental role in several industries' most critical business processes. Availability of valuable datasets is key for unleashing the potential of AI added value, especially in key industries like business services, manufacturing, wholesale, retail, and infrastructure providers (e.g., 5G operators). Emerging data marketplace infrastructures enable data to be shared and traded in a trusted, secure, and transparent manner that respects ownership. Blockchain technology enables data sharing at scale. Federated analytics on distributed data sources enable the sharing of derived insights without sharing the original data, thus balancing privacy, autonomy, and IP protection. An example of this approach is behind the Collaborative Cancer Cloud. These developments will make data available for AI development in new ways, offering unprecedented opportunities to scale the ecosystem.
- **OB3. Innovative data-driven business models enabled by new value ecosystems:** Moving from "data for business intelligence" to "data for AI" also means transitioning from internal processes to more collaborative and participative cross-domain ecosystems. The most innovative data-driven business models show a wide variety of value creation possibilities, from direct data monetization to access-based valorization of data assets on sharing platforms. Mobilizing data from business processes and services can offer a myriad of new opportunities, where data-driven often also implies engagement with strategic partners and the development of data ecosystems around these opportunities. A prominent example is a drive by the Mobility Open Blockchain Initiative's (MOBI, www.dlt.mobi/) consortium to accelerate the data sharing for autonomous vehicle technology.
- **OB4. Opportunities to tap into "safe" personal data:** The GDPR sets precise requirements for the safe use of de-personalized data outside its original purpose. Personal data will increasingly be considered for cross-sectoral applications following explicit consent and true anonymization (a technical challenge). Driven by the private customers' use of digital services, cross-sectoral services will continue to emerge. The banking industry, for example, was forced to rethink

its market position following the Payment Services Directive, and based on transaction data, new services are being provided across insurance, banking, and health service providers.

5.2 Opportunities for Citizens

European citizens will benefit from easy and secure data sharing in various ways:

- **OC1. Full control over personal data:** Under the GDPR, data platforms must guarantee legally compliant data privacy and sovereignty, affording individuals higher control and traceability of their data. In this ideal scenario, individuals will be able to monitor what data they share, which space it is stored in, and who can access or use it while retaining the right to alter these decisions. However, in addition to the management of inversely private data, which remains a challenge in itself, the need for individuals to fully comprehend the implications of these rights remains. Therefore, the BDVA is active in gathering and disseminating cutting-edge developments in the area of privacy-preserving technologies for Big Data, contributing to a better digital single market and increased end user data protection.
- **OC2. Well-being and quality of life benefits from personal data sharing in key sectors:** GDPR-compliant data sharing platforms enable emerging European technology to perform data analysis for a clear personal (or social) benefit in important sectors such as health. Furthermore, the safe management of legally compliant personal health data records allow for broader analysis (see "data for research" opportunity, next section) of health, wellness, and life data for improved understanding, risk identification, and prevention of diseases directly benefiting private users.
- **OC3. Access to personalized and cross-sectoral B2C services:** Digitization brings production closer to consumers, offering more choice and personalization independent of geographic location. Interoperable data ecosystems are required to enable bundling services during digital transactions. The business opportunity introduced by tapping into personal data will benefit private citizens, e.g., banking and insurance services offering consumers alternative product sales and choices between the most cost-effective options.
- **OC4. Increased opportunities for personal data monetization:** New European legislation incentivizes individuals to share their data, introducing C2B business models that allow them to remain in control of their data while directly receiving fair monetary or economic benefits.
- **OC5. New professional opportunities:** Further innovation will define new career pathways and generate additional jobs whose education, re-skilling, and up-skilling will continue to be supported by national and regional authorities that understand their long-term value.

5.3 Opportunities for Science

Academia is expected to benefit from the following opportunities:

- **OS1. Increasing socioeconomic impact of research data across domains and borders:** Converging standards for data and metadata representation, sharing models, licensing, and exchange protocols will make it increasingly easier to discover, integrate, or otherwise jointly process and analyze data in other scientific domains. This broadens collaboration opportunities between different scientific fields and promotes the value of making generated data available, at least partially, as open data, for the broader good. In addition, the establishment of appropriate guidelines can promote collaboration between scientific and national bodies to address societal challenges better, effectively reducing data access bottlenecks faced by European researchers.
- **OS2. Advancing science and open innovation through data availability:** Access to data for research purposes remains limited since companies need to invest effort in preparing data for little or no apparent gain. After clear business incentives for data exchange and digital infrastructure that removes friction in the process are in place, little additional effort will be required to make the data available (different license agreements) to academia. In return, it will be easier to involve researchers with access to this data in open innovation activities, thus accelerating innovation in companies. Due to the data access conditions, data challenges remain common in academia. However, while platforms such as Kaggle are very successful, they tend to be limited to specific challenges with little flexibility in the evaluation metrics. An increase in data availability enables academics to design and run more complex challenges, thus improving scientific crowdsourcing to advance science and identify solutions benefitting industry. This can help eliminate the imbalance between European and non-European industries when providing data to scientists as a core part of their business; examples from the USA and China include Google, Amazon, and Alibaba. Aside from industry data, science will also benefit from the GDPR-enabled availability of depersonalized "personal" data from millions of European citizens, offering new research opportunities in societal challenges such as healthcare and transport.
- **OS3. Monetization opportunities brought about by emerging data-driven business models:** Providing controlled access to research data will enable scientists, universities, and research institutes to have the opportunity to exchange or monetize their research data by making it available in a controlled way to other institutes and companies. This will strengthen cooperation between research and industry, enable different data to be integrated and analyzed, and thus introduce new revenue opportunities for academia.

5.4   Opportunities for Government and Public Bodies

These opportunities will benefit all levels of government, national, and European public authorities:

- **OG1. Data commons for better government services:** Opening public domain datasets and systems offer opportunities to improve existing services or develop new ones. Moreover, it can increase accessibility and simplification of e-Services. The Single Digital Gateway Regulation (SDGR) promises to make company and citizen data available across Europe in a one-stop-shop manner. The once-only principle makes it easy for companies to settle and set up businesses across Europe. Harmonizing and opening up government data also enables governments to act as data platforms offering digital services to both companies and citizens.

- **OG2. AI-enhanced digital services:** AI-boosted digital services can help predict and analyze national and European data in a privacy-preserving and ethical manner. Collaboration, piloting, and information sharing between government agencies support data platform exploitation. Based on working national examples, EU government bodies can form information sharing Data Spaces to support emerging EU-wide data service management. For example, Finland has a network of government agencies to share best practices of introducing AI to organizations.

- **OG3. Real-time European statistics:** An integrated European data sharing space can provide real-time monitoring across key sectors at both national and EU levels. Examples include economy, security, and health: customs statistics can monitor imports and exports and help with cross-border security, whereas health data can make disease outbreaks visible to all relevant healthcare authorities.

- **OG4. Lean business environment enabled by access to government services:** Public governmental services can be connected with industrial data for leaner business planning. For example, financial planning can be based on real-time information on the effects of rules and taxation regulations. In addition, better-integrated information systems enable automation of taxation, leading to cost-saving and predictable national budgets.

- **OG5. Evidence-based policymaking:** Data for policy, sometimes also referred to as evidence-based policymaking, uses Big Data in the policymaking process. It allows policymakers and governmental bodies to use public sector data repositories and collaborate with private actors to improve and speed up policy cycles and explore new areas of policymaking in a data-driven way.

- **OG6. Data as evidence of policy compliance:** Policymakers and regulators will increasingly depend on data-driven policy compliance solutions. In a data-driven society, traditional compliance mechanisms are challenged due to the increasing velocity and complexity of regulating cyber-physical systems. Advances in open data, the Internet of Things, and edge computing create a wide array of new data to be used by public organizations, smart cities, manufacturing and production lines, and logistics. These data can serve as evidence for validation of whether specific policy-related conditions, implied by European-wide regulation and policies, are met. This can open new horizons on how certifications on organizational procedures can be provided. Automating compliance and sticky policies [6] can already be witnessed in APIs or blockchain for smart contracting in data markets.

6   Call to Action

BDVA has identified five recommended preconditions for successfully developing, implementing, and adopting a European data sharing space [1]. Following widespread consultation with all involved stakeholders, the recommendations have been translated into 12 concrete actions. These can effectively be implemented alongside the Horizon Europe and Digital Europe Programmes [7]. This call for action is aligned with the European Commission's latest data strategy [5]. The recommended actions are categorized under five independent goals, convergence, experimentation, standardization, deployment, and awareness, each targeted toward specific stakeholders in the data sharing ecosystem. The implementation of the five goals should take place within the timeframe shown in Fig. 4. Assuming the convergence of initiatives required over the next 3 years will yield satisfactory outcomes, deployment efforts can be scaled up with experimentation acting as a further catalyst. Other deployment efforts need to go hand in hand with intensified standardization activities, which are key to a successful European-governed data sharing space. Activities targeted at greater awareness for all end users can initially target organizations, entities, and individuals that can act as data providers and then extend to all potential consumers as solid progress is achieved.

To catalyze the convergence of existing national and regional concepts, efforts, priorities, and strategies:

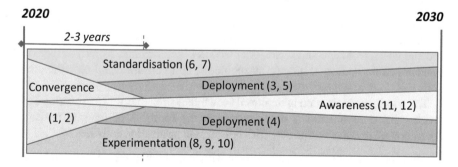

Fig. 4 A suggested timeframe for implementing the recommended actions over the next decade (x-axis). The y-axis illustrates the relative, rather than absolute, effort distribution between the five phases. The absolute effort value is expected to fluctuate, reaching a peak around the 3–5-year mark [1]

- **A1:** Launch coordination actions to map existing initiatives at EU, member state, regional, and municipal level (leveraging existing EC CSAs) and identify the common building blocks to create an impact and foster standardization actions (see A6). Efforts should particularly seek to collect, publish, and systematically analyze use cases (see A8) and align legal (including tax treatments) and governance structures necessary for safe and fair level European-wide data sharing (see A5).
- **A2:** Further invest in a Euro-wide skills strategy to identify major skill and expertise gaps in the European workforce, devise Euro-wide re-skilling and up-skilling roadmaps, advise education and training bodies with remedial actions, and offer further scholarships and fellowships in crucial areas of study.

Given the nature of the above actions, they are intended for all stakeholders, but particularly the EC (and the CSA instrument), consortia behind existing EC CSAs, government at member state, regional and municipal level, industrial and research associations, standardization bodies, consumer organizations, as well as (in the case of A2) educators and industry as employers.

Assuming an acceptable degree of convergence is achieved, the deployment of successful trusted data sharing framework and governance structures can be achieved by:

- **A3:** Funding the development of rules and guidelines for iterative, multi-phase design, creation, scale-out, or merging of existing Data Spaces.
- **A4:** Launching research and innovation actions (including EC R/IAs) to evolve high-impact technology (privacy-, interoperability-, security-, quality-, and ethics-by-design) emphasizing an interoperability initiative across existing spaces and the deployment of trust solutions for data sharing assurance.
- **A5:** Developing EU-wide Data Governance practices to clearly define rules of conduct for the fair use of exchanged data after testing different approaches to assessing the impact of data sovereignty, campaign for the revision of relevant legislation, and explore means for conformity assessment (e.g., voluntary, or licensed certification, data sharing assurance agreements) that guarantee trust.

The above actions can be primarily implemented by the EC (RIA instruments), governmental organizations, regulators, industry associations, direct data suppliers and end users, technical and sector-oriented R&D associations, as well as (see A5) standardization bodies, assurance stakeholders, and consumer/end user organizations.

The widespread adoption of deployed accountable, reliable, and trustworthy Data Spaces will only succeed if organizations at the forefront of relevant research, development, and innovation activities bolster standardization activities by:

- **A6:** Proactive coordinated engagement with international standardization bodies to establish key performance indicators (KPIs) that drive standardization of common building blocks (see action no. 1), successful technology (see action no. 4), methods (e.g., data rights management as ownership solution), and best practices (e.g., guidelines for international data valuation, privacy preservation, and FAIR principles for non-research data) behind trust-enabling solutions.

- **A7:** Assurance of sufficient conditions (assurance, market demand, government procurement) to enable widespread adherence to established standards, rules, and guidelines and promotion within networks.

The above actions are primarily intended for technology and infrastructure providers (e.g., BDVA i-Spaces), standardization bodies in the information and communications technology (ICT) domain, sector-oriented (vertical) associations, data marketplace operators, direct data suppliers, and consumers.

Deployment activities should be supported and guided to exploit various experimentation instruments to make maximum use of safe environments for testing evolutionary steps of a European data sharing space by:

- **A8:** Investing in piloting to test-drive identified data sharing use cases in safe and dynamic regional and European-wide cross-sectoral scenarios (pairing member states with different levels of progress in data sharing infrastructures).
- **A9:** Engaging with major initiatives offering experimentation activities that rely on data sharing (EDIH future network, BDVA i-Spaces network) to jointly explore market capabilities for sustainable new business, cross-sectoral opportunities, and organizational models (e.g., for data valuation and organizational transformation) and create innovative solutions.
- **A10:** To encourage innovation and motivate data sharing, setting up European regulatory sandboxes for trusted data sharing and safe exploration of risks.

The joint realization of the actions requires the attention of funding bodies within member states, industrial targets across Europe, start-ups, entrepreneurs and technological providers, direct data suppliers, consumers, investors, venture capitalists, and incubators in coordination with governmental organizations and regulators. In parallel to all actions, greater awareness of the opportunities offered by an open, fair, and ethical data economy needs to be achieved. To this end, we call for the following supplementary actions:

- **A11:** Launch a campaign that influences organizations to revisit their data strategy and lifecycles to ensure sharing-ready data by design.
- **A12:** Launch an EU-wide citizen-oriented campaign for an open, democratic, and fair data economy and the right to the free flow of safe and trusted data.

Drivers of activities for greater awareness include the EC, government at member state, regional and municipal level, sector-based industrial associations, entrepreneurs and technology providers, and consumer/end user organizations.

7   Conclusion

This chapter described initial challenges, opportunities, and calls to action to set the scene for common European Data Spaces. Setting a conceptual basis using a community vision is a necessary first step. However, progress toward actionable Data Spaces will require an iterative learning curve. Therefore, action and learning from action is essential, creating a feedback loop for development. Our proposed roadmap delineates and estimates progress, while framing the roadmap further can encompass more elaborate timeframes of technological architectures laid out by the key stakeholders.

There are many critical points which the proposed action items mitigate as the objective is very ambitious. Ensuring the EU's pole position in data sharing and Data Spaces requires investment and strategic cooperation between European organizations in addition to technical competence development. The digital transformation will be a reality. There is no turning back. The competitiveness of the EU depends on the successful implementation of common European Data Spaces.

Acknowledgments We would like to thank the contributors from the BDVA Task Force on Data Sharing Spaces, including Simon Scerri (Fraunhofer), Tuomo Tuikka (VTT), Irene López de Vallejo (DisCO.coop), Martina Barbero (BDVA), Arne Berre (SINTEF), Davide dalle Carbonare (Engineering), Oscar Corcho (UPM), Edward Curry (Insight Centre for Data Analytics), Valerio Frascolla (Intel), Ana García Robles (BDVA), Robert Ginthör (Know-Center), Sergio Gusmeroli (POLIMI), Allan Hanbury (TU Wien), Jim Kenneally (INTEL), Antonio Kung (TRIALOG), Till Christopher Lech (SINTEF), Antonis Litke (NTUA), Brian Quinn (INTEL), Dumitru Roman (SINTEF), Harald Schöning (Software AG), Tjerk Timan (TNO), Theodora Varvarigou (NTUA), Ray Walshe (DCU), and Walter Weigel (HUAWEI). We would like to acknowledge the comments from BDVA members and external communities received.

References

1. Scerri, S., Tuikka, T., & Lopez de Vallejoan, I. (Eds.) (2020). *Towards a European data sharing space*, Big Data Value Association, Brussels.
2. Curry, E. (2020). *Real-time linked dataspaces*. Springer International Publishing. https://doi.org/10.1007/978-3-030-29665-0
3. Curry, E., Derguech, W., Hasan, S., Kouroupetroglou, C., & ul Hassan, U. (2019). A real-time linked dataspace for the Internet of Things: Enabling 'pay-as-you-go' data management in smart environments. *Future Generation Computer Systems, 90*, 405–422. https://doi.org/10.1016/j.future.2018.07.019
4. Zillner, S., Curry, E., Metzger, A., Auer, S., & Seidl, R. (Eds.) (2017) *European big data value strategic research & innovation agenda*. Big Data Value Association. Available: http://www.edwardcurry.org/publications/BDVA_SRIA_v4_Ed1.1.pdf
5. *Communication: A European strategy for data*. (2020). Available: https://ec.europa.eu/info/sites/info/files/communication-european-strategy-data-19feb2020_en.pdf

6. Pearson, S., & Casassa-Mont, M. (2011). *Sticky policies: An approach for managing privacy across multiple parties. Computer, 44*(9). doi:https://doi.org/10.1109/MC.2011.225.
7. Zillner, S., Bisset, D., Milano, M., Curry, E., García Robles, A., Hahn, T., Irgens, M., Lafrenz, R., Liepert, B., O'Sullivan, B. and Smeulders, A., (eds) (2020) *Strategic Research, Innovation and Deployment Agenda - AI, Data and Robotics Partnership. Third Release.* September 2020, Brussels. BDVA, euRobotics, ELLIS, EurAI and CLAIRE.

Printed in the United States
by Baker & Taylor Publisher Services